T0295386

VERTEBRATE
PALEOBIOLOGY

LIFE OF THE PAST Thomas Holtz, editor | James Farlow, founding editor

VERTEBRATE
PALEOBIOLOGY

A Form and Function Approach

Sergio F. Vizcaíno, M. Susana Bargo,
Guillermo H. Cassini, Néstor Toledo,
and Gerardo De Iuliis

INDIANA UNIVERSITY PRESS

This book is a publication of

Indiana University Press
Office of Scholarly Publishing
Herman B Wells Library 350
1320 East 10th Street
Bloomington, Indiana 47405 USA

iupress.org

© 2024 by Sergio F. Vizcaíno, M.
Susana Bargo, Guillermo H. Cassini,
Néstor Toledo, and Gerardo De Iuliis

All rights reserved
No part of this book may be reproduced
or utilized in any form or by any
means, electronic or mechanical,
including photocopying and recording,
or by any information storage and
retrieval system, without permission
in writing from the publisher.

First printing 2024

Library of Congress
Cataloging-in-Publication Data

Names: Vizcaíno, Sergio F., author.
Title: Vertebrate paleobiology : a form and
    function approach / Sergio F. Vizcaíno,
    M. Susana Bargo, Guillermo H. Cassini,
    Néstor Toledo, and Gerardo De Iuliis.
Description: Bloomington, Indiana :
    Indiana University Press, [2024] |
    Series: Life of the past | Includes
    bibliographical references.
Identifiers: LCCN 2023059567 (print) |
    LCCN 2023059568 (ebook) | ISBN
    9780253070470 (hdbk.) | ISBN
    9780253070487 (web PDF)
Subjects: LCSH: Evolutionary
    paleobiology. | Extinct vertebrates. |
    Paleobiology. | BISAC: SCIENCE /
    Paleontology | NATURE / Animals /
    Dinosaurs & Prehistoric Creatures
Classification: LCC QE721.2.E85 V59 2024
    (print) | LCC QE721.2.E85 (ebook) |
    DDC 560—dc23/eng/20240416
LC record available at https://
    lccn.loc.gov/2023059567
LC ebook record available at https://
    lccn.loc.gov/2023059568

To the memory of
*Rosendo Pascual (1925–2012), Leonard B. Radinsky (1937–1985),
and Robert McNeill Alexander (1934–2016)*

*In loving memory of Dito and Luna (GHC)
To Bárbara and Lumen (NT)*

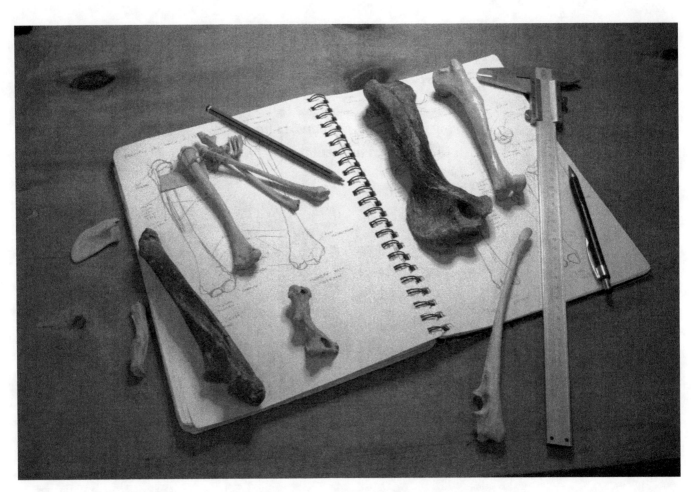

Paleobiological hypotheses based on a form-function approach are strongly rooted in a deep knowledge of the anatomy of the forms under study. As noted in chapter 2, it is important to work directly with specimens, including both the fossils under study and the extant organisms used for comparative purposes. Drawings, diagrams, and personal notes are practically irreplaceable tools in helping to understand the functional anatomy of skeletal elements. Composition and photography by Néstor Toledo.

Contents

Prologue

At the beginning of the 1980s, the renowned Argentinean biologist and paleontologist Osvaldo Alfredo Reig (1929–1992; fig. 1) argued that Argentinean paleontology should move beyond mainly "empirical" pursuits, such as classification and description, toward the adoption and development of novel theoretical approaches. Among other ideas, he especially encouraged the study of the "functional anatomy" of extinct vertebrates as a means to interpret their paleobiology.

In several memorable academic confrontations, such as the one that occurred in 1985 during the I Jornadas Argentinas de Mastozoología (Mendoza, Mendoza Province), Reig criticized Rosendo Pascual (1925–2012; fig. 2), a veritable giant in the development of modern South American mammalian paleontology and perhaps its most important modern practitioner, as the prime example of the sort of empiricism that Argentinean paleontology had to leave behind. Those who were PhD students in the División Paleontología Vertebrados of the Museo de La Plata, like the no-longer-young authors of this book (SFV and MSB), can attest that Pascual was much more than merely an empiricist in Reig's sense, and he encouraged biologists within his sphere of influence to pursue research interests other than classification and description. In the mid-1980s, Pascual remarked to one of us (SFV) on the importance of studying the variations in the morphology and histology of the placenta of extant eutherian mammals and, more relevant for this book, the importance of "functional anatomy" for investigating the different feeding types of the great diversity of extinct mammals of the continent. By that time, vertebrate paleontology in other countries had begun to draw on functional morphology and biomechanics as tools for interpreting the way of life of extinct vertebrates. However, and unfortunately, in Argentina there were no trained paleontologists or biologists capable of advising doctoral students on such investigative tools, and access to publications and the great reference collections of modern specimens from abroad was very limited. Thus, vertebrate paleontologists from Argentina and other South American countries largely took up Reig's

0.1. Osvaldo Alfredo Reig (1929–1992) was a versatile Argentinean evolutionary biologist, born in Buenos Aires. He worked at the Universidad de Buenos Aires, Universidad Nacional de Tucumán, and Universidad Nacional de Mar del Plata (Argentina); Universidad Central, Universidad de Los Andes, and Universidad Simón Bolívar (Venezuela); Universidad de Chile and Universidad Austral de Chile (Chile); Universidad de Barcelona and Museo Nacional de Ciencias Naturales of Madrid (Spain). Concerned about the relationship between science and society, he lived in exile. Photo courtesy of Carlos Quintana.

call, although mostly through investigations into biostratigraphy, phylogeny, and historical biogeography. Indeed, these three themes figure prominently in the titles of many articles published by South American paleontologists of that generation.

It was not until the 1990s that the functional perspective got underway in Argentinean paleontology. In 1992, our unorthodox Uruguayan colleague Richard A. Fariña (born in 1957; fig. 3) burst onto the scene, with the orthodox IX Jornadas Argentinas de Paleontología de Vertebrados (Trelew, Chubut Province) as backdrop, and

0.2. Rosendo Pascual (1925–2012) was an Argentinean paleontologist, born in Mendoza. He undertook all his scientific activity at the Universidad Nacional de La Plata, where he assembled one of the largest research groups of vertebrate paleontologists worldwide and achieved recognition as one of the most renowned international scientists of his time. He proposed the timescale for South American fossil mammals, using the Argentinean fossil record as the pattern for South America. Photo owned by authors.

0.3. Richard Alfredo Fariña Tosar is an Uruguayan paleontologist, born in Montevideo in 1957. He is the pioneer in applying biomechanics and considering metabolism in the interpretation of megafaunal paleobiology and paleoecology of the South American Pleistocene. Never one to follow an orthodox path, Fariña generates and promotes intellectually challenging hypotheses, and his articles and presentations at scientific meetings catch the attention of every paleontologist. He trains researchers interested in the more biological aspects of paleontology at both the undergraduate and graduate levels. Among his several talents, Fariña is also an outstanding communicator of science for the general public. Photo courtesy of Sebastián Tambusso.

presented several decidedly challenging hypotheses on the Pleistocene megafauna. Richard, who had been trained in biomechanics under the guidance of Professor Robert Mc-Neill Alexander (1934–2016, Leeds University, England; see chap. 2), proposed glyptodont bipedalism, and that some giant terrestrial sloths, forms that had traditionally been considered to be exclusively herbivorous, should be included in the carnivore guild. Soon after, Richard and two of us (SFV and MSB) began a collaborative project that deployed the paleobiological approach and promoted the creation of working groups on either side of the Río de La Plata. In Argentina, the core of our group is in the División Paleontología Vertebrados, which Pascual supervised for so long, but there now are extensions in other institutions, such as the Museo Argentino de Ciencias Naturales "Bernardino Rivadavia" (Buenos Aires), Facultad de Ciencias Naturales e Instituto Miguel Lillo of the Universidad Nacional de Tucumán, Centro Regional de Investigaciones Científicas y Transferencia Tecnológica (CRILAR) of La

Rioja, Instituto de Bio y Geociencias del NOA (IBIGEO) and Instituto de Ecorregiones Andinas (INECOA) of Salta, and Centro de Investigaciones en Ciencias de la Tierra (CICTERRA) of the Universidad Nacional de Córdoba, among others.

We have, clearly, been influenced and guided by many other researchers, be it through all or some combination of their publications, conversations, collaborations, or data sharing. We refrain from attempting to list them all for fear of omitting someone, but we would be remiss in not highlighting the efforts of Leonard B. Radinsky, Christine Janis, and Richard F. Kay. As we noted (Cassini et al. 2021a) in a volume dedicated to Leonard, his clear, concise, and rigorous *The Evolution of Vertebrate Design* has inspired many young biology students to frame their doctoral studies within a form-function paradigm. In Christine, whose large databases of extant species and methodological approaches served us as models, we discovered a muse to guide our early efforts. Our collaboration with Richard, a renowned specialist in the functional anatomy, adaptations, and evolution of primates, has been among the most fruitful of our careers. We shared with Richard 20 years of fieldwork in Patagonia with the aim of reconstructing the paleoecology of the Miocene Santacrucian fauna. Beyond these, the list of distinguished characters who have contributed to the reconstruction of extinct vertebrates is lengthy indeed. Most of them are identified throughout the course of this book as authors of the various works we cite. Among those who have also had a notable influence on us include Robert McNeill Alexander, Charles Oxnard, Martin Rudwick, Nikos Solounias, and Walter Greaves.

Our own development in the study of form and function has been mostly a process of trial and error, with much material to analyze, many issues to study, and considerable conceptual framework and methodology to learn and, indeed, develop. The contributions of editors and reviewers to our work have been crucial to this process. The manuscripts sent to journals and repeatedly returned, with corrections and suggestions or perspectives that had not occurred to us, have enriched us intellectually. The interaction within the group and with other researchers interested in the topic has expanded the scope of the issues to be dealt with and the methodologies to be used, and therefore have created a critical body for the exploration of new methodologies and discussion of conceptual frameworks. Thus, we believe that we are furthering development and innovation of the functional aspect of Reig's and Pascual's expectations.

An earlier version of this text, *Forma y Función en Paleobiología de Vertebrados*, published in 2016 by the Editorial de la Universidad Nacional de La Plata (fig. 4), Argentina, offered our first efforts in bringing together the many developments, both conceptual and methodological, that have been employed and developed by paleobiologists over the past several decades to address questions on and reconstruct the lives of extinct vertebrates. For this new and updated edition, we have been joined by a longtime colleague (GDI), whose expertise and interests lie mainly in intraspecific variation and the systematics of extinct South American mammals, particularly xenarthrans, although he has also contributed to analyses of form and function. While such aspects of paleontology as alpha taxonomy are not, strictly speaking, directly involved in paleobiological investigations, they help provide finer-grained analyses by clarifying the taxonomic units employed and thus lend greater resolution to the paleobiological and, particularly, paleoecological investigations.

We hope that this expanded version of the book provides a point of entry for those who intend to begin studying fossil vertebrate form and function. By no means do

EDITORIAL DE LA UNLP

0.4. Logo of the Editorial de la Universidad Nacional de la Plata, Argentina. The oak leaf was chosen as the university's emblem by its founder Joaquín V. Gonzalez (1863–1939) because the oak is a tree consecrated to Zeus and linked to Pallas Athena, goddess of genius and intelligence, and symbolizes strength, toughness, firmness, and perenniality.

we intend this to be a definitive text; it is just a brief summary of the many paths we have traveled, where we are now, and where we believe we are headed. Our narrative tracks the more reliable paths we have taken. That is why we refer mostly to our own work, summarizing the methodological, conceptual, and epistemological discussions that we have faced and pointing out the alternatives. Our perspective is that, due to the singularity of the extinct vertebrate faunas from South America, functional studies will help decipher processes at the organismic level that will contribute to understanding supraorganismic approaches (including environmental reconstruction) in evolutionary analyses.

The text consists of nine chapters. In chapter 1, we contextualize the use of the term *design* in biology within an engineering framework, define paleobiology and its objectives, and outline our approach in the application of the principle of form-function correlation; further, we define terms used throughout the text and describe the protocol implemented for our paleobiological studies. This protocol is based on three aspects: body size, substrate preference and use, and feeding. In subsequent chapters, we initially treat each of them independently and then together as a whole. In chapter 2, we detail the methodological tools used in our studies: biomechanics, functional morphology, and ecomorphology. Chapter 3 reviews the mechanical properties of the most important biomaterials, such as skeletal tissues and muscles, encountered in the study of vertebrate paleobiology. Other anatomical aspects are summarized in the appendix. In chapter 4, we consider the first biological character listed above, describing the effects of size on different aspects of vertebrate biology, and methods of assessing body size in extinct vertebrates. Chapters 5 and 6 are devoted to substrate preference and use, the former dealing with the diversity of vertebrate locomotory systems and locomotion in fluids (aquatic and aerial) and the latter with locomotion in terrestrial environments. Chapters 7 and 8 examine feeding. Chapter 7 discusses food classification and types of feeding, and briefly surveys the diversity of the vertebrate cephalic feeding system, whereas chapter 8 specifically analyzes this system, including the more commonly studied skeletal and dental elements, as well as soft-part components such as the lips and tongue. In chapter 9, we demonstrate how the application of the form-function system principle, complemented by information from other disciplines, allows more precise and comprehensive paleoecological reconstructions.

Acknowledgments

We are deeply grateful to the many people and institutions that have, in one way or another, influenced us and our work, made it more enjoyable, and, in many ways, even possible. SFV, MSB, GHC, and NT are particularly indebted to the Argentine public educational system, from our primary institutions to the Universidad Nacional de La Plata and the Universidad Nacional de Luján, which helped shape and nourish us personally and intellectually. Further, we acknowledge the valuable contributions of the governmental organizations that support science in Argentina: the Consejo Nacional de Investigaciones Científicas y Técnicas (CONICET), the Comisión de Investigaciones Científicas de la Provincia de Buenos Aires (CIC), and the Agencia Nacional de Promoción Científica y Tecnológica (ANPCyT). We all recognize and appreciate the privilege of conducting our professional lives largely within natural history museums: the Museo de La Plata (SFV, MSB, and NT), the Museo Argentino de Ciencias Naturales "Bernardino Rivadavia" (GHC), and the Royal Ontario Museum (GDI).

We are particularly grateful to Indiana University Press for allowing us to publish this book, to James Farlow (former editor of the IU Press series Life of the Past) for originally inviting us and reviewing the first version, and to Gary Dunham (director of the IU Press and Digital Publishing) for his continued attention and guidance. We sincerely appreciate the efforts of the reviewers of this second version of our book. Although they did not agree with all of our views, their comments and suggestions provided us the opportunity to detect weaknesses, necessary updates, and aspects that required further development to convey our messages more comprehensively to our readers. The deficiencies that persist in this text are attributable solely to us.

The subject matter of this book and our treatment of it have been deeply enriched by discussions and collaborations with and critical readings and reviews by numerous people throughout our careers. We refrain from attempting to provide a complete list here, because we are sure to commit more than one injustice by neglecting to mention several of them. However, we permit ourselves license to mention a few who have had a special impact on some or all of us, including Richard A. Fariña, Richard F. Kay, Charles S. Churcher, Robert McNeill Alexander, M. Laura Guichón, and David A. Flores. We are thankful also to the many individuals who contributed to the production of the illustrations. Their contributions are acknowledged in the illustration captions.

Last, we are grateful to each of our families for having been fundamental supports in our lives, including our careers: Miriam, Josefina, and Julieta (SFV); Pablo and Julia (MSB); Eduardo, Norma, Patricia, Silvita, Florencia, Carolina, and Adolfo (GHC); Bárbara and Lumen (NT); Virginia, Theodore, and Jacob (GDI).

Early morning along the Atlantic coast of Argentine Patagonia at the Estancia La Costa locality (see fig. 9.11). The tide has largely receded, exposing the beach platform from which many of the Santacrucian fossils discussed in this book were recovered. Members of the Museo de La Plata–Duke University team may be seen searching for fossils. Photography by the authors.

VERTEBRATE
PALEOBIOLOGY

Facing left, **1.1.** Georges Léopold Chrétien Frédéric Dagobert Cuvier (1769–1832), known as Baron Cuvier. The French naturalist, born in Montbéliard, is considered one of the founders of comparative anatomy. Engraving by George T. Doo (1840), based on a painting by W. H. Pickersgill (1831).

Facing right, **1.2.** Étienne Geoffroy Saint-Hilaire (1772–1844). The French naturalist, born in Étampes, was the main opponent to Cuvier's ideas. He established the principle of the unified plan of corporeal composition. Saint-Hilaire and Cuvier participated in a famous debate in 1830. This nineteenth-century engraving was published by Rosselin Editorial, 21 Quai Voltaire, Paris.

The Design of Fossil Vertebrates as a Tool for Interpreting Their Biology

Geoffroy S.t Hilaire.

Form-Function and Design

The word *design* is commonly defined as a plan or drawing (e.g., a blueprint) produced to demonstrate the appearance and function or workings of an object, such as a building or garment, before it is made. Therefore, design implies functional, as well as aesthetic and symbolic, considerations.

But what do we mean when talking about design in biology? In using such terminology, we must, first and foremost, be entirely clear that we are not in any way suggesting or supporting the existence of a designer and hence a plan, concepts encapsulated in their current guise under

intelligent design. The latter refers to the ideology (as opposed to a scientific hypothesis) that holds that the origin and evolution of organisms and the Universe are products of the rational and deliberate actions of supernatural intelligent agents (see Scott and Matzke 2007). In this context, Bock (2009) has strongly urged that the concept of design is inappropriate in biology and should be omitted from all biological explanations. We fully appreciate Bock's view, but because in this book we will, to a great extent, consider organisms as living machinery, we concur with Rudwick (1964) that machines can only be described for what they are by referring to the way their design enables them to function for their intended purpose and that such purposive language (including *mechanisms*, *devices*, and *design*, among others) cannot be avoided and should not be disguised by "pseudosubstitutions" of apparently non-teleological terms.

Indeed, the term has frequently and for a long time been used by biologists without major conflict (e.g., van der Klaauw 1948; Schaefer and Lauder 1986; Wainwright 2007; Lieber and Ward 2011; Shanahan 2011; Kilbourne and Hoffman 2013; Bertram 2016a, 2016b, 2016c), especially among morphologists who, consciously or unconsciously, share the principle of the correlation of parts that the French anatomist Georges Cuvier (1769–1832; fig. 1.1) enunciated in his *Leçons d'anatomie comparée* between 1800 and 1805. According to Cuvier, the anatomical structure of each organ is functionally related to all the other organs of an animal, and the functional and structural characteristics of organs result from their interaction with the environment. Thus, an animal's functions and habits determine its anatomy. By contrast, Cuvier's coeval Geoffroy Saint-Hilaire (1772–1844; fig. 1.2) stated that anatomical structure precedes, and therefore requires, a particular way of life. This began a great debate that continues today. It is advisable to read the summaries of the history of the discussion of form versus function by Padian (1995), Schwenk (2000), and Huneman (2006), and Lister's (2014) review on the value of the fossil record to evaluate whether behavioral changes precede morphological evolution.

When discussing design, biological and otherwise, it is difficult to avoid reference to the great Italian polymath Leonardo da Vinci (1452–1519; fig. 1.3), who had great respect and veneration for nature's complexity and diversity. In notebooks kept in the Windsor Castle library, Leonardo praises the wonderful works of nature ("opere mirabili della natura"): "Human ingenuity will never devise an invention more beautiful, more simple or more direct than does nature because in her inventions nothing is missing, and nothing is superfluous." We will have the opportunity presently to delve more deeply into the relevance of these concepts.

Regarding nature, da Vinci studied, drew, and painted the geological forms created by water, the growth of plants due to their metabolism, and the anatomy of the animal body in movement (fig. 1.4). According to Capra (2007), da Vinci may have been an unrecognized founder of modern science, given that he developed what then was a new empirical approach that implied the systematic observation of nature, reasoning, and mathematics; that is to say, important features of what today is known as the scientific method. Capra (2007) explained that da Vinci's approach

1.3. Leonardo di ser Piero da Vinci (1452–1519), or simply Leonardo da Vinci, was born in Vinci, Florence. He was a remarkable polymath, described as the paradigm of the Renaissance man, who was devoted to multiple scientific and artistic interests. He defined himself as an engineer and architect, and even a sculptor, rather than as a painter. Leonardo was a restless observer of nature and strove to decipher its laws. Self-portrait in chalk on paper made after 1510; housed in the Royal Collection of Windsor, London Castle, London.

to science was visual; he frequently asserted that painting implies the study of natural forms and drew attention to the intimate connection between the artistic representation of those forms and the intellectual understanding of their intrinsic nature and underlying principles. In many of his creative efforts, Leonardo paid particular attention to the human body, its beauty and proportions, and the mechanisms that produce its movement. He also studied animal proportions and movements, as demonstrated by the drawings he produced as studies for his design of an equestrian statue venerating the father of Ludovico Sforza, the Duke of Milan. The famous unfinished project, which represented Francesco I Sforza on a rearing horse with a fallen foe at its feet, was to be a monument of colossal dimensions, surpassing all other similar works, which demanded many years of empirical and intellectual effort.

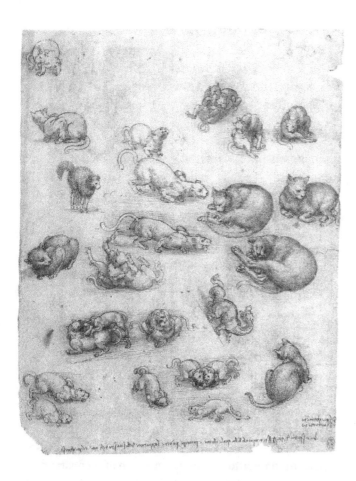

1.4. Drawing made by Leonardo da Vinci for studying animals in movement. This piece includes mainly domestic animals (cats) and a fantastical creature (a dragon). This classic Renaissance piece, created in 1513 in ink and chalk, is housed in the Royal Collection of Windsor, London Castle, London.

To understand the movements of the horse's body and transfer them to a statue, da Vinci needed to explore the mechanical principles thoroughly and systematically. In addition, he had to overcome the technical difficulties of posing such a large structure supported only by the horse's rear legs.

Returning to our initial question of what is meant by design in biology, it is blindingly obvious that most modern biologists are fully aware that there need not be any plan (Padian 1995) or designer behind biological design. In referring to design, biologists mean the form, often recurrent, present in different organisms that resolves particular functional needs. According to Muñoz (2019), as evidenced by repeated patterns from the level of populations to whole lineages, many aspects of form-function evolution are highly deterministic, and when form-function relationships exhibit one-to-one mapping (see chap. 2), natural selection often finds the same solutions to similar problems. Even in systems exhibiting many-to-one

mapping (Wainwright et al. 2005; Wainwright 2007), mechanical sensitivity can bias morphological evolution to a few traits that are mechanically highly effective. Thus, natural selection retains solutions resulting from the appearance of accidental variability, and these can be studied from an engineering perspective. Indeed, as postulated by Rudwick (1964, 34), this very perspective reaffirms the application of the term *design*, but (to reiterate) it does so independently of the existence of a designer and, consequently, of a predetermined plan. Nowadays, morphological studies based on understanding the underlying engineering principles of design are gaining ground in theoretical biology, whether at the cellular level, organismal level, or even that of multiple organisms.

Biology has joined the debate process via optimization theory (Alexander 1996), according to which design operates to obtain better results at lower costs (da Vinci's "nothing is missing, nothing is superfluous"). Thus, animal structures, movements, behavior, and life histories in general are features that are optimized through multiple evolutionary routes (e.g., adaptation, exaptation, stochastic change). In *Principles of Animal Design: The Optimization and Symmorphosis Debate*, Weibel et al. (1998) presented an extended discussion on whether animals are effectively designed according to the same rules that engineers use for building machines; that is, by maximizing performance levels through economical use of materials and energy.

Symmorphosis is a theory of biological optimization that proposes that structure responds effectively to functional demands as a result of regulated morphogenesis during growth. A primary component of this theory is that differences in functional demands on an organ require quantitative adjustments of its structural parameters so that functional capacity matches maximum demands. For example, Weibel et al. (1998) observed that the architecture of blood vessels ensures blood-flow distribution with minimum energy expenditure, and the internal bone structure is patterned according to stress distribution and, at the same time, is quantitatively adapted to total stress. With training, athletes can precisely adjust the structure of the muscular, skeletal, and cardiovascular systems to accommodate a higher functional demand, but these modifications quickly revert when training is interrupted. Symmorphosis theory postulates an ideal adaptation of structural design to functional capacity (which, when taken to the extreme, may be summarized by da Vinci's phrase "nothing is superfluous"). However, engineered designs of complex systems incorporate redundancies as security factors against potential imperfections in functional performance and variable conditions, and therefore systems do not necessarily reflect the most economical designs. By using the symmorphosis concept, such deviations from an ideal economical design can be detected.

Although there are examples that challenge the applicability of this concept to human physiology under different conditions (trained vs. untrained, healthy vs. unhealthy; e.g., Gifford et al. 2016) or to animal weaponry (e.g., structures used in intraspecific competition that raise the question of the circumstances under which the benefits of investing in large structures outweigh the associated costs; see Emlen 2014; Shuker and Simmons 2014), the general principles and application of symmorphosis in several distantly related lineages of animals have been recently considered valid. For example, the respiratory system and flight and hopping muscles in insects (Snelling et al. 2012), the respiratory structures and functional needs in mice (Canals et al. 2010), and the evolution and functional differentiation of the diaphragm muscle of mammals (Fogarty and Sieck 2019). Further, van der Zwaard et al. (2016), in a reply to Gifford et al. (2016), stated that their conclusions contradict those of the latter authors.

A few paragraphs earlier, in discussing biological design, we noted that we do not support intelligent design as it relates to the origin and evolution of organisms. However, we recognize that, apart from this context, intentional and intelligent biological design is possible. The example of the athletes previously mentioned and any kind of artificial selection, involving direct or indirect genetic manipulation, may be considered as instances of such design, certainly more so than the products of natural selection, as behind them there is often a predetermined plan.

Paleobiology

So far, we have focused on the interpretation of design, with the understanding that in addition to the possibility of studying form, we can know function by observing it in living organisms. After all, as Wainwright (2007, 394) claimed, "The capacity of organisms to perform the tasks of their daily lives is rooted in the design of the mechanical, physiological, and biochemical systems that make up the body." Therefore, in one way or another, from the time of Cuvier onward, paleontologists have pinned their hopes on using design to assign biological roles to extinct organisms (i.e., formulating paleobiological hypotheses).

The term *paleobiology* (as palaeo-biology) was first used by Buckman (1893; see Sepkoski 2012), but its eventual widespread use is due to the Austrian paleontologist Othenio Abel (1875–1946; fig. 1.5). In *Grundzüge der*

Palaeobiologie der Wirbeltiere (Principles of the Paleobiology of Vertebrates), Abel (1911, 15) stated, "Ich führe für jenen Zweig der Naturwissenschaften, der sich die Erforschung der Anpassungen der fossilen Organismen und die Ermittlung ihrer Lebensweise zur Aufgabe stellt, die Bezeichnung 'Paläobiologie' ein." (For the branch of the natural sciences that aims to elucidate the adaptations of fossil organisms and their way of life, I introduce the term *palaeobiology*; English translation, Kutschera 2007, 172).

Following from Abel's definition, the main objective of paleobiological approaches, then, is to reconstruct the lives of extinct organisms from the available fossils, including interactions with other organisms and the environment. Some five decades ago, however, the term began to be applied to a broader range of studies involving fossil organisms, including those focusing on phylogenetic relationships, geological time frames, extinction and speciation rates, paleobiogeography, fossil preservation biases, and macroevolutionary patterns such as taxonomic diversity through time (Jackson and Erwin 2006). The objectives of paleobiology in this broader sense may be summarized as reconstructing both the lives and evolutionary histories of extinct organisms, effectively rendering the term a synonym of paleontology and thus leaving us without a specific term to refer to the narrower scope of reconstructing the lives of extinct organisms. Our aim in this book is to examine methods by which the lives of extinct organisms may be reconstructed and so to apply paleobiology to this realm of research. We are well aware that other avenues of research, some of which were just noted, are often considered as falling under the umbrella of paleobiology; some are only superficially covered, whereas others are not considered at all here.

Like other historical scientists, paleobiologists face incomplete data and lack direct observation and experimental access to their objects of study. This has led some philosophers and scientists to be pessimistic about the epistemic potential of the historical sciences (including paleobiology, of course). On the contrary, analysis of an article on the paleobiology of the extinct marsupial lion *Thylacoleo carnifex* led the philosopher of science Adrian Currie to claim that historical science often produces plausible, sophisticated hypotheses based on two important methodological features (Currie 2015). First, he considered historical scientists to be methodological omnivores who construct purpose-built epistemic tools tailored to generate evidence about highly specific research subjects. This allows them to marshal multiple lines of independent evidence and thus maximize their epistemic reach. Second, their approach resembles investigative scaffolding; research proceeds in a piecemeal fashion, information

1.5. Othenio Lothar Franz Anton Louis Abel (1875–1946), an Austrian paleontologist and evolutionary biologist who also studied law, was born in Vienna. He taught paleontology at the University of Vienna from 1917 to 1934. In 1914, Abel proposed that fossils of extinct miniature elephants inspired the myth of the mythical giant Cyclops, with the central nasal opening having been interpreted as an eye socket. Photo from Universitätsbibliothek der Universität Wien, Archiv der Universität Wien, Forum Zeitgeschichte.

only gaining evidential relevance once certain hypotheses are well supported.

In contrast to most extant animals, direct observation of behavior and measurement of physiological variables are impossible for extinct forms. A standard method used in interpreting paleobiological aspects involves the application of actualism, through which past events are inferred by analogy with observable processes that occur in the present. In an engaging philosophical review of convergence as evidence for evolution and adaptation, Currie (2013) defined *analogy* as a similarity between two lineages that meets two conditions: (1) the trait must be present in the two lineages, but not in their common ancestor (it must be homoplastic); and (2) the trait must have evolved in the two lineages nonaccidentally. This author focused on what he termed *analogous inferences*; that is, inferences that take a trait-environment dyad from one lineage and project it onto another. Currie (2013, 771) concluded that analogy plays a central role in the confirmation of adaptive hypotheses, providing important "evidence for the construction and testing of historical hypotheses about biological form and function."

Although it is in general use, there are cases in which analogy is not entirely appropriate when applied to lineages with morphologies not represented within the range of variation of their extant relatives, or in cases of distant phylogenetic affinity. For instance, Norman and Weishampel (1985) and Nabavizadeh (2014) noted that although ornithopod dinosaurs have been compared to large extant ungulates due to their dental morphology with flat occlusal surfaces, they are more appropriately viewed as representing alternative pathways to achieving a comparable transverse grinding function. Indeed, there are so many particularities in other dental features and jaw morphology (e.g., placement of the craniomandibular joint, jaw curvature, muscular attachment sites, and the mandibular symphysis) that there are no modern analogs with which to compare them sufficiently and, as a consequence, mechanisms for oral food processing have been proposed for these animals that are unrecognized in modern herbivores (Nabavizadeh 2014). Among mammals, extant xenarthrans represent a severely restricted sample of the total diversity preserved in the fossil record of their extinct relatives. Given their shared history, the extant representatives of the three major groups (Cingulata, Folivora, and Vermilingua) may provide a valuable basis for paleobiological inference. However, many extinct taxa are morphologically so dissimilar from their extant relatives that they suggest very different ways of life. Vizcaíno et al. (2018) claimed that in such cases, extinct forms do not have modern models (or analogs) within their phylogenetic group, and so the application of a simplistic and strict actualistic approach can produce nonsensical reconstructions. Analyzing linear dimensions of the appendicular skeleton of extant and extinct xenarthrans, these authors evaluated the limitations of the use of extant xenarthrans as morphological models for paleobiological reconstructions. They found that many extinct xenarthrans are more like extant mammals from groups other than their closest relatives and that many of the equations derived from extant xenarthrans predict unrealistic results as a product of dimensional and shape differences between most of the extinct and extant xenarthrans. This does not invalidate actualism and the use of analogs, but it suggests the need to apply other approaches, such as those that address form-function relationships but are not necessarily drawn from closely related extant species. Although coarse-grained "analogies supporting wide-scope, highly specific

1.6. Leonard Burton Radinsky (1937–1985), an American paleontologist born in Staten Island, NY, was mainly interested in the functional anatomy of fossil mammals, including ancestors of primates, horses, and modern carnivores, and was an ardent advocate of paleoneurology. A political activist, he was also committed to social justice in the world beyond his laboratory. For a scientific and personal account of Radinsky, see Kay (2021). Photo from Hanna Holborn Gray Special Collections Research Center, University of Chicago Library.

of its design based on analogous forms that are not close kin may be used instead. Radinsky (1987) strongly asserted that data must be carefully and rigorously gathered to establish form-function relationships and that the behaviors or functions correlated with a particular anatomical form in extant species must be sought (see also Vizcaíno and Bargo 2021). A recent example of this is the volume published in honor of Radinsky, edited by Cassini et al. (2021b), which deals with the form-function relationship in South American extant mammal lineages from morphofunctional and ecomorphological perspectives. For instance, Vizcaíno and Bargo (2021) reinforced Radinsky's view that testing form-function correlation should be addressed as a prerequisite for developing hypotheses on adaptation. This is in line with Currie's claims that homoplastic (e.g., convergent) traits, though uninformative in a phylogenetic perspective, are instead highly significant from a functional and adaptive framework.

If there are no biological analogs available at all, mechanical analogies can be employed (see Rudwick's paradigmatic analysis, 1964; chap. 2, this text). Understanding an animal's biological design (i.e., how an animal's specific biological attributes functioned) may allow interpretation of the roles it played in a particular ecosystem of the geological past, providing crucial information for understanding ecosystem evolution and, therefore, corresponding paleoenvironments.

An outstanding contribution that cemented the modern status of functional studies of extinct vertebrates in the context of similar work on living animals, together with a broad philosophical view of the subject's development within the framework of phylogenetic analysis, is the already classic *Functional Morphology in Vertebrate Paleontology* edited by Jeffrey J. Thomason (1995). Two recent efforts, by Vizcaíno et al. (2016) and Croft et al. (2018), emulated Thomason's effort in scope, but both provided updates on methodological tools, with the work by Vizcaíno et al. (2016) delving more deeply into the conceptual and philosophical frameworks relevant to this research field. Another interesting publication is *Issues in Paleobiology: A Global View*, edited by Sánchez-Villagra and MacLeod (2014). In this contribution, Vizcaíno (2014) pondered the limitations of actualism and analogy in reconstructing life modes in extinct organisms and called for further expansion of tools applicable to paleobiology (see further in this chapter).

Despite paleobiology probably being the branch of paleontology that has developed most rapidly over the past three or four decades, there are very few comprehensive texts focusing directly on the paleobiology of the bizarre

inferences are problematic," Currie (2013, 783) suggested that by using the methodological modeling tools he proposed, "as well as incorporating analogies into integrated explanations, such worries can be partly mitigated."

In vertebrate paleontology, the main source of information is fossilized bones and teeth. Therefore, although indirect sources can be used, such as ichnites (fossil tracks) and taphonomic evidence, most paleobiological information comes from skeletal material. A way of approaching paleobiology is to apply the principle of form-function correlation (Radinsky 1987; fig. 1.6), which contends that there is a close relationship between the two, such that the latter can be inferred from the former. If there are no comparable extant relatives of the animal of interest, analysis extinct vertebrates (mainly mammals) of South America, and even these provide very little (e.g., Croft et al. 2018) or limited discussion of the conceptual framework for the understanding of their biological design (see Fariña et al. 2013). This dearth of information encouraged Vizcaíno et al. (2016) to compile what was then known on the subject, and these themes are expanded in this updated version with the hope of stimulating further development of the intellectual curiosity and practices of this discipline in students and young researchers.

Our aim is, then, to help new paleontologists understand that a fossil is not merely a product of fossilization useful for phylogenetic analysis or for assigning an age to the stratigraphic level where it was found. A fossil also retains information about the creature's habits, its environment, and its role in the ecosystem where it lived, and this paleobiological interpretation requires knowledge of a combination of biological and physical concepts. For that purpose, throughout the following chapters we will attempt to familiarize the reader with the basic methodologies for interpreting vertebrate paleobiology from biological form, to facilitate understanding of necessary basic concepts of biology and physics, and to introduce the attributes of extinct forms that allow the generation of hypotheses about their past biological role.

Definition of Terms Regarding Form and Function

One of the most important initial aspects of the communication process in science is the clear and unambiguous expression of the cognitive constructs or images produced by visualizing an object of study, given that any researcher carries conceptual baggage. Or, more simply: What do we mean to say when we say . . . ? Often, we believe that we capture the essence of what our colleagues want to express, until we realize that they operate in a conceptual frame slightly or markedly different from ours, and this influences the final configuration of the mental image about a particular event or concept. The conclusions they arrive at, after applying a determined methodology, are shaped by their specific epistemological framework. On the whole, in every process of scientific communication, clearly defining the terms and basic concepts of research is an advisable practice. Therefore, before going further, in this section we provide definitions of general biological concepts that will be used throughout the text. Other, more specific concepts will be defined in other chapters.

The most relevant term that is necessary to define is *morphology*, by which we mean the study of form (including size) and structure of organisms. As defined by Koehl

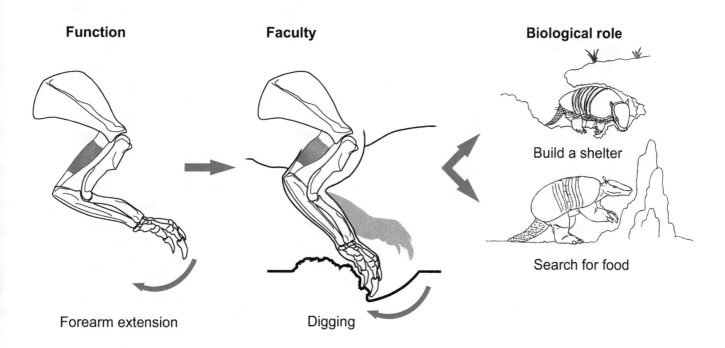

1.7. Example of the link between the morphology of a feature and its biological meaning. In armadillos, as in other mammals, the function of the triceps muscle is to extend the forearm, which confers (for example) the faculty of digging, the biological role of which may be to seek or obtain food and build a shelter, among other possibilities.

(1996), it makes reference to an organism and its structure, at whatever organizational level, from molecular to organismic. It involves aspects of the organism's organization (anatomy) that may be coupled to a functional outcome (e.g., form-function complex, Cuvier's organic-functional correlation law, or Currie's trait-environment dyad). Hence, it is necessary to link morphology with the function that structures can (or could) perform and then infer its possible biological meaning (or biological role, defined below). In this text, we will use the following definitions, based on those originally supplied by Bock and von Wahlert (1965) and Plotnick and Baumiller (2000):

Feature: any part, attribute, or character of an organism; these may be, for example, morphological, behavioral, or physiological. The structures of an organism are its morphological features.

Form: the appearance, configuration, and composition and arrangement of material of a feature, including size.

Shape: refers to form but excludes size.

Function: what a feature does or how it works. It includes physical and chemical properties arising from its form. A given feature can have multiple functions (see chap. 2). Note that this definition differs from the ecological view of function, which addresses how an organism operates in an ecosystem (Orlandi Laureto et al. 2015).

Faculty: the form-function complex; that is, the combination of a given form and a particular function. Faculty is any particular activity that the form-function complex allows an animal to do.

Biological role: how the organism uses the faculty during its lifetime, in the context of its environment. The same faculty can have multiple biological roles. According to Bock and von Wahlert (1965), the biological role cannot be predicted with certainty from the study of form and function and must be directly observed. Biological role generally corresponds to the concept of life habit.

The differences among the last three definitions become more evident with an example. In mammals in general, the function of the triceps muscle is to produce the extension of the forearm that allows, among other things, the faculty of digging, the biological role of which may be, for example, to build a shelter, seek or obtain food, or dig a burrow (fig. 1.7).

Basic Protocol for Paleobiological Studies Based on Form-Function Correlation

According to the definitions provided in the previous section and following the form-function correlation paradigm, a comprehensive paleobiological study implies an analytical sequence that includes describing form, modeling or inferring the function, formulating hypotheses on faculties, and speculating on biological role (speculating, because an important distinction between extant and extinct organisms is that biological role can be observed directly for the former but not the latter).

Bock and von Wahlert (1965) identified several methodological approaches applicable in the functional analyses of extinct organisms, which they grouped in three conceptual categories: phylogenetic, analogical, and biomechanical.

Given that we frame our design definition in the context of biological evolution, the inferences about function and behavior in extinct organisms should be interpreted through their phylogenetic relationships. Organisms reflect a compromise between adaptation to environmental demands and the tendency for relative stasis (which may indeed have been preserved by selection; see Shanahan 2011), generally termed *phylogenetic inertia* (that could be driven by historical contingency) or *evolutionary constraint* (as a result of genetic, developmental, or even functional constraints). That is, organismal morphology is limited or moderated by its evolutionary history, and it is not structured or designed specifically or only in response to its adjustment to the habitat. In Shanahan's words, "No organism can be re-designed from scratch, but must always begin with the characteristics inherited from its ancestors—characteristics that may limit its species' subsequent evolution" (2011, 67).

From the perspective of design alone, evolutionary constraint can result in suboptimal solutions in response to selection pressures. Thus, the phylogenetic component represents a variable that potentially causes confusion (or "noise") when inferring habits from functional morphology, and its critical and detailed knowledge is essential to understand and measure a functional morphological signal. According to Savazzi (1999), phylogenetic approaches are comparative and inductive ways for interpreting function in extinct organisms that are carried out by comparing homologous structures in extinct and extant organisms (see also Currie 2013). This requires the use of optimization algorithms that map features of the taxa on phylogenies and reconstruct their ancestral state. When the structures and functions are unique (autapomorphies) in extinct organisms, there is no available information about their functions from the phylogenetic frame of reference. Furthermore, in many cases homologous structures (that is to say, shared due to common ancestry) in closely related taxa have different functions. In general, the more distantly related the taxa are, the more likely that the phylogenetically determined relationship between structure and function will break down (Lauder 1995). Today, there are several ways of excluding the effects of phylogeny on our analyses, such as those that seek anatomical characters that covary with habit in nonrelated lineages (taxon free), or methodologies that intend either to identify and remove the effects of the phylogenetic component or to calculate the percentage of variation explained by phylogeny (comparative methods; see chap. 2). According to Blomberg and Garland (2002), the variation explained by phylogeny is termed *phylogenetic signal* (PS) and defined as a tendency for related species to resemble each other more than they resemble species drawn at random from the tree (i.e., an explicit result of an analysis). Nevertheless, according to Shanahan (2011), adaptation and phylogenetic inertia should not be treated as alternatives, because phylogenetic legacy includes previous adaptations.

Similarly, if an extinct lineage develops a phenotype adapted to a particular environmental condition, the phenotype need not necessarily be identical to that developed among extant animals, and nonphylogenetic approaches are therefore required. When suitable counterparts are not found, we look for analogs, preferably biological analogs (Radinsky's form-function correlation); should this fail, then we seek out mechanical analogies. Finally, biomechanics examines the interrelation among biological structures and physical processes, allowing the quantification of functional properties of the former and the evaluation of their effects on the faculty and the biological role. Biomechanics does not require extant homologues or mechanical or biological analogs (Radinsky 1987), but one of them could be suggested as a starting point.

More recently, a basic protocol for paleobiological studies based on the form-function correlation has been described (Vizcaíno et al. 2008, 2016) that identifies three essential biological attributes for each taxon: body size, substrate preference and use (including locomotory mode), and feeding (fig. 1.8). Although these principles have been used for the past four decades (e.g., Andrews et al. 1979; van Couvering 1980; Reed 1998), they have not been fully applied toward understanding the great paleobiological diversity in South America.

Body size is a feature that exerts a notable influence on an animal's life history and processes and can be estimated for extinct vertebrates (Peters 1983; Schmidt-Nielsen 1984; Damuth and MacFadden 1990; Smith and Lyons 2013). Nearly all biological variables are decisively influenced by or correlated with body size, such as metabolism (Kleiber 1932), dimensions of limb bones and locomotion biomechanics (Alexander et al. 1979; Alexander 1985, 1989; Fariña et al. 1997), population density and home range (Damuth 1981a, 1981b, 1987, 1991, 1993; Lindstedt

1.8. Protocol for paleobiological studies: body size, substrate preference and use, and diet or feeding, combined, are the supports permitting our comprehensive vision of organisms in their paleoecological context.

et al. 1986; Reiss 1988; Swihart et al. 1988; Nee et al. 1991), organization and social behavior (Jarman 1974), and tendency for extinction (Flessa et al. 1986; Lessa and Fariña 1996; Lessa et al. 1997).

The morphological study of mandibular and locomotory systems allows interpretation of the movements for which they are optimized. Jaw analyses are useful for formulating a hypothesis about the animal's diet, while analyses of the locomotory system allow inferences on the kind of preferred substrate and the way it is used. Obviously, these two aspects, together with body size, are considered some of the most relevant data for interpreting an animal in its paleoecological context.

The methodological tools we will apply in the following chapters of this book to carry out this protocol are functional morphology, biomechanics, and ecomorphology. Functional morphology analyzes how form causes, allows, or limits the functions that an organism can perform; biomechanics analyzes the relationships between form and function of organisms using physical and engineering principles; and ecomorphology analyzes the variation of form in relation to environment, ontogeny, and phylogeny (Plotnick and Baumiller 2000). For descriptive purposes, we present them in the next chapter as three distinct topics, but in reality there is considerable overlap among these three concepts, and they are all interrelated. An important aspect of these methodological tools is their intersection with morphometry, the measurement or quantification of anatomical features, including bones and teeth. As explained further in chapter 2, there are two main categories of morphometrics: linear morphometrics, which involves lengths (and derivatives such as area) between homologous anatomical points; and geometric morphometrics, in which the coordinates of a set of anatomical landmarks are considered in a bi- or tridimensional space.

Methodological Tools

Biomechanics

Mechanics is the branch of physics that studies and analyzes bodies in movement and at rest under the action of forces. As noted in chapter 1, biomechanics is the application of the principles of mechanics, including concepts and methods of physics and engineering, to organisms. It allows quantification of the functional properties of structures and the assessment of their effects on the biology (i.e., faculty and biological role) of organisms.

Biomechanics has had a long-standing tradition in the history of knowledge. The earliest known treatise on the topic is *De Motu Animalium* (On the movement of animals) by the Greek polymath Aristotle (384–322 BC; fig. 2.1), who viewed animal bodies as mechanical systems. During the Renaissance, the topic was taken up by the Neapolitan intellectual Giovanni Alfonso Borelli (1608–1679; fig. 2.2), who applied his knowledge of physiology, physics, and mathematics to biomechanics. Inspired by Aristotle, he produced his own *De Motu Animalium I* and *De Motu Animalium II*, published posthumously in two parts, in which animals were compared with machines and mathematics was used in testing hypotheses (fig. 2.3). Borelli was the first to suggest that muscles produce movement through contraction of their fibers, and he recognized that forward movement first involves the forward displacement of a body's center of gravity, followed by swinging of the legs to keep balance. In addition, he studied the mechanics of the heart and arteries and used microscopy to investigate

Above, **2.1.** The Greek philosopher Aristotle (384–322 BC) was a logician and scientist born in the city of Estagira in ancient Greece. Plato's disciple and Alexander the Great's teacher, he transformed each area of knowledge with which he dealt, influencing the intellectual thinking of the Western world. Bust engraving in cap and gown with the inscription, "The great Aristotle enquired into the cause of things." National Library of Medicine, Bethesda, MD.

Right, **2.2.** Giovanni Alfonso Borelli (1608–1679) was a Neapolitan physicist and mathematician and is considered a father of biomechanics. Following the advice of Marcello Malpighi, who founded the study of histology, Borelli focused on medicine, in which he applied physical and mathematics laws to biological processes. Lithograph, circa 1803–1839, by Godefroy Engelmann is based on Vigneron's drawing. National Library of Medicine, Bethesda, MD.

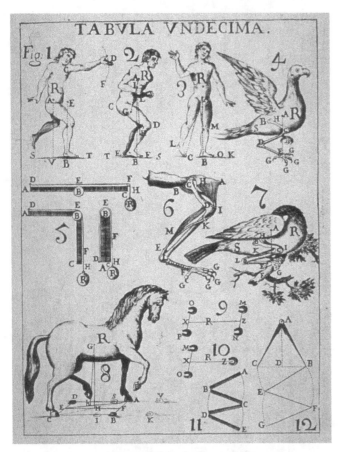

2.3. Print reproduction of a plate from Borelli's *De Motu Animalium*, a work that deals with organisms as compared with machines. It includes humans, birds, the various gaits of horses, and levers and pulleys among its subject matter.

2.4. Galileo Galilei (1564–1642), Italian astronomer, philosopher, engineer, mathematician, and physicist, was born in Pisa, which then belonged to the Grand Duchy of Tuscany. This engraving, *Effigies incognita* by Doménico Cunego (1769), represents Galileo in old age.

the movement of the guard cells regulating the stomata of plants.

Borelli contributed to the modern notion of scientific research, sharing Galileo Galilei's (1564–1642; fig. 2.4) principle of testing hypotheses through observation. Due to his discoveries, Borelli has been recognized as the "father of modern biomechanics" by the American Society of Biomechanics, which established the Borelli Award in 1984 as its highest honor for research in this field.

Biomechanics was established as a recognized discipline and as an area of independent research in the late twentieth century, thanks largely to the efforts of Y. C. Fung (1919–2019; fig. 2.5), who, like Borelli, is credited as a founder of biomechanics based on his research, especially on humans. Biomechanics applied to humans is manifest in several disciplines, such as medical biomechanics (which seeks to minimize the physical effects of pathologies; Lu and Chang 2012), sports biomechanics (which seeks to enhance performance in sports; Blazevich 2007), and occupational biomechanics (which studies the interaction of the human body with elements of working environments; Chaffin et al. 2006).

Below, **2.5.** Yuan-Cheng "Bert" Fung (1919–2019), a naturalized US citizen, was born in Changzhou, China. A bioengineer, he was a pioneer in applying principles of analytical and quantitative engineering to studies of the human body and its diseases, and he contributed, among other things, to the development of artificial skin. Photo courtesy of Jacobs School of Engineering, University of California, San Diego.

2.6. Robert McNeill Alexander (1934–2016), a British zoologist, was born in the city of Lisburn, Northern Ireland. He specialized in the mechanics of organisms, particularly of human and animal locomotion, including the role of elastic mechanisms in running and jumping, scaling of structure and movement in animals, and the mathematical modeling of walking, running, and jumping. Photo courtesy of Professor McNeill Alexander (2015).

In parallel, biomechanics found its place in the huge field of animal biology in 1968, with the publication of the book *Animal Mechanics* by the outstanding British zoologist Robert McNeill Alexander (1934–2016; fig. 2.6), who also contributed to the application of biomechanics in the interpretation of fossil vertebrate paleobiology with his book *Dynamics of Dinosaurs and Other Extinct Giants*, published in 1989. In botany, Karl Niklas (1992) described multiple aspects of the biomechanics of plants in the book *Plant Biomechanics: An Engineering Approach to Plant Form and Function.*

Much of the knowledge derived from biomechanics is based on what are known as biomechanical models, which allow prediction of the behavior, resistance, fatigue, and other aspects of different parts of the body when subjected to certain conditions. The application of biomechanics to paleobiology is similar to the procedure known in engineering design as reverse or inverse engineering, by which each component of an object is analyzed separately to discover how it functions, with the goal of duplicating or improving it. Reverse engineering is applied for inferring the scope of functions that a certain structure could perform.

Some mathematical tools of biomechanics are linear algebra, differential equations, vector algebra, tensor calculus, and numerical methods. Nowadays, it is possible to study very complex phenomena and test and refine theoretical models through computerized simulations (computational biomechanics), with control of many parameters or with the repetition of their behavior.

Some elementary examples of biomechanics research include the study of the forces applied on limbs; the flight aerodynamics of birds, bats, and insects; the swimming hydrodynamics of fish and other aquatic vertebrates; and locomotion in general, in all forms of life, from individual cells to whole organisms. In such cases, straightforward application of Newtonian mechanics provides suitable approaches at each level and is particularly useful in paleobiological reconstruction. However, more sophisticated approaches are also available, such as continuum mechanics (fig. 2.7). The latter proposes a unified model for materials (solids or fluids) that are considered continuous masses; their mathematical description is achieved by application of differential equations, such as in the method of finite elements.

In discussing form and function in chapter 1, we emphasized the need for precision and clarity for the terms used in science, as this is an essential aspect in ensuring unambiguous communication among scientists. In a similar vein, we provide below definitions of terms related to biomechanics used in the examples in this chapter as well as in other chapters of this book.

Definition of Terms and Concepts Relating to Biomechanics

MECHANICS AND RELATED TERMS

Mechanics: A branch of physics that studies and analyses the body in movement and at rest under the action of forces. Two branches of mechanics include:

Statics: that part of mechanics that deals with the forces applied on a body in the absence of changes in motion.

Dynamics: that part of mechanics that deals with the effect that forces applied to objects have on their motion.

MEASUREMENTS AND UNITS

The state of any material or physical system can be described through its physical quantities—that is, physical

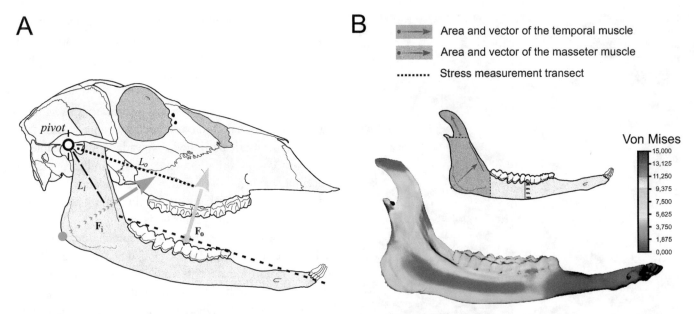

2.7. Methods used in the investigation of the mechanics of the masticatory apparatus. *A*, illustration of the skull and dentary of the Pampas deer (*Ozotoceros bezoarticus*) depicting the application of classical Newtonian mechanics on the masticatory apparatus; **F**$_i$, *vector* along the line of action of the input force of the masseter; F$_o$, output force; L$_i$, input arm; L$_o$, output arm. *B*, illustrations of the dentary of the Pampas deer (*Ozotoceros bezoarticus*) indicating results as modeled by typical continuum mechanics analysis; larger illustration indicates the distortions in the modeled material (i.e., bovine Haversian bone), with *color patterns* resulting from finite elements analysis depicting von Mises stress distribution (describing deformation of the modeled material), according to the methodology of Fletcher et al. (2010).

Within the figure B legend:
- Area and vector of the temporal muscle
- Area and vector of the masseter muscle
- Stress measurement transect

Von Mises scale: 15,000 / 13,125 / 11,250 / 9,375 / 7,500 / 5,625 / 3,750 / 1,875 / 0,000

properties that can be quantified by measurements. Physical quantities can be divided into scalars and vectors.

Scalars have only magnitude (i.e., defined completely by a unique number), are expressed in a determined unit, and lack other attributes (e.g., direction); common examples are time and temperature. In general, scalars are invariant; that is, they do not depend on the chosen system of reference (a convention of space-time coordinates, such as the Earth's surface today). Their abbreviation is usually represented as an unbolded symbol in italics (e.g., t is the abbreviation for time).

Vectors are physical quantities that have at least two descriptive attributes: magnitude (or module) and direction. Force and velocity are examples. They are often abbreviated in bold font or in italicized unbolded font with an arrow printed above the vector symbol (e.g., **v** and $\vec{v}$ are vector symbols for velocity; here, they are represented in bold and unitalicized font). Some vectors, such as force, require a third attribute, called *line of action*, which is the line through the point at which the force is applied and in the same direction as the vector. Vectors depend on the system of reference; thus, it is crucial that the system of reference be clearly defined before any calculations with vector quantities are carried out.

A base quantity is one that is distinct in nature (i.e., it cannot be defined by other quantities). There are seven base quantities, of which length, time, mass, and temperature are those most commonly used in paleobiological studies. Derived quantities, on the other hand, are built using combinations of these base quantities, such as speed, acceleration, force, and momentum, as noted below. The units for these quantities (and others discussed in this book) follow the International System of Units (SI), which is the system used in science. Some relevant physical quantities are presented in the text box "Units" and in figure 2.8.

NEWTON'S LAWS

Newton's laws of motion are three principles that explain most problems related to body movement (fig. 2.9). In Newtonian mechanics (i.e., at scales larger than atoms), force is mathematically defined as the product of the mass of the object and its acceleration (**F** = M**a**). Note that the force units result from the interaction between mass units (kg) and acceleration units (m/s²).

Law of inertia. Each object tends to stay in the same state until a new force acts to cause it to move, stop, or change direction. Inertia is the object's tendency to resist a change from its initial state.

Law of motion. The rate of change in the movement of an object is proportional to the force that acts on it.

Law of action and reaction. For each action between two objects, there exists a reaction of the same magnitude but of opposite direction.

The limitations imposed on these laws by the theory of relativity are mathematically significant only when an

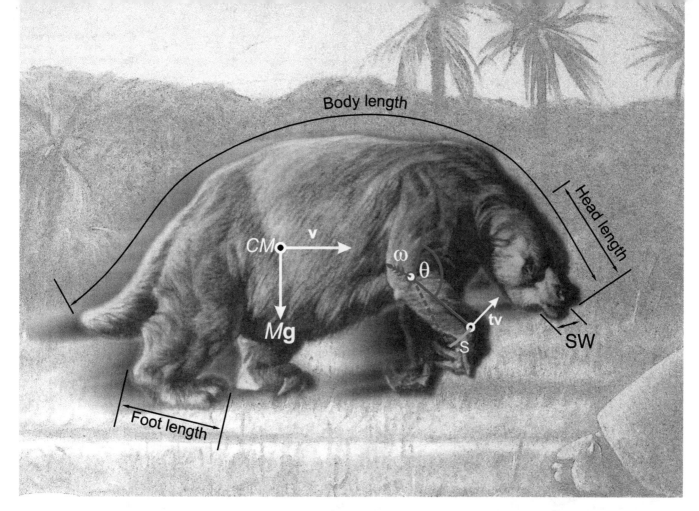

2.8. Physical quantities. Scalar and vector quantities relevant to the displacement of the giant sloth *Megatherium americanum*: *SW*, snout width; head and foot length are measured as rectilinear distances between points; body length is measured along the animal's dorsal profile; **v**, velocity of displacement is measured in meters per second; *M*, mass is measured in kilograms; **g**, acceleration of gravity (9.8 m/s²); *Mg*, resultant of weight acting on *CM*, the center of mass; ω, angular velocity (perpendicular vector projecting out from the plane of the illustration); **tv**, tangential velocity; θ, angular displacement; *S*, length of arc displacement. Drawing based on an old illustration exhibited at the División Paleontología Vertebrados of the Museo de La Plata, Argentina.

UNITS

Length (*l*) is a base physical quantity. It is defined as a concept of distance between points (see, e.g., size and allometry in chap. 5). Biology involves scalar linear measurements such as length, width, height, depth, and diameter. The unit of length is the meter (m).

Time (*t*) is the base quantity referring to the concept of the flow of events. The unit of time is the second (s).

Mass (*M*) is another base quantity. It defines the amount of matter in an object and is an intrinsic property of matter. The unit of mass is the kilogram (kg); like length and time, mass is a scalar.

Density (ρ, the Greek letter rho) is the amount of matter per unit of volume of a given substance (M/V). Its unit of measure is kg/m³.

Force (*F*) is a concept that describes the effect of a body acting on another and is derived from quantities describing mass, distance, and time. The unit of force is the newton (N; $1\ N = 1\ kg\ m/s^2$).

Weight (*W*) is a force measurement and, therefore, a vector quantity. It represents the continuous action of gravity accelerating a body toward the center of the Earth. Being a force, its unit of measure is the N (or kg m/s²).

Center of mass (*CM*) is a virtual point where the total mass of an object could be concentrated for analytical purposes. At rest, it is located at the position where the object is balanced. The center of mass changes its relative position with the changes in form of the body as the animal moves.

Velocity (*v*) is a vector quantity that describes the rate of change in the position of an object over time. It is expressed in meters per second (m/s) and includes magnitude and direction. A related concept is speed (*v*), which is a scalar quantity describing only the magnitude of the rate of change in position over time. It is also expressed in meters per second (m/s).

Acceleration (*a*) is the rate of change in velocity. Its unit of measure is meters per second per second (m/s²).

Law of Inertia

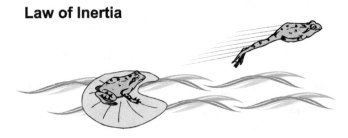

Law of Motion

Law of Action and Reaction

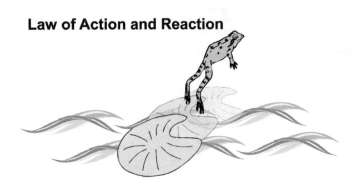

Left, **2.9.** Newton's laws. Law of inertia: inertia is an object's tendency to resist a change in its initial state. A new force can cause it to move, stop, or change direction. On the left, the frog remains static in the absence of any new forces; on the right, once in motion, the frog's movement continues, unless a force causes it to change. Law of motion: the rate of change in the movement of an object is proportional to the forces that act on it; in the frog's case, its movement is proportional to the force applied by its leg muscles and the force of gravity. Law of action and reaction: for each action between two objects, there exists a reaction of the same magnitude but of opposite direction. The force exerted by the frog on the water lily (action force) results in a force of the same magnitude but of opposite direction exerted by the combination of the water and water lily (reaction force) on the frog.

Below, **2.10.** Free-body diagram. Example of a tiger frog (*Hoplobatrachus tigerinus*) descending along an inclined plane with an angle θ. The forces acting in this system are concentrated at the frog's center of mass (*white dot*). For the frog to progress, the force of friction between the adhesive disks of the frog's feet and the plane must be lower than the normal component of the force of the frog's weight.

Facing, **2.11.** Physical systems. *A,* mechanism: with one part (its lower element) fixed, the rest of the components move in a determined direction from the initial position (*white*) to the final position (*dark gray*). *B,* machine: a mechanism that transfers an applied force; in this case, the force produced by the piston is transferred by the linked elements of the machine. *C,* idealized diagram of the mechanism formed by the set of movable elements of the ophidian craniomandibular system; contraction of the pterygoid protractor and elevator muscles results in the application of a force that is transferred through the linked elements and results in outward rotation of the fangs. Based on Liem et al. (2001) and Diogo and Abdala (2010).

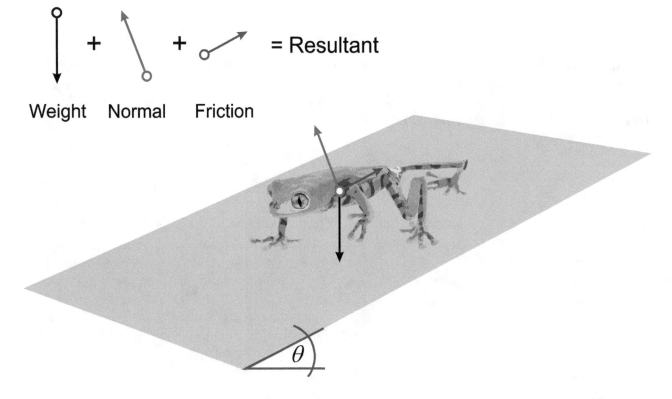

Weight   Normal   Friction   = Resultant

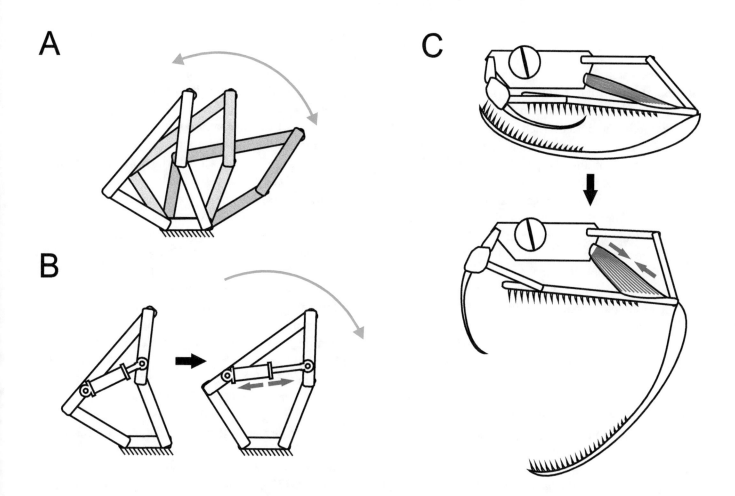

object's velocity approaches that of light. As we may be quite confident that no extant or extinct vertebrate has ever approached this velocity, we may rest assured of the appropriateness of applying these laws to our paleobiological interpretations.

In classical Newtonian mechanics, there is also a concept variably termed *linear momentum*, *translational momentum*, or, more simply, *momentum*. This is a physical quantity that describes the state of an object in movement and is calculated as the product of the object's mass and its velocity at a determined instant (e.g., when a bird flies in a straight line). There is also an *angular momentum* in rotational mechanics (e.g., when a bird circles in a thermal).

FREE-BODY DIAGRAMS

The forces that act on a body can be studied by using free-body diagrams, which indicate all the external forces applied on the body (fig. 2.10). In such diagrams, an object or body is chosen and isolated, with the forces acting on it (such as the rope that suspends a weight or the horse that pulls a wagon) represented by arrows that indicate their direction. It is essential that the forces in a free-body diagram be correctly represented before applying Newton's law of motion, and that the sum of all applied (extrinsic)

forces on the object equal its mass multiplied by its acceleration:

$$\mathbf{F}_{ext} = M\mathbf{a}$$

Indeed, the forces due to gravity and friction must be also represented. If more than one body is involved, a separate diagram is made for each one.

PHYSICAL SYSTEMS

A physical system is a portion of the Universe that is modeled for analysis and consists of an aggregate of objects or material entities where there is a causal link or interaction among its components (fig. 2.11). Among the terms relevant to physical systems are:

Kinematic chain: a representation of a series of linked components or parts that move.

Mechanism: a kinematic chain with restricted movements, so that the movement of an element produces a defined and predictable movement on the other elements of the chain.

Machine: a mechanism for applying or transferring forces.

Levers are machines consisting of a rigid bar with a point of support (fulcrum or pivot) and at least two forces present: one to be overcome, the load or resistance (usually a weight to be supported, lifted, or moved), and one, the effort, exerted to perform the desired action. The distance between the pivot and the position along the rigid bar where each force is applied is called a lever arm. Thus, a given arm corresponds to each force. The perpendicular distance from the direction of force (termed the *line of action* of the force) to the pivot is called the moment arm. There is an input and output moment arm (fig. 2.12). When the force is orthogonal to the rigid bar, then the lever arm and the moment arm are the same. But when the force is applied obliquely with respect to the bar, the lever arm and the moment arm are different. In this case, the moment arm must be calculated from known data (bar length and angles) using trigonometric equations. Each of the forces applied to the lever produces a rotational component known as torque or moment of force (or simply moment). This is defined mathematically as the product of the force (F) and length (l) of the moment arm; it is a vector quantity (because force is a vector) and measured in Newton meters (Nm).

Mechanical advantage: is the ratio (F_o/F_i) of the output force (F_o) to the input force (F_i) acting on the system. A lever that is optimized to provide an output force greater than the input force has a mechanical advantage greater than 1.

Velocity quotient: is the relationship ratio (l_o/l_i) of the input moment arm to the output moment arm. A lever optimized for providing higher velocity has a velocity quotient greater than 1.

When the system is at equilibrium, the torques of the input force and the output force are equivalent in magnitude but opposite in direction; the lever, therefore, does not rotate. Mathematically, it corresponds to the expression $F_o \, l_o = F_i \, l_i$ or its equivalent $F_o/F_i = l_i/l_o$. It is evident from this that the mechanical advantage is equivalent to the inverse of the velocity quotient. This last expression is known as the law of the lever, which was demonstrated geometrically by Archimedes and later mathematically by Galileo. Because the velocity is always measured as distance over time and the arc traveled by the bar depends on the arm length, long output moment arms and short input moment arms favor velocity, whereas short output moment arms and long input arms favor force.

There are different kinds of levers (fig. 2.12):

First-class levers: the pivot is located between the input force and the output force (load or resistance). The mechanical advantage can be any value greater than 0. This kind of lever can be optimized to the degree desired by lengthening the moment arm of the input force with respect to that of the output force (at the expense of a decrease in the angular velocity of the system). Examples are the systems formed by the head and nuchal muscles, with the first cervical vertebra as pivot, and by the forearm and triceps, with the elbow joint as pivot.

Second-class lever: the load or resistance is located between the input force and the pivot. This kind of lever always produces a mechanical advantage greater than 1, and so is very efficient for force enhancement, though not for velocity. An example is the system formed by the foot and the gastrocnemius muscle, with the ankle as pivot.

Third-class lever: the input force is located between the load (resistance) and pivot. The mechanical advantage is always less than 1 and the velocity quotient greater than 1.

FORCES ACTING ON MATERIALS

Compression occurs when two forces act on the object from opposite directions toward the object's center. The particles of the object approach each other, tending to make the material more compact.

Tension occurs when two forces act in opposite directions away from the object's center. The particles move away from each other, resulting in stretching of the material.

Flexion occurs when forces applied on the object tend to bend it. This produces a combination of compression on one side of the object and tension on the other: while one side of the material is compressed, the other is stretched. A beam that is bent undergoes compression and tension on opposite sides.

Shear occurs when forces are applied in opposite directions at different planes of the object, so that the material's particles tend to move over each other, which often produces a break.

Torsion occurs when rotational forces in opposite directions at each end of the object tend to twist it on its central axis.

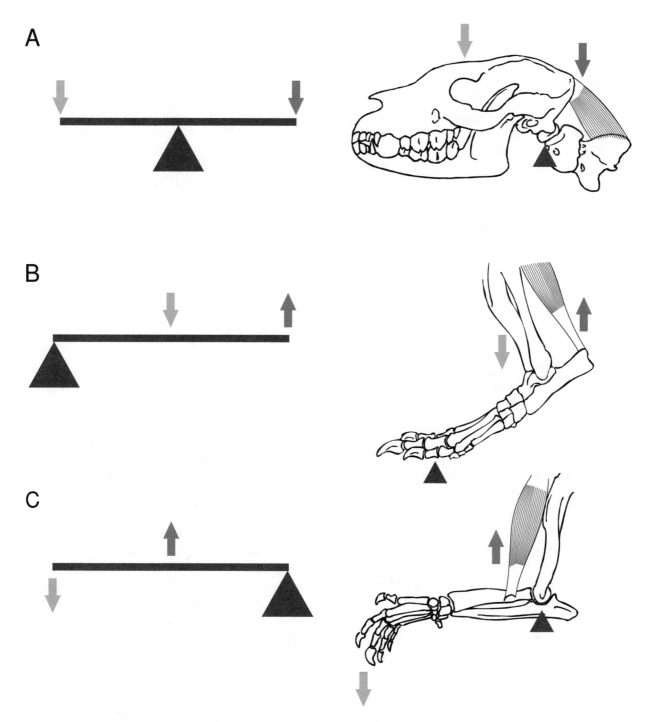

2.12. *A*, first-class lever, with the pivot (*black triangle*) situated between the input force (*red arrow*) and the resistance (*blue arrow*); example, system formed by the skull and first cervical vertebra, with the pivot between the skull and vertebra, skull weight as resistance, and force produced by the contraction of the nuchal muscles as the input force. *B*, second-class lever, with the resistance located between the input force and pivot; example, system formed by toe support on the ground as pivot, the force of body weight as the resistance, and the force applied by the gastrocnemius muscle as the input force. *C*, third-class lever, the input force applied between the pivot and resistance; example, system formed by the elbow joint as pivot, the weight of the forearm as resistance, and the force applied by the biceps muscle as the input force.

This kind of lever optimizes velocity and is less efficient for force. Examples include systems formed by the forearm and the biceps muscle, with the elbow as pivot, and by the mandible and the set of masseter, pterygoid, and temporal muscles, with the articular condyle as pivot.

FORCES ACTING ON MATERIALS

The characteristics of the materials that constitute the objects may allow them to be deformed or even broken in response to applied forces. The kinds of stresses (fig. 2.13; the Greek letter sigma, σ, is the symbol for stress)

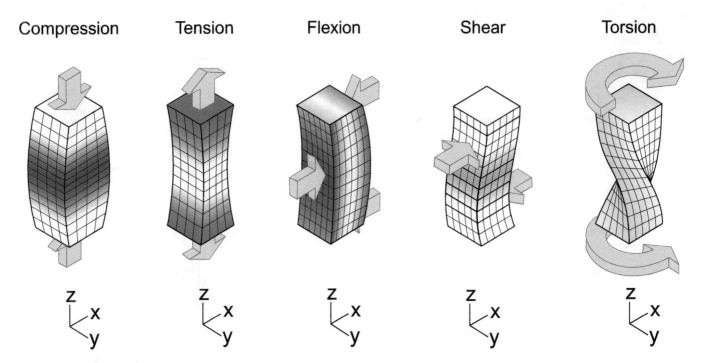

| Compression | Tension | Flexion | Shear | Torsion |

2.13. Forces acting on materials. The areas in *blue* indicate the approach of particles to each other and concentration of compressive forces; in *red*, areas of particle separation and concentration of tensile forces; in *yellow*, displacement of particles over each other; the *yellow-green gradient* in torsion indicates areas of displacement and a combination of forces.

that the elements of the structures must withstand are compression, tension, flexion, shear, and torsion, defined in the text box "Forces Acting on Materials" and in figure 2.13. Each of these forces reaches a maximum value in a given material just before it ruptures. These maxima are experimentally calculated for each type of material. Most materials are better able to withstand compression than tension and shear. The repeated and prolonged use of a material can cause localized structural damage known as *fatigue*. Under such circumstances, the stress values that produce a fracture are much lower than is usual for a given material, and a spontaneous failure (i.e., rupture) occurs due to the appearance and propagation of microfractures.

THEORY OF TRAJECTORIES

When an object bears a load, the material that composes it transports the resulting internal stress along trajectories or stress paths, from molecule to molecule. A beam attached by its base to a wall (i.e., a cantilever) bends under its own weight; while the beam's lower surface withstands compressive forces, its upper surface withstands tensile forces. They are greatest at the surfaces and decrease to the center of the beam (i.e., the neutral surface). The resulting compressive and tensile forces are transported, within the substance of the beam, across trajectories that cross at right angles and are concentrated near the beam surfaces (fig. 2.14).

Examples of the Application of Newtonian Mechanics and Continuum Mechanics

MECHANICAL ADVANTAGE

In the paleobiology of tetrapods, it is often quite straightforward to apply Newtonian mechanics to those parts of the skeleton that work as lever systems, provided that the musculature has been reliably reconstructed. For instance, among mammals, armadillos (Xenarthra) are characterized as being good diggers. Digging is performed by scratching the soil with their strong claws by means of powerful forearm extension. Thus, the forelimb acts as a machine that is useful for transferring the force produced by its musculature to the substrate. Although the complete movement also involves other limb segments, we may analyze the action of the forearm through a free-body diagram that considers the ulna (the longest bone of the forearm) as a lever with its fulcrum at the articulation with the humerus. The input arm is represented by the olecranon, the part of the ulna onto which the triceps muscle inserts. This muscle pulls (input force) on the olecranon, producing a torque (moment) that results in forearm extension. The output arm is represented by the portion of the ulna distal to the articulation, which transmits the output force to the hand that will act on the substrate. Within the diversity of extant armadillos, the ratio of the length of the olecranon to that of the ulna distal to the

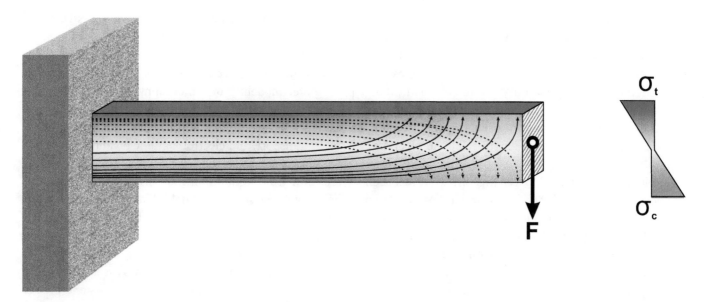

2.14. Theory of trajectories. The force **F** applied on the end of the beam is the load due to the beam's weight, producing flexion. The *lines* show the component distribution of symmetrical stresses. *Solid lines* indicate arcs of compressive stress (σ_c, in *blue*) and *dotted lines* indicate arcs of tensile stress (σ_t, in *red*). The trajectories of resulting internal compressive and tensile forces cross at the beam's end. The beam axis (*yellow*), where the material is not subjected to stress, is the neutral axis. The stress intensity that the material must withstand increases linearly from the neutral axis in upward and downward directions.

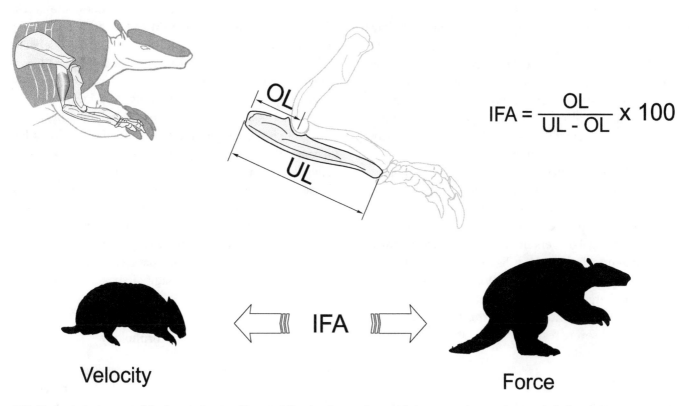

$$IFA = \frac{OL}{UL - OL} \times 100$$

2.15. Mechanical advantage of the forearm in armadillos, modeling the ulna as a lever, with the pivot at the articulation with the humerus (*top* images). The triceps muscle provides the input force that extends the forearm. The index of fossorial ability (IFA; also referred to as the olecranon index; see text) indicates the mechanical advantage of the triceps: OL, olecranon length; UL, ulna length; UL-OL, the length of the ulna distal to the articulation. The illustrations at the *bottom* indicate a range of IFA values; low IFA values (to the *left*) occur in an ambulatory armadillo (three-banded armadillo, *Tolypeutes matacus*) that digs occasionally; high IFA values (to the *right*) improve the output force in the more powerful digging armadillos (such as the giant armadillo, *Priodontes maximus*).

articulation (i.e., the mechanical advantage) is higher in species that dig more frequently and/or intensely. An index of fossorial ability (IFA; Vizcaíno et al. 1999; Vizcaíno and Milne 2002) may be calculated as a means of assessing differences among mammals in the mechanical advantage of the triceps muscle during forearm extension. IFA is expressed as the olecranon length (OL) divided by the difference between ulna length (UL) and olecranon length (UL-OL is the length of the ulna distal to the articulation) or OL / (UL-OL) × 100 (fig. 2.15). IFA has been applied to Miocene and Pleistocene armadillos (Vizcaíno et al. 2003, 2006b, 2012b). This relationship, however, was first introduced by Vizcaíno et al. (1999) to investigate the proficiency of digging among armadillos by considering the force applied at the end of the forelimbs, but the relationship may also be used to explore additional functional trade-offs—that is, its performance in faculties other than digging (e.g., locomotion in general). For example, a relatively shorter olecranon may increase the forelimb's velocity for improved running speed or flexibility for improved climbing ability; a relatively longer ulna is found in gliding mammals compared to closely related nongliding arborealists, likely increasing the surface area of attached patagia (Thorington and Santana 2007; Chen and Wilson 2015; Meng et al. 2017; Grossnickle et al. 2020). The index thus has implications beyond digging proficiency and may be more appropriately referred to as the olecranon index (OI).

It is important to point out here that two aspects already noted in chapter 1, body size (i.e., allometry; see chap. 4) and evolutionary constraint, should be considered in the formulation of paleobiological hypotheses. The studies performed on armadillos allowed the influence of body size on the index to be ignored (Vizcaíno and Milne 2002). This index is also effective in identifying good diggers in mammalian groups other than armadillos. However, even though the most specialized diggers within each group have the highest IFA values for a particular group, the most specialized diggers in these groups (e.g., Carnivora, Rodentia) have lower values than seen in armadillos that are less specialized for digging (Vizcaíno et al. 1999; Vizcaíno and Bargo 2021). This indicates, or at least strongly suggests, that the ancestrality of this habit in armadillos has imposed severe restrictions on subsequent modifications (Vizcaíno et al. 1999).

FINITE ELEMENTS

Finite element analysis (FEA) is an engineering approach initially developed to simulate and predict stress/strain distributions in human-made objects with relatively simple geometry. With advances in computer technology, FEA has emerged as a powerful tool in the investigation of mechanical behavior in complex biological structures, its nondestructive nature making it of particular value in paleontological studies. FEA is based on the method of subdividing a digitized solid (obtained, for instance, by computer modeling or tomography) into cubic units or "blocks." The calculation of the physical variables of interest, such as structural analysis, heat transfer, fluid flow, mass transport, and electromagnetic potential, is carried out separately for each of these blocks. Then, the variables are unified in a quantitative description of the physical behavior of the solid as a whole. The set of parameters to which the model is subjected (e.g., force, stress, acceleration) is determined a priori by the researcher based on clearly defined criteria. An approach of finite elements in tridimensional models of skulls allows assessment of the distribution of zones of the skull that withstand greater stress and the formulation of explanatory hypotheses for this distribution. Degrange et al. (2010) applied this procedure and investigated strategies of prey capture of an extinct bird, the phorusrhacid *Andalgalornis* (body mass estimated at 40 kg) (fig. 2.16).

Based on regression analysis, Degrange et al. (2010) estimated bite force at the beak tip as 133 N. The results obtained from analysis of finite elements of a red-legged seriema (*Cariama cristata*, a gruiform bird, one of the closest living relatives of *Andalgalornis*) and an eagle (*Haliaeetus albicilla*) showed that the *Andalgalornis* skull underwent relatively high tensions when subjected to lateral loads, and low tensions when the force was applied dorsoventrally (sagittally) and in simulations of backward pulling. Given that an animal, for its own physical integrity, would not be expected to partake in potentially dangerous behaviors, Degrange et al. (2010) considered it unlikely that *Andalgalornis* fought with its beak in subduing large prey, because this would subject the beak to varied forces, including lateral loads that would have caused high tensile strain. Therefore, they proposed that either it consumed smaller prey that could be killed and consumed safely (e.g., by being swallowed whole) or that it used the beak to deliver multiple well-targeted sagittal strikes to reduce the prey to ingestible portions.

Functional Morphology

In this book, we define *functional morphology* as a causal, comparative, and inferential approach that operates at the interface between form and biology, where function is established. Functional morphology, then, involves the study of relationships between the form or anatomical design of a biological structure and the group of functions that it can perform. We have already mentioned that this

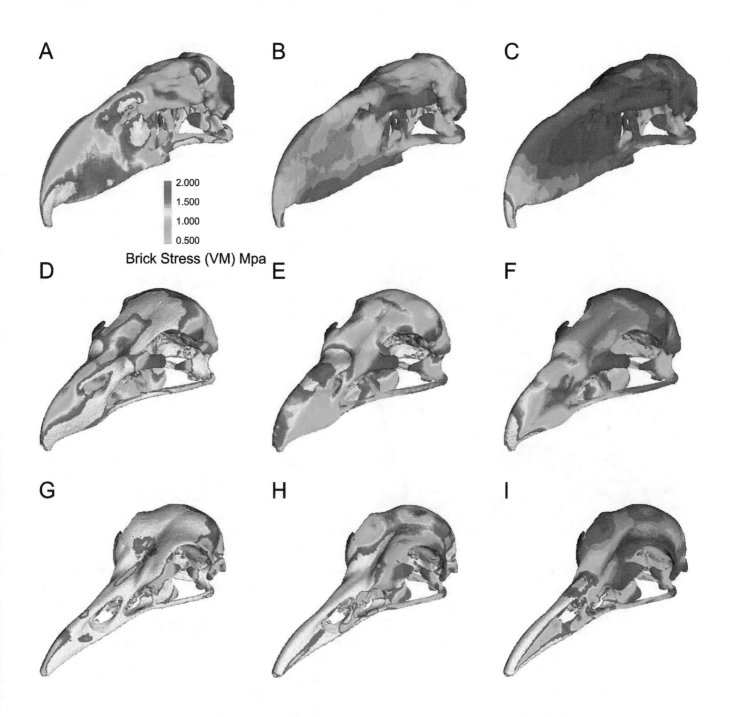

2.16. Finite elements. *A–C*, the phorusrhacid *Andalgalornis steulleti*. *D–F*, the eagle *Haliaetus albicilla*. *G–I*, the seriema *Cariama cristata*. Stress distribution under three simulated load conditions during prey capture: lateral shakes (*A, D, G*), normal bite (*B, E, H*), and backward pulling (*C, F, I*). The *colors* indicate von Mises stress distribution (VM), a magnitude proportional to material distortion, in megaPascals (MPa, an SI unit of pressure for stress). According to the provided color scale, *blue* indicates the lowest deformation values and *red* the highest. The areas in *white* indicate VM values that exceed the maximum of 2 MPa. Modified from Degrange et al. (2010).

was conceptualized by Radinsky (1987, 8–10) in what may be defined as a "paradigm of form-function relationship" (Vizcaíno and Bargo 2021). That is, functional morphology involves developing explanatory hypotheses of how form allows and/or restricts the function or functions relevant to the biology of the organism. For many authors, reconstructing the relationships between form and function is a basic step in understanding the way of life of extinct and extant organisms (Betz 2006; Benton 2010; Vizcaíno and Bargo 2021).

In 1984, the British primatologist Charles E. Oxnard (fig. 2.17) pointed out advantages of this kind of inferential approach to structure and function, of which the most relevant for us are:

1. It allows much wider analyses of different anatomical regions and diversity of animal groups than could be achieved with an experimental design.
2. It is able to deal with rare, incomplete, and fragmentary specimens, for which an experimental approach is impossible (especially important in paleontology).
3. It requires only the study of museum specimens already collected, without interfering with living species, many of which are currently threatened.

2.17. Charles Oxnard, a British primatologist living in Australia, was born in Durham, United Kingdom, in 1933. His research ranges from anatomical dissection and the mathematical modeling and engineering of anatomical structures, to Fourier transformations, Lagrange's analysis, and finite elements applied to bone architecture and biomechanics. Photo courtesy of Charles Oxnard (2015).

In paleobiology, as noted in chapter 1, function is not verifiable through direct observations, and form is but the starting point for morphofunctional inference. In most cases, only the form of the preserved hard parts, such as bones and teeth, is available. However, it is possible to obtain additional information about the form of soft parts through the morphology of hard parts (we will refer further to this in this chapter).

Being a comparative discipline, functional morphology in paleobiology compares extinct taxa with extant models with known functional attributes. Sometimes, however, no extant organism is sufficiently similar to be used as a model; in such cases, mechanical analogs may be employed. As such, functional morphology as practiced by paleobiologists incorporates elements and concepts of

biomechanics and ecomorphology, differing from these disciplines in its mainly qualitative approach. But this does not mean that quantitative data are not used in paleobiological analyses. Nonetheless, the inference of function from form generally follows a qualitative approach.

The relationship between form and the complex of functions that the form can accomplish can be summarized through an "averaged biomechanical situation," which was defined by Oxnard (1984) as the integration of the functional information in a mechanical profile of each component (e.g., a limb). The loads that a given part of an organism can be subjected to may differ not only in magnitude but also in duration, depending on the activity and its intensity and frequency. Activities that are infrequent but of high intensity can have a greater impact on the design of this body part than activities that are more frequent but of lower intensity. Such a mechanical profile accounts for the functional capacities of the object under study, be it an isolated skeletal element, limb, body segment, or even the entire organism (fig. 2.18).

Practical Manual and Technological Approaches

Functional morphology is strongly rooted in a deep knowledge of the anatomy of forms under study. We emphasize here the value of working directly with specimens, including both the fossils under study and the extant organisms used for comparative purposes (fig. 2.19). Specimens must often be observed carefully under different types of lighting and with the aid of hand and binocular magnifiers. In some cases, the use of raking light helps to reveal surface topography otherwise difficult to discern. The specimens should be handled (to the degree that their preservational state permits), because touch may reveal details that are difficult to distinguish visually. Drawings, diagrams, and personal notes are practically irreplaceable tools in helping to understand the anatomy of a structure, as is the use of clear and agreed-upon terminology for unambiguous communication with other researchers.

There are alternative ways of approaching form, especially when the aim is to distinguish structures and internal details that are difficult to see or when dealing with specimens whose fragility discourages handling. Latex or silicone casts can be made, for example, of internal spaces such as the cranial cavity (Paulina Carabajal and Canale 2010), as these materials are sufficiently flexible to be removed readily from such internal cavities. Bargo et al. (2009) reconstructed the masticatory movements of the Miocene sloth *Eucholoeops* by means of qualitative analysis of the wear stages of the teeth. As part of their study, they used resin casts of the teeth (made from molds of

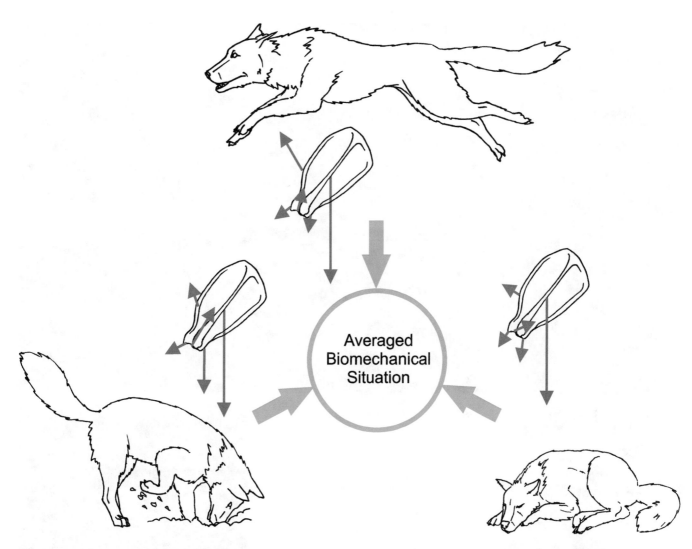

2.18. Averaged biomechanical situation. The *vectors* reflect the different combination of loads (forces, in this case exerted by muscles and body weight) to which the analyzed structure (in this case the scapula) is subjected during the diverse activities the organism performs.

the original specimens) for visualizing more clearly the possible masticatory movements. As just noted, it is often necessary to incorporate direct psychomotor experience, manipulating the elements (or their casts) to better understand their possible movements and restrictions. Bargo et al. (2009) also used morphological information of soft-part structures (i.e., muscles) as inferred from the skeletal elements to complement understanding of function, as well as quantitative estimates of body mass to contextualize the functional hypotheses.

In addition, devices more sophisticated than hand magnifiers and gauges are available for understanding and describing form; for example, X-rays and, more recently, computerized tomography, which are particularly suitable in the reconstruction of the brain and cranial nerves. Tomographic digitization and editing allow the creation of "virtual models" of the brain and nerves and of other internal skull structures (Fernández and Herrera

2009; Fernicola et al. 2012; Bona et al. 2013; Tambusso and Fariña 2015; Boscaini et al. 2020), and they can complement the information supplied by internal natural molds (Herrera et al. 2013). The laser scanning of surfaces is also being used to produce digital casts (Muñoz et al. 2017, 2019; Muñoz 2021) and virtual models of elements or their parts can be 3-D printed into polymer casts. Also, photogrammetry has been demonstrated to be an affordable alternative technique (see Otero et al. 2020 and references therein).

Inferring Function from Form: Comparative Models and Functional Analogs

There are several ways of inferring function or functions from the analysis of form. One approach is to make use of morphofunctional knowledge of extant organisms, as suggested by Radinsky (1987), analyzing the extent to which

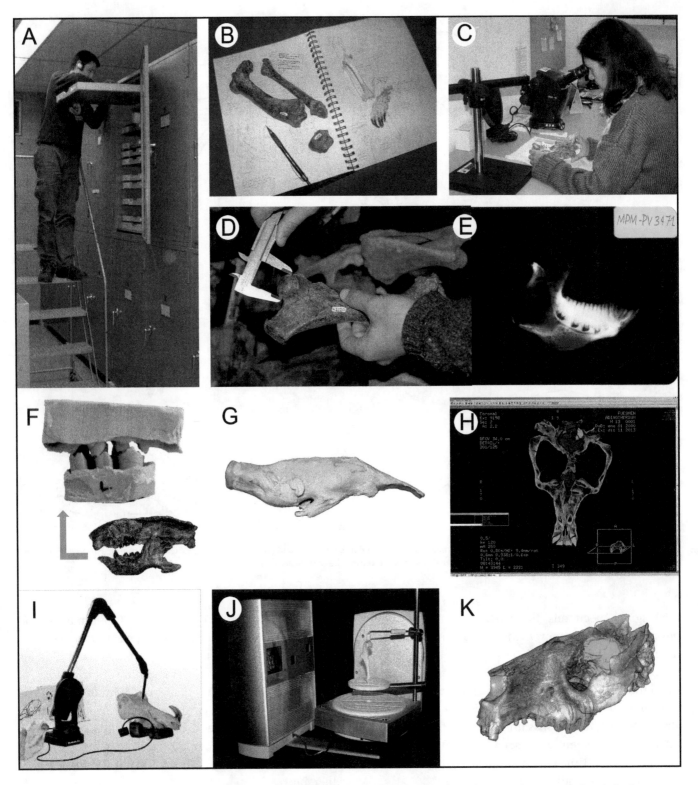

2.19. Practical approaches to the study of morphology. *A*, study of specimens in collections. *B*, traditional comparative anatomical analysis. *C*, specimen observation with a binocular magnifier. *D*, using calipers to measure a glyptodont femur. *E*, use of X-ray imaging to study tooth characters of *Interatherium*, a Miocene notoungulate. *F*, use of a tooth cast of the extinct sloth *Eucholoeops* for studying masticatory mechanics. *G*, silicon mold of the cranial cavity of the extant crocodilian *Gavialis gangeticus* (courtesy of Paula Bona). *H*, CT scan to reveal the cranial cavity of the Miocene notoungulate *Adinotherium*. *I*, MicroScribe G2L digitizer arm used for acquiring 3-D coordinates. *J*, surface laser scanner NextEngine Desktop 3D. *K*, in *green*, digital mold of the cranial cavity of *Adinotherium* obtained through CT scan.

there is morphological similarity between the extinct taxon under study and the extant taxa for which functional properties are known. The premise is that a similarity in form implies a similarity in function. For instance, in a study on two Paleocene fossil marsupials, *Pucadelphys* and *Mayulestes*, Argot (2001) looked for causal associations between functional-anatomic patterns of the forelimb and habits in extant didelphids (arboreal, generalized, and terrestrial opossums) taken as models to understand the functional meaning of the main bony morphological features. This knowledge was then applied to make functional inferences for the extinct taxa based on their morphology. This study considered the greatest muscular differences among living species with different locomotory habits to functionally explain several morphological features observed in fossil skeletons.

Quantitative data can also be employed (e.g., body size, indices, proportions) to facilitate comparisons (Fernández et al. 2000) and to apply statistical tests for distinguishing more clearly the degree of morphological difference among taxa (Sargis 2002a, 2002b). This approach is based mainly on the correlation of function between homologous structures. In this regard, functional morphology replicates the logical basis of quantitative ecomorphology. The functional models should be extant organisms chosen by the researcher based on clearly established criteria such as phylogenetic kinship, equivalent body size, and general morphological similarity, among others. For instance, Candela and Picasso (2008), when studying the extinct Miocene South American porcupine *Steiromys*, used not only its closest extant relatives (both South and North American tree porcupines) as models for functional inference, but also other, more distantly related taxa (e.g., the Brazilian guinea pig, the Patagonian mara, and the plains viscacha) that represent a wider functional diversity for enhancing the scope of the comparison.

In other situations, the function of a feature of an extinct organism is directly inferred from its material attributes, based on the knowledge of well-known extant organisms. For instance, it can be hypothesized that the femur of an extinct rhinoceros will have the same mechanical properties as the femur of an extant rhinoceros, and these inferred attributes can be used as the basis for elucidating femoral functions in the extinct animal. In this regard, functional morphology adopts a more experimental format, closer to biomechanics.

When there are no homologous structures for comparison (e.g., in the case of organisms belonging to a lineage without extant representatives) or in cases where weak morphological resemblance casts doubt on any functional similitude, it is necessary to search for alternative functional models (Vizcaíno et al. 2018). These will generally be extant organisms that are sufficiently similar structurally to the extinct taxon under study to allow proposing a hypothesis of functional similarity. For instance, Rudwick (1964) detailed the sequence of logical inferences that might stem from the analysis of the anterior limb of a pterosaur (fig. 2.20) with that of crocodiles, which were then considered their closest living representatives (see below). Little could be inferred about its function using crocodiles as functional models. What functional inferences, after all, could be made, given the vast differences in structure of the pterosaur and crocodile forelimbs? Using birds and bats as functional models might prove more fruitful; Rudwick could not use a homology criterion for inferring function, but he was able to make headway based on general physical principles common to all flying animals. This example is nowadays outdated because more recent phylogenetic hypotheses place archosaurs, to which birds belong, and pterosaurs as sister groups, with crocodiles as outgroup to both of them (see Nesbitt 2011). However, this does not alter the heuristic value of Rudwick's argumentation, because birds and pterosaurs evolved flight independently, and the comparison between them is still based on functional similarities related to flight rather than on homology. In proposing hypotheses of functional similarity, Gould (1993) echoed the notion of the usefulness of considering extant organisms that are sufficiently similar structurally to the extinct taxon under study. This author reasoned that functional assumptions on ichthyosaur fins based on the limbs of lizards, ichthyosaurs' closest extant relatives, would be of very limited heuristic value. By contrast, however, analogy-based comparisons with whale flippers result in more explanatory functional hypotheses.

Restricting comparison by resorting purely to homology and actualism runs the risk that a reconstructed extinct organism results in a kind of chimera, an entity embodied from an averaged agglomeration of features from extant relatives, which Pagel (1991) called an "average animal." Vizcaíno et al. (2018) performed comparative exploratory analyses to assess the morphometric similarity between extant and fossil xenarthrans, concluding that the use of extant species as the only models could lead to biased estimates of functional features of extinct forms.

Mechanical Analogs: Rudwick's Paradigmatic Analysis

In special cases, when morphology is so radically different from that of any extant organism that it renders its use as a plausible model highly suspect, it is necessary

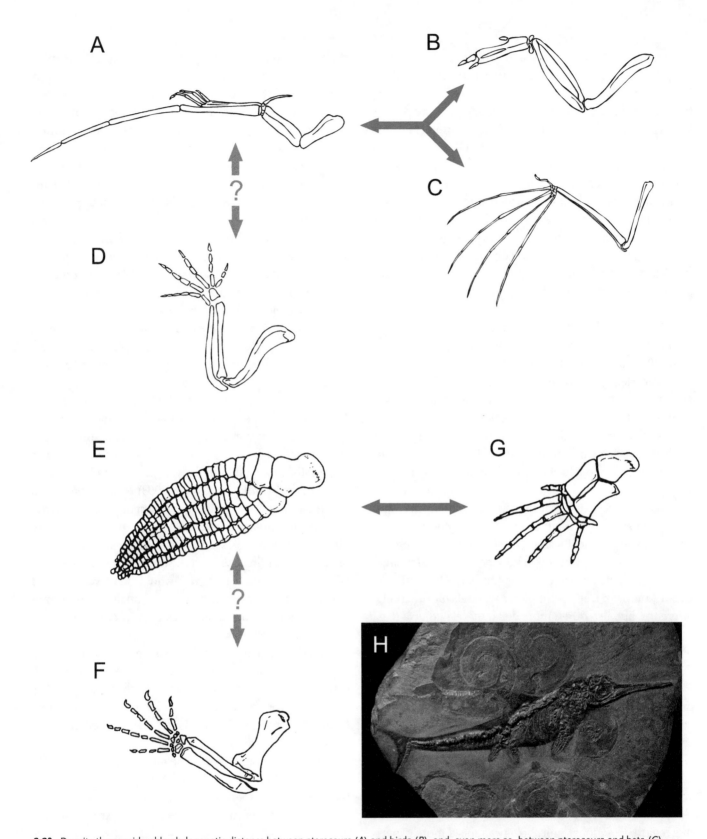

2.20. Despite the considerable phylogenetic distance between pterosaurs (*A*) and birds (*B*), and, even more so, between pterosaurs and bats (*C*), the forelimbs of birds and bats serve as more useful functional models. The structural similarities among the three kinds of wings reflect the basic mechanical requirements of flight, rather than common ancestry. Although pterosaurs evolved a winglike structure independently from that of birds and bats, the forelimb of these last two vertebrates is still more suitable as a functional model than the forelimb of the common ancestor of birds and, *D*, crocodiles (birds' more recent living relatives). Another example of functional analogs using ichthyosaurs: despite being much more phylogenetically distant from ichthyosaurs (*E*) than lizards (*F*), the forelimb, or flipper, of modern cetaceans (*G*) is more useful as a functional model, showing analogous (and not homologous) specializations. *H*, an exceptionally well-preserved *Temnodontosaurus* ichthyosaur from the Lower Jurassic of Hölzmaden (Germany) showing the forelimb, as well as the outline of the body with clear dorsal and caudal fins (figure modified from https:// commons.wikimedia.org/wiki/File:Harpoceras_%26_Ichtyosaure_(p).jpg).

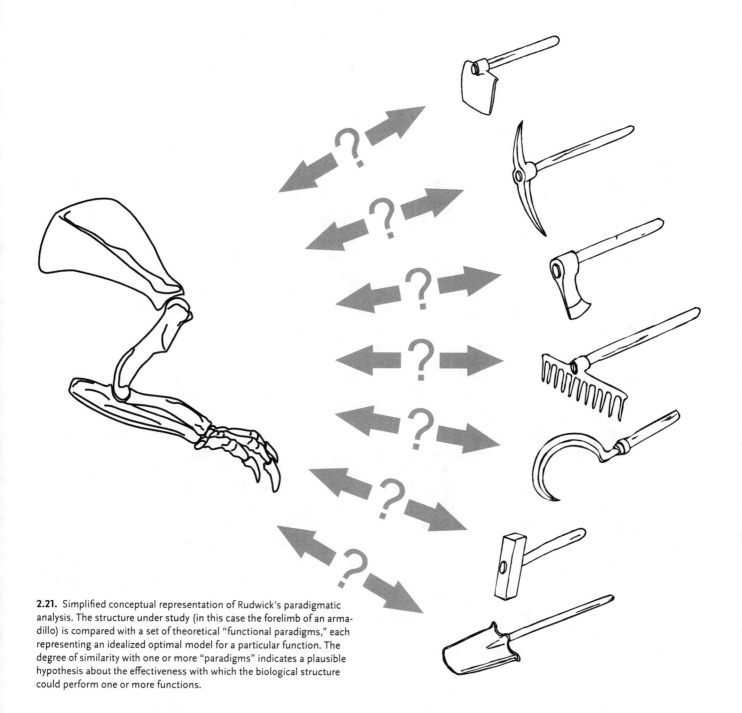

2.21. Simplified conceptual representation of Rudwick's paradigmatic analysis. The structure under study (in this case the forelimb of an armadillo) is compared with a set of theoretical "functional paradigms," each representing an idealized optimal model for a particular function. The degree of similarity with one or more "paradigms" indicates a plausible hypothesis about the effectiveness with which the biological structure could perform one or more functions.

to appeal to mechanical analogs, as in the paradigmatic analysis by Rudwick (1964). This author developed a protocol of comparative functional analysis in which several functional hypotheses for a particular structure of an extinct organism under study are proposed. The next step involves developing ideal hypothetical structures that are capable of carrying out the proposed functions. These model structures are called "functional paradigms," and they are created by the researcher on the basis of physical principles (fig. 2.21).

It is worth remembering here that living organisms may be described as chemical machines as well as mechanical

machines, and thus the function of a structure is not necessarily tied only to its physical form. Amundson and Lauder (1994), for example, added thermoregulation to the list of functional hypotheses for pterosaur wings. The last step of paradigmatic analysis is the comparison between these ideal analogs and the real structure. The existence, however, of an exact analogy with only one of the "paradigms" should not be expected, because structures in living creatures generally perform several functions, and their form is a compromise solution between the demands of those functions and the restrictions imposed by heritage and life history. For example, the forelimbs of large felids could not become specialized for running to the same extent as in other mammals, such as the antelopes and deer upon which they prey, because felids must also perform critical functions in prey capture and manipulation, as well as in climbing, with their forelimbs. Thus, the forelimbs of felids are suboptimal designs from a perspective focused exclusively on running performance. In addition to this, as analyzed by Wainwright et al. (2005, 256) for fish skulls, different designs can render similar performances for a given function ("many-to-one mapping of form to function"), which can be especially true for complex biomechanical systems; hence there can be more than one "paradigm" for any given function. For this reason, similar functions (and faculties enabled by them, such as flying or digging) can be performed by structures with different designs. However, Rudwick's (1964) method is an interesting and useful approach, because the exercise itself of hypothesizing ideal structures for a given function can be fruitful in that it may help to broaden the spectrum of hypotheses by revealing design solutions not evident in our initial understanding of the problem.

Often, although without explicit acknowledgement of the use of Rudwick's protocol, a researcher may analyze a particular feature of a fossil through biomechanical methods combined with a comparative search for functional hypotheses that would best fit the performance of the structure. This methodology is especially evident in the morphofunctional analyses of articulations in mammals by, for example, Argot (2001, 2002), Sargis (2002a, 2002b), Candela and Picasso (2008), and Toledo et al. (2013).

Research can also refer to mechanical analogs, simply by comparing a particular structure, say the pterosaur forelimb, with a series of functional paradigms; that is, structures ideal for the promotion of, for example, flapping flight, gliding, propelled swimming, and capture. Rudwick (1964) pointed out that the limit to the number of functional inferences we could make is not so much in the capabilities of the extinct organisms under study as in the limits of our understanding of the operational principles of physics.

Phylogenetic Brackets for Reconstructing Soft Tissues

Carrano and Hutchinson (2002) performed a detailed reconstruction of the pelvic and hind-limb musculature of the theropod dinosaur *Tyrannosaurus* based on a protocol proposed by Witmer (1995; although Bryant and Russell 1992 had already proposed the same idea) called

EPB (extant phylogenetic brackets; fig. 2.22). The muscular reconstructions allowed them to carry out a series of functional inferences, such as action lines of muscles and relative mechanical advantages, through the qualitative application of biomechanical concepts. Here, the importance of the reconstruction of soft tissues is highlighted as the source of necessary information for the functional inferences.

The EPB method uses extant taxa, those phylogenetically closest to the fossil under study (the "brackets" within which the fossil is encompassed; see fig. 2.22), for analyzing the relationship between soft tissue and an associated feature of a skeletal element, using the parsimony criterion on a cladogram (a tree representing phylogenetic relationships). First, the relationships between a given osteological correlate and the associated soft feature to be reconstructed (a muscle scar and the muscle that attaches to it, for instance) are determined in extant taxa used as "phylogenetic brackets." It is assumed that these bone feature–soft feature relationships have been inherited from a common ancestor; that is, they are homologous. Then, the condition for the bony correlate is evaluated in the extinct taxon under study. Afterward, the associated soft tissue optimization by parsimony is carried out along the tips and nodes of the cladogram, with assessment of three possible states: present, absent, or ambiguous (in general, optimizing only three taxa is easy, without need to resort to cladistic software). During optimization, the hypothesis is formulated during the tip-to-root (i.e., working our way down the cladogram) assignment of states to nodes, and tested during the root-to-tip (i.e., working our way up the cladogram) assignment of states.

A particularly interesting aspect of the method is that it allows testing hypotheses of soft-tissue reconstruction with definite levels of certainty that can be used to discriminate among hypotheses with greater or lesser degrees of support. The three main alternatives described by Witmer (1995) are detailed as follows.

Inferential Level I. If in both extant brackets the bony correlate and its associated soft feature are both present, and the bony correlate is also present in the extinct form, the presence of the associated (homologous) soft feature in the fossil can be inferred with a high degree of certainty.

Inferential Level II. If one of the extant brackets lacks the bony correlate or its associated soft feature, inferring the presence of the soft tissue in question in the extinct organism would not be supported by evidence. The assignment would be ambiguous or equivocal.

Inferential Level III. Finally, if the associated soft feature–bony correlate relationship is not present in any of the

Inferential Level I

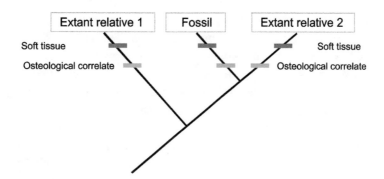

Inferential Level II   Inferential Level III

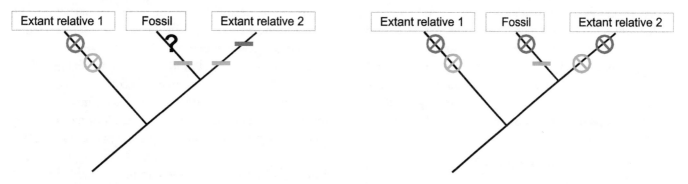

2.22. Extant phylogenetic brackets (Witmer 1995). At inferential Level I, the presence of the causal dyad *osteological correlate–soft tissue* in both extant relatives allows hypothesizing the reconstruction of the soft tissue in the extinct organism with better support than at Levels II and III, where the *osteological correlate–soft tissue* dyad is absent in one or both extant relatives.

extant brackets, the assignment in the extinct organism is negative. According to Witmer, this does not mean that the soft feature in the extinct organism cannot be inferred, but simply that such an inference is more speculative because it is not supported by the parsimony criterion. On the other hand, in cases when the morphological evidence for the presence of the soft feature is very strong ("compelling" sensu Witmer), rejecting its presence based solely on its absence in extant relatives could be incorrect. For example, returning to our example of ichthyosaurs, both the dorsal fin and tail fin and their osteological correlates are absent in extant lizards, but the fossilized evidence of well-developed dorsal and tail fins in many exquisitely preserved ichthyosaurs indicate they were present beyond any doubt (Gould 1993; see fig. 2.20G).

In situations where the soft tissue under consideration does not have an associated bony correlate, there are alternative grading levels of inference degree. These levels are called I', II', and III', and the assignment procedure is the same. Although it is obvious that a level I inference implies more certainty than a level I' inference, the hierarchy of certainty among all the alternative inference

degrees and the original ones are not yet clear (Carrano and Hutchinson 2002).

As frequently happens with any methodology, EPB is not trouble-free. For instance, many soft structures extend beyond the limits of their skeletal correlates, and it therefore becomes difficult to know which homology hypothesis is being tested: Is it that of the soft tissue or that of the bone that is its correlate? Witmer's (1995) original work suggested that the method could be used not only for testing the robustness of soft-tissue reconstruction, but also inferences on function and biological role. The use of parsimony for inferring function is debatable, however, as previously noted in Rudwick's (1964) example of pterosaur wings and Gould's (1993) example of ichthyosaur fins. Besides presence or absence, other muscle features that can be estimated with confidence in fossil taxa include length (which does not imply fiber length; see chap. 3), volume, and cross-sectional area (Perry and Prufrock 2018). For example, relatively great development of muscle scars (entheses), the dimensions of which are correlated to the amount and direction of mechanical load exerted by muscle on the bone tissue, is a clear indication of a powerful muscle.

In most cases, the reconstruction of soft parts is carried out by analyzing specimens qualitatively and relying on as thorough an understanding of the anatomy of extant forms as possible; we reiterate the importance of knowing the soft anatomy of the related extant forms. This knowledge is derived mainly from dissections, which should be performed whenever it is possible to do so, because for many groups the available knowledge of soft anatomy is, at the very least, out of date. Indeed, the rate in the updating of anatomical knowledge based on dissections has slowed due to ethical concerns over the sampling of endangered species (Perry and Prufrock 2018). However, as noted above, it is not unusual that the morphology of an extinct form under study differs radically from that of its closely related extant forms, or that there are no extant forms that are closely related. This requires comparison with more distantly related forms as models for soft anatomy. It is often possible that the taxon under study may not have had soft-tissue structures with precise homologies among extant forms; such cases necessitate a thorough and careful explanation of anatomical and functional hypotheses.

Ecomorphology

Van der Klaauw (1948) formalized the ecomorphology concept under the term *ecological morphology*, defining it as the study of relationships between the morphology of an organism and its environment. The term was originally used to explain variation in design (morphology and physiology). A series of works in the 1950s, wherein this author developed the functional cranial theory, made it possible to establish closer links in the relationships among form, function, and environmental pressures (Dressino and Lamas 2003). In the 1960s, ecologists began to explore the relationships between species' morphology and the ecological niche that they occupy within communities (Wainwright and Reilly 1994). Even though, at that time, studies addressing morphological disparity (or design differences among species or clades) and its correlation with environmental characteristics were emphasized, by the end of the 1970s the causal consequences of anatomical differences between species (or between individuals in a species) began to be studied to explain the ecology of organisms. Bock (1977, 1980, 1988, 1990) made the distinction we adopt in this book between function (the action, or how structures work) and biological role (how faculties enabled by the function of the structure are used in the organism's life). As stated by the Spanish poet Antonio Machado (1875–1939), "Wayfarer, there is no path, the path is made as you go" (Caminante no hay camino, se hace camino al andar); so too does ecomorphology continue in making its path toward becoming formally recognized as a research program. This has been made possible through the availability of large databases and technological advances in analysis. The latter allowed application of multivariate statistical tools, such as discriminant analysis developed by Fisher (1936), phylogenetic flexible discriminant analysis developed by Motani and Schmitz (2011), and other comparative methods (e.g., Brooks and McLennan 1991; Harvey and Pagel 1991; Giannini 2003; Ollier et al. 2006; Pennel and Harmon 2013) that evaluate phyletic diversity and allow assessment of whether a character of particular design can be considered as an adaptation.

Over time, ecomorphological studies converged on quantifying the variation in form of organisms and evaluating the patterns of such variation. In particular, the goal is to evaluate or explain the proportion of variation that can be attributed to the environment (adjustment to a functional task imposed by it) and to history (ontogeny and phylogeny). The basic premise of ecomorphology stipulates that the covariation between present morphological and environmental features is the result of natural selection and adaptive evolution (Klingenberg and Ekau 1996). This hypothesis coincides, then, with Cuvier's correlation principle and Currie's trait-environment dyad that were mentioned in chapter 1. Both assume that variation of an organism's form is correlated with the action of different environmental and biological pressures (Wainwright 1991; Feilich and López-Fernández 2019), but ecomorphology relies on natural selection and adaptation for explaining the adjustment to the environment (Schwenk 2000). When the environment imposes restrictions on the morphology and ecology of organisms, the causal relationship between form and function can be predicted (Karr and James 1975). However, it is necessary to evaluate the proportion of variation due to history (phylogeny) to avoid the risk of falling into a naive adaptationism (sensu Gould and Lewontin 1979) that discovers that organisms are adapted to their environment.

Ecomorphology involves a great part of Liem's (1991b, 764) vision of a new synthesis in morphology, the mission of which aims "to explain diversity and adaptation, both in historical and functional terms, and to discover the causes of richness of unique phenomena that characterizes biology." If we return again to Radinsky's (1987, 8) form-function paradigm and consider the first part of one of the approaches for interpreting functional aspects of extinct species—"Form-function correlation involves looking for the behaviors or functions that are correlated with a particular anatomical form in living species"—we may, as noted by Cassini et al. (2021a), trace the principles and fundamentals of ecomorphology. In summary, ecomorphology deals with making connections between how an organism is constructed and the ecological and evolutionary consequences of its construction or design.

Levels of Analysis

Arnold (1983) proposed that testing hypotheses of adaptation requires an analysis of levels (fig. 2.23) that allows the assessment of variation in form (the product of hereditary genetic variation) that results in variation of biological efficiency (fitness).

At the beginning of this book, some of the terms involved in this analysis in levels were defined; here we intend to contextualize them when dealing with each level, following Reilly and Wainwright (1994).

Level 1 – Morphology. This is the most basic level and represents the description of any aspect of structure(s) or phenotypical feature(s) that could be related with the interactions among organisms, and between these and the environment. It can be any level of organismic design, from molecular and cellular structures to different components of complex morphological systems or behavioral mechanisms.

Level 2 – Function. The rigorous analysis of function and its morphological bases is the following step for a well-based understanding of subsequent ecological and evolutionary consequences of form. However, the functional analysis does not need to be carried out in the context of the biological role.

Level 3 – Fundamental niche. This level of analysis assesses performance as a measure of how well a particular behavior or function is performed; it is the link between the design of the organism and its effects on its relationships with other organisms, the environment, and evolution. This level can be divided in two sublevels, referred to as "potential" and "actual" use of resources. The performance capacity, either the actual measurement or that predicted from morphology, generally reflects the maximum range of abilities of a morphological system in performing a function. For instance, the potential prey range of a fish that eats molluscs will be limited to consuming those with shells that the fish is able to grind (Reilly and Wainwright 1994). Therefore, the performance capacity of such a fish can be used to infer the fundamental niche or potential prey that could be exploited. It is the relationship between morphology and function that determines the performance capacity of an individual and, thus, its potential use of resources.

Level 4 – Realized niche. This is the performance measurement or real resource use under natural conditions. The realized niche space determined by morphology and

the particular function(s) under study involves quantification in the field of behavioral ranges (performance) or resource use patterns. In the case of the fish that grinds shells, the realized niche is the range in the kinds of prey consumed by individuals in nature. The comparison between the fundamental niche and the realized niche identifies which portion of performance capacity is used by the organism in nature and how well the performance capacity equals the resource use. When the predicted resource use approaches the actual resource use in natural populations, it is reasonable to propose that performance capacity represents the maximum range of resources that the individual will use. This relationship should be considered beforehand so the performance capacity can be related to the following level: reproductive success or fitness.

Level 5 – Fitness. The final evaluation of the adaptive meaning of morphological characters through their functional roles and potential and real performance is its effect

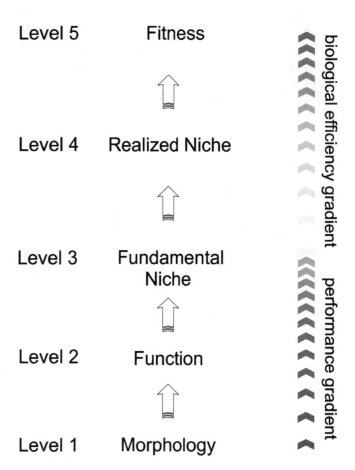

2.23. Hierarchical relationships of interconnected analysis levels in ecomorphological studies. The function and performance provide the key links for relating the patterns of morphological and ecological variation. Integration of information from ecology (realized niche, actual resource use, and reproductive fitness), performance capacity (and hence the inferred fundamental niche), and functional morphology is necessary for interpreting the ecological role of organisms (modified from Reilly and Wainwright 1994).

on the individual's reproductive success. This implies demonstration of whether or not the features under consideration are heritable, and the characteristics of selection acting on their performance through the analysis of quantitative genetics, survival rates, and number of surviving descendants. In other words, it implies studies of natural selection in natural populations.

Integration of Ecomorphological Studies

In reality, there are very few studies that deal with all the levels described above, with most studies including only two or three. Inferences about the effects of one level on the next higher level depend on the features under study being causally relevant for that following level of analysis.

One of the difficulties in ecomorphological inference is that the function of many morphological features is insufficiently understood to allow inference of biological role, whether in the fundamental or realized niche; for this reason, it is necessary to begin with functional studies. Furthermore, the adaptive meaning of many important ecological features is not always clear, therefore requiring greater integration among the data supplied by ecologists, functional biologists, and morphologists.

Hence, the results of the paleoecological studies based on ecomorphology should be considered as hypotheses to be validated. To this end, functional morphology and, especially, biomechanics are applied to assess the relationship between performance and variation in form so that those ecologically relevant functions from design conflicts might be differentiated (Huey and Kingsolver 1989).

Thus, ecomorphological approaches are mainly based on assessing patterns of association between morphology and ecology in a group of extant species and considering also their shared evolutionary history. According to Feilich and López-Fernández (2019), all types of data can be useful, and their choice will depend, among other factors, on the biological issues to be addressed, the availability of specimens for study, and the existing knowledge about the organisms under study and those of the comparison sample.

The first steps in ecomorphological studies are, therefore, selection of one or more species for study, determination of the morphological aspects to analyze, and quantification of its or their morphology through methodological approaches such as linear or geometric morphometrics (fig. 2.24), as explained in chapter 1. Linear morphometrics quantifies form by using dimensions such as lengths,

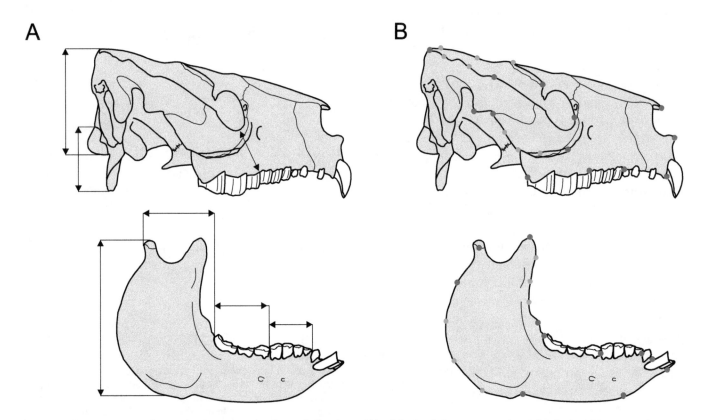

2.24. Morphometric approaches on the same research subject (skull and mandible of the South American ungulate *Adinotherium*). Linear morphometrics (*A*) quantifies form by means of distances such as lengths and heights (*arrows*), while geometric morphometrics (*B*) proceeds by collecting coordinates of landmarks (points related to anatomical features, in *red*) and semilandmarks (points that are located along a curve or contour without being referred to site-specific anatomical features, in *blue*). Modified from Cassini et al. (2011) and Cassini and Vizcaíno (2012).

widths, and heights; that is, distances between specific features measured by calipers directly on specimens, or images or 3-D models of them. These measurements can be analyzed together or separately, or combined to produce ratios or indices. In most cases, these variables are mixed bags of size and shape information, which can be problematic to separate for analytical purposes (Zelditch et al. 2012). There are several protocols to account for size variation (including standardization and size transformation); however, how and whether to apply them depends on the questions or hypotheses being addressed. Angular data, which have the advantage of being invariant to size, may also be used (Slice 2006).

Geometric morphometrics, by contrast, proceeds by characterizing specific morphological features via the collection of the Cartesian coordinates (bi- or tridimensional) of landmarks and semilandmarks. The resulting set of coordinates is called a landmark configuration (Zelditch et al. 2012). There are different protocols to process the orientation, scale, and location information encoded into the coordinates, and to separate its shape and size information (the centroid size; Kendall 1986; Bookstein 1989); among them, Procrustes superimposition methods (i.e., based on minimizing squared distances) are the most common (Goodall 1991; Rohlf 1999). See Zelditch et al. (2012) for further information about this topic.

Even though there are some exceptions (e.g., caviomorph rodents, for which digging is also evaluated), in general, cranial aspects are assessed in relation to diet, while postcranial features are considered in relation to locomotion and other uses of the substrate. The ecological variables of the chosen species should be well known and can be assigned to either discrete or continuous categories. Examples of discrete categories based on diet include Spencer's (1995) classification of African bovids into grass feeders, mixed feeders preferring grass, mixed feeders preferring dicots, dicot feeders, and frugivores, based on the proportion of the kind of food included in the diet; and Samuels' (2009) characterization of rodents as vertebrate eaters, insectivores, omnivores, specialist herbivores, and generalist herbivores. An example of discrete categories of substrate use was employed by Elissamburu and Vizcaíno (2004) to classify caviomorph rodents into diggers, generalized, cursorial, swimmers, and jumpers. Examples of continuous-type variables include maximum bite strength (Herrel et al. 2000), food item hardness (Herrel et al. 2008), diet item composition (Cassini and Toledo 2021), and maximum speed (Garland and Janis 1993).

The relationship between ecological and morphological variables is assessed by means of a diversity of statistical analyses. In this sense, the growth of available computational power and the rise of open-source software have prompted an outburst of analytical tools, especially for multivariate statistics. Generally, principal components analysis (PCA) is performed first. PCA defines new axes that are linear combinations of the original variables and aligned to the axis of greatest variance. These new axes are called principal components (PCs), the first of which accounts for the greatest proportion of variation. Variation decreases in the following axes until the sum of all of the axes equals the total variance in the sample. Indeed, we expect that PCA condenses the major proportion of variance into a few (usually three to five) PCs that allow inspection of the morphospaces (i.e., mathematical spaces describing and relating phenotypes of organisms; Mitteroecker and Huttegger 2009) depicted by them. It has the advantage of reflecting the sample's intrinsic variation (i.e., without criteria defined a priori), where the researcher seeks patterns that correlate morphospaces with ecological characteristics. Although PCA is not a hypothesis-testing technique, it is a variable-reduction method that permits exploratory analysis (see, e.g., Gingerich 2005) as well as testing of the first or first few PCs, which account for the most of variance, and usually better fit the statistical assumptions. By contrast, methods such as partial least squares (PLS; Bookstein et al. 1990) allow assessment of the association between two or more groups of variables (i.e., morphological and ecological) that, unlike regression analyses, do not assume that one of the groups predicts or determines the other(s).

MANOVA (multivariate analysis of variance) and MANCOVA (multivariate analysis of covariance) are statistical techniques that test hypotheses on the association between morphological and ecological features in a multivariate environment (and the respective ANOVA and ANCOVA in bivariate contexts and their nonparametric alternatives, such as the Mann-Whitney U test). PCAs are often very useful for inferring ecological features in those species of unknown habits that share the same morphospace, but, unlike PLS, they do not provide information on the degree of correlation.

Discriminant analysis (DA; also termed *discriminant function analysis*, DFA, or *linear discriminant analysis*, LDA) is a classical technique of multivariate statistics that allows classifying new samples within predefined groups using discriminant functions (see below) adjusted for maximizing the between-groups to within-groups ratio of variance. This technique is often used in paleobiological studies that aim to predict assignment to a particular category (e.g., fossorial, herbivore) with a high degree of confidence. A data set of variables from extant specimens with known habits is used as a reference sample (also known as a learning sample) with established ecological categories for building mathematical equations, technically termed *discriminant functions*. These equations are linear combinations of the variables under study that allow the clearest separation (discrimination) of the sample into these categories. However, not all the morphological variables considered may turn out to be informative or meaningful in discriminating among categories. Therefore, we can use a stepwise procedure to discover which of a series of equations maximize the discrimination of categories by adding and removing the variables one at a time until we discover which combination of variables is most useful in discriminating groups, thus generating an equation (i.e., discriminant function) based on these variables that best separates the categories. Such an equation must then be evaluated against a test sample (i.e., extant animals of known ecological habits, which were not included in the reference or learning sample) or by means of cross-validation tests (see Hair et al. 2019 for a holdout sample cross-validation, and subsequent modifications applied to characterize morphological distinctiveness by Cardini et al. 2009, Abba et al. 2015, and Campo et al. 2020). Finally, the equation is applied to datasets of extinct organisms to assign them to a category. We should be aware, though, that if we use a reference sample with limited ecological categories, then extinct organisms will be classified into one of them, even if they have morphological features that depart from the reference sample. For example, Cassini et al. (2011), using DA in combination with classification trees to predict the feeding behavior in Santacrucian notoungulates, found that rodent-like typotheres exhibit morphological features that exaggerate the patterns exhibited in modern grazing ungulates. Although the typotheres were classified as grazers, these authors proposed alternative hypotheses for these particular morphological features (e.g., putative digging habits) and the need for seeking better herbivorous analogs to evaluate these inferences.

Motani and Schmitz (2011), based on the protocols by Hastie et al. (1994) and combined with a phylogenetic generalized least squares (GLS) regression (Martins and Hansen 1997), developed phylogenetic flexible discriminant analysis, which accounts for phylogenetic covariance when predicting group membership. In the Motani and Schmitz (2011) analysis, identifying an optimal Pagel's lambda value (Pagel 1999) is required, as it provides the strongest correlation between morphology (variables or indices) and ecology (substrate preference categories). Lambda describes the transformation necessary to convert the phylogenetic covariance matrix so that it conforms to a Brownian motion evolutionary model (i.e., a random

variation of traits not produced by an assumption of evolution; see Blomberg et al. 2020). A lambda value close to 1 does not modify the covariance matrix, and the model fits an evolutionary pattern (i.e., the variance can be explained as a result of phylogeny), whereas a lambda of 0 implies that the tree becomes a star phylogeny and represents a model entirely independent of evolutionary pattern.

In either classical or phylogenetic flexible discriminant analyses, if reliable equations (i.e., discriminant functions) are obtained, they can be used to confidently assign the extinct taxa under study to a predefined category. These methods, in other words, always result in a classification that is determined by the variety of categories in the reference sample (as noted above).

Many paleobiological studies are performed following two steps. First, analyses that quantify the relationship between morphological features (e.g., craniodental variables) and ecological features (e.g., composition of diet and habitat preference) that are observable in extant communities are carried out. Second, the patterns detected are used for inferring ecological aspects of the extinct species.

As previously mentioned, ecomorphological approaches require evaluation of the proportion of variation due to the history of organisms (ontogeny and phylogeny). Two major approaches can be applied: lineage or community ecomorphology. In lineage ecomorphology, aspects of the chronological distribution and phylogeny in the morphospace created by the members of a clade or lineage can be mapped and interpreted. This approach is particularly useful in the case of extinct lineages without extant representatives, because it allows consistent patterns of successive taxa replacement to be detected (e.g., in endemic South American ungulates of Notoungulata; Giannini and García-López 2014). However, these approaches require a second level of analysis for proposing or testing paleobiological hypotheses. Community ecomorphology is intended to characterize the communities by identifying patterns that link morphological features and ecological features and their interactions, independently from the taxonomic composition (taxon-free) and the time period under consideration (Mendoza et al. 2005). This approach has been often applied to artiodactyls and perissodactyls to characterize past communities from a classical morphometric perspective (Mendoza 2005; Mendoza et al. 2006; Mendoza and Palmqvist 2006, 2008; De Esteban-Trivigno et al. 2008).

In ecomorphological studies where form variations due to the phylogenetic history of the group can be distinguished, the proportion of the variation that can be explained by common ancestry (i.e., the phylogenetic signal) should be considered. Several methodological approaches from the field of comparative biology for achieving this and other goals have been proposed (Felsenstein 1985; Garland et al. 1993, 2005; Grafen 1995; Ackerly 2000; Paradis and Claude 2002; Giannini 2003; Grossnickle 2020a; Ollier et al. 2006; Mahler et al. 2013; Pennel and Harmon 2013; Slater 2013; Slater and Harmon 2013). They are collectively known as phylogenetic comparative methods (PCMs). It is expected that these methods complement the classical parametric statistical analyses, the assumptions of which cannot be met due to lack of sample independence, because the biological species share an evolutionary history rather than being independent events. However, other authors point out that the need for adding phylogenetic information to achieve a successful comparative analysis depends on the nature of the intended inferences (Sanford et al. 2002; Quader et al. 2004). According to Sanford et al. (2002), the approximations intended to identify and explain the variations related to phylogenetic relationships should form part of a framework distinct from that of comparative biology, as in the latter the knowledge of one organism is used for guiding the study of another.

One of the difficulties of the comparative methods is that intraspecific variation in design is often reduced to an average. Although it may be challenging, some analyses have been proposed to address this issue (e.g., phylogenetic ANOVA, Garland et al. 1993; canonical phylogenetic ordination, Giannini 2003; Bayesian multilevel models, Nations et al. 2019). For example, canonical phylogenetic ordination analyses allow testing the covariations between matrices of morphological and ecological features, and a third matrix reconstructs the phylogenetic relationships. In addition, many of the available methods cannot be performed if any of the ecological categories is represented by a single species. Others require calibrated (time-reconstructed) phylogenetic trees for their application, which in turn are burdened with their own algebraic and evolutionary assumptions (see Münkemüller et al. 2012 and Cornwell and Nakagawa 2017 for a review). Notwithstanding, the development and availability of free statistical and morphometric software (e.g., the R suite and its libraries) allow applying a large diversity of last generation analyses with reasonable computing times.

3   Biomaterials

The performance of materials, and thus of the structures made from them, depends in large degree on the mechanical properties of the materials. For example, for much of the last several thousand years burnt clay brick has been a nearly ubiquitous component of human-made structures because it combines solidity with relatively little mass, a result of the water loss that occurs during the intense heating of the brick-mold phase. However, while a brick undergoes heavy loads (forces that are transmitted upward and downward from the upper part of a column or wall to the ground), it is not very resistant to horizontal movement such as that commonly occurring during earthquakes. By contrast, traditional Japanese houses are designed to be flexible, which increases people's survival during earthquakes and typhoons. Almost all traditional Japanese building materials are organic: soft wood of perennial trees for the structure, mulberry paper for the walls, and bamboo for elements that ensure structural integrity. When choosing materials for a certain product, the designer searches for a specific combination of properties suitable for the function the product is expected to fulfill; such properties include density, elastic modulus, breakage modulus, tensile strength, and coefficient of thermal expansion.

In like manner, in studies of form-function correlation in biology (especially in biomechanics), it is essential to know the basic properties of biomaterials. The different tissues forming body organs, such as skin, bones, and blood vessels, have unique material properties. The passive mechanical response of a particular tissue can be attributed to the characteristics of the molecular composition of extracellular components, such as that of diverse proteins (e.g., elastin and collagen), as well as their orientation and organization within the tissue (e.g., dense regular fibrous connective tissue vs. dense irregular fibrous connective tissue), and any mineralized components, as in bone.

To determine the scope of biomechanics' application to vertebrate paleobiology, it is important to understand the properties of different skeletal elements involving connective tissues (especially bones, but also cartilages, tendons, and ligaments), the muscles moving them, and teeth. The appendix at the end of this book provides information on the most relevant tissues and organ systems for studying form-function correlation in extinct vertebrates. In this chapter, we deal with several functional and mechanical aspects of these biomaterials.

Tissue Response to Mechanical Stresses

As in any type of material, the forces applied (loads) and the deformations (changes in shape) they produce can affect properties of living tissue, modifying its growth and remodeling. In general, the forces that act on a biological material and can cause deformation in its structure, mainly in its form and the distribution of its components, are compression, tension, and torsion. Analyses of deformation behavior determine the "resistance" of a biological material to the application of a load upon it. In mechanics, stress is defined as the internal force per unit area that develops in the material due to the applied load, and the deformation induced by this stress is described by the term *strain*. Different materials respond to equivalent loads by experiencing different stresses and exhibiting different degrees of deformation before eventually failing. For example, steel can withstand a load that will deform rubber but will itself deform under a greater load. Mathematical rules that describe deformation in response to load allow the classification of the behavior of each material (elastic, plastic, viscoelastic, fluid) under specific conditions.

Two related concepts are important in considering the behavior of different materials: toughness and strength. Both are defined based on stress versus strain curves at the value of stress at failure (see Alexander 2017). Toughness describes the total amount of energy (as stress) that a material can absorb before total breakage, including elastic and plastic deformation. The former refers to deformation that is not permanent, so the material returns to its original shape; the latter involves deformation beyond elastic deformation and is permanent (the material cannot return to its original shape). On the other hand, the strength of a material can be defined as the maximum stress that it can withstand before it yields to permanent breakage (i.e., the limit of its plastic behavior). Steel will deform just a little before breaking, while rubber will deform easily and extensively before breaking, but both materials can show similar strength (maximum stress before breaking). In biology, tough materials (such as collagen) tend to have low strength.

Biomaterials are challenging to study because they are complex in structure and composition (i.e., made up of two or more components; structures often combine two or more materials that differ in their mechanical properties) and are inhomogeneous (i.e., are not isotropic). Most tissues are composed of several types of complexly arranged biomaterials, which often renders general rules described for isolated biomaterials inapplicable to tissues. For example, bone exhibits different strength in its longitudinal and transverse sections (i.e., it is anisotropic); a more familiar example is wood, which can withstand considerably more loading applied along than across its grain. In addition, tissues can modify or adapt the structure and behavior of their constituent biomaterials in response to new demands throughout the organism's life (see symmorphosis, chap. 1). Examples of such responses include atrophy, referring to a decrease in tissue volume due to absence of loads, and hypertrophy, referring to an increase in volume under the influence of a steady load (e.g., in athletes). We may also note hyperplasia, which denotes cellular division and proliferation under stress, and metaplasia, generally a pathological response in which a particular tissue type changes into another.

One of the most evident examples of the ability to adapt to new demands is observed in bone (fig. 3.1). To play their role in support and protection, bones must undergo

sudden deformation. However, bone is dynamic and changes gradually during an individual's life as a response to environmental factors. Four kinds of environmental factors can alter a bone's basic form: infectious diseases, nutrition, hormones, and mechanical loads. We are here concerned with the latter.

The forces produced by gravity and muscular contractions determine a bone's final form, but over an individual's life these loads on the bone also change. Furthermore, a bone's response to mechanical stress depends on the duration of force. A young animal is very active (i.e., it runs, jumps, and so on), but in adulthood it changes its activities: it may travel long distances, fight for territory, put on extra mass as part of caring for progeny, or produce larger forces when digging. In turn, as the animal increases in size, changes in proportions become a determining factor (see chap. 4). Increase in body mass puts greater mechanical demands on the elements supporting the body, which must thus withstand the concomitantly increased reaction forces without structural failure (Vassallo et al. 2021). By contrast, increasingly older individuals tend to reduce their activity, which reduces loads and hence mechanical stress. In short, for several reasons, the forces to which bones are subjected change over an individual's life history.

Collagen fibers and calcium salts help strengthen the bone. In fact, collagen fibers are very resistant to tension, while the mineral salts (calcium phosphate) are resistant to compression. This combination of properties makes bones extremely resistant (i.e., they can absorb considerable energy before failing) to both tensile and compressive stresses.

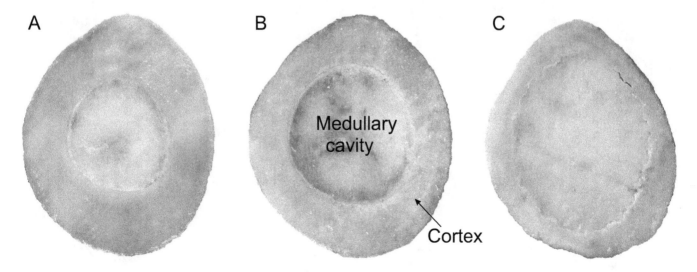

3.1. Physiological responses of bone tissue. Cross section of the same bone of the posterior limb (femur) of three different individuals of cattle. *A,* bone with cortical mass increase (hypertrophy). *B,* normal bone. *C,* bone with cortical mass loss (atrophy).

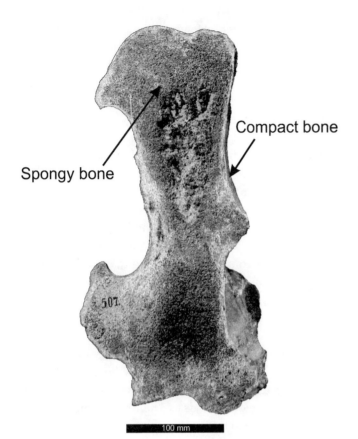

Compact bone

Spongy bone

Above, **3.2.** Longitudinal section through a left humerus of a Pleistocene scelidotheriine sloth (Museo de La Plata; MLP 507). Note the peripheral distribution of cortical (compact) bone tissue and the extensive development of spongy bone tissue internally. Photo by the authors.

Bones are formed by compact (cortical) and spongy (cancellous) bone tissues (fig. 3.2; see appendix), and the distribution of both types is directly related to their response to mechanical factors.

An entire bone or some portion of it can be loaded in compression, tension, flexion, shear, torsion, or some combination thereof (fig. 3.3). Muscular action during movement (including fetal movements) also produces mechanical loads applied to the bone, causing a dynamic reaction of the bone tissue: the bone thickens or reorganizes its trabecular meshwork (see below).

Bones are anisotropic (i.e., their structure and properties change in three-dimensional space), but are approximately isotropic transversely (i.e., their properties remain nearly constant); they are stronger along their longitudinal axis than perpendicular to this axis. In very large terrestrial animals (elephants, sauropod dinosaurs; fig. 3.4), limbs are arranged as simple columns that support compressive forces generated by weight (see chap. 5). In

Below, **3.3.** Forces acting on bones. The color code corresponds to that in figure 2.13: areas shaded in *blue* indicate regions where particles approach each other and compressive forces dominate; in *red*, where they separate and tension dominates; in *yellow*, where particles are displaced or twisted over each other; the *yellow-green gradient* in torsion indicates zones with displacements and a combination of forces.

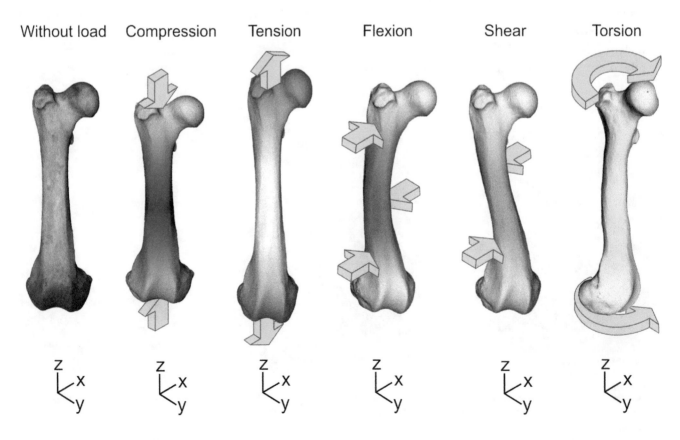

Without load   Compression   Tension   Flexion   Shear   Torsion

A

B

Above, **3.4.** Extant and extinct very large, graviportal vertebrates with columnar limbs; see chapter 6. Bones of the appendicular skeleton aligned columnarly are indicated in *white*. In the African elephant (*A*), both the scapula and pelvis have a columnar arrangement, in alignment with the limbs, whereas in the sauropod dinosaur (*B*), only the pectoral girdle is so aligned.

Below, **3.5.** Extant and extinct vertebrates with flexed appendicular joints. The appendicular bones, indicated in *white*, are angled more obliquely with respect to each other. A, red panda (*Ailurus*). B, Miocene notoungulate *Adinotherium* (modification of Robert Bruce Horsfall's illustration published in Scott 1913). Scale = 50 cm.

A   B

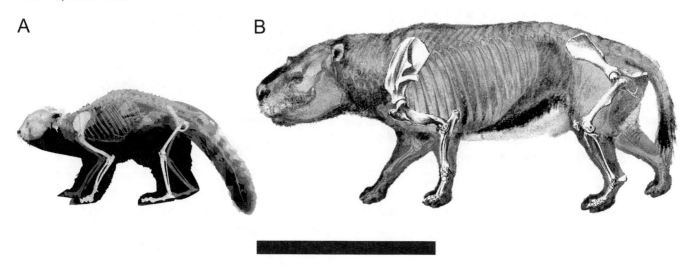

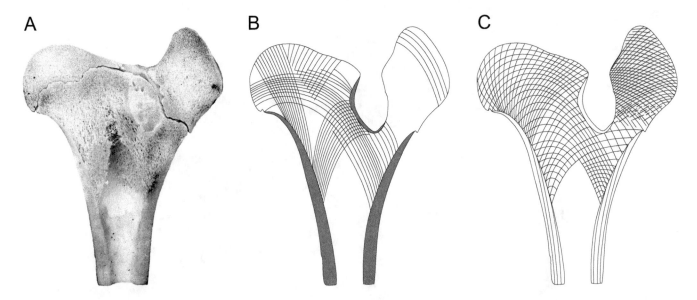

3.6. Wolff's law of bone transformation exemplified in a bovine femur, where spongy bone trabeculae extend along trajectories of load dissipation, and the bone density reflects stress intensity. *A*, longitudinal section of the proximal end of the bone showing cortical and trabecular tissue. *B*, representation of arc patterns that follow trajectories of trabecular bone, according to von Meyer (1867). *C*, representation of trajectories on which the loads would act on the trabecular bone, according to Wolff's (1870) approximation where, unlike *B*, arc intersections are orthogonal.

these animals, bones are solid rather than hollow, optimizing the role of limbs in weight support. By contrast, many other terrestrial vertebrates are specialized for running. In such vertebrates, a limb's upper part (humerus or femur) is oriented less vertically in lateral view (in anterior and posterior view they are vertical). These bones function as angled beams and are subject to flexion, which tends to stress the dorsal surface in tension and the ventral surface in compression (fig. 3.5). They are tubular, with the wall of the diaphysis (the shaft-like portion of a bone) formed by a relatively thin layer of compact (or cortical) bone, making it well suited to withstand loads from all directions.

The load system on epiphyses (the expanded ends of a bone, where it typically articulates or forms a joint with another bone) is different from that on the diaphysis, because epiphyses must transmit loads from an articular surface to the diaphysis. This transmission is the role of spongy bone. A longitudinal section through a humerus or femur of a running animal reveals that spongy bone is a lattice of fine laminae and rods, composing the trabeculae (or "small beams"). The laminae extend from the thin cap of compact bone that forms the articular surface and mainly converge toward the compact bone forming the diaphyseal wall. The organized arrangement of trabeculae has been compared with the distribution of stress trajectories in loaded beams.

The theory of trajectories, considered in chapter 2, also applies to bone, except for the fact that bone is a living tissue that reacts dynamically to mechanical loads. This is known as Wolff's law of bone transformation (Wolff 1869,

1892, 1986), which establishes that bone remodeling occurs in proportion to the mechanical demands imposed on it. According to the theory of trajectories, the laminae of trabeculae extend along trajectories of load dissipation, and bone density reflects the stress intensity. This explains why the diaphysis of a long bone, where the trajectories are effectively aligned along the long axis of the diaphysis, is formed by compact bone. If stress patterns change, bone responds by altering trabecular orientation to optimize resistance to load. This idea arose in Germany by the mid-nineteenth century, when the engineer Kart Culmann suggested to the anatomist Hermann von Meyer that the trabecular bone trajectories of the human femur, metatarsus, and calcaneum are aligned with the main stress directions (von Meyer 1867; Rüttimann 1992). However, the trajectories depicted in von Meyer's illustration (1867) were not orthogonal, in contrast to what would be expected, and it was Julius Wolff (1869) who correctly drew the trajectories, following a mathematical rule (Skedros and Baucom 2007). Although the illustrations of von Meyer (1867) and Wolff (1869) are different (fig. 3.6), they are often mistaken for each other in much of the literature.

Bone is also involved in different forms of armor. Protection is its commonly accepted function, but armor might serve multiple functions and be shaped by trade-offs among them (e.g., as a compromise between strength and thermal capacity in some lizards; Broeckhoven et al. 2017). Among extinct mammals, glyptodonts have a conspicuous carapace that has been proposed as protection against predators and/or tail-club blows of conspecific individuals (Alexander et al. 1999; Blanco et al. 2009; Arbour and Zanno 2018). The combination of dense compact layers and a porous lattice core in glyptodont osteoderms might provide an optimized combination of high strength and toughness, and thus high energy absorption (du Plessis et al. 2018).

Tendons, Ligaments, and Cartilage

Soft tissues, such as tendons, ligaments (fibrous connective tissue), and cartilage (special connective tissue), are combinations of matrix and fluid proteins (see appendix). In each of these tissues, the main element that resists deformation is collagen, though its quantity and type vary according to the function that each tissue must perform. The typical function of tendons is to connect muscles with bones (in some cases, tendons connect to each other, as in the tendon sheets, or aponeuroses, involved in several abdominal muscles), and so they are subjected to tensile loads. Tendons must be strong to permit body movement and at the same time be able to accumulate elastic energy. Ligaments (most commonly) connect bones and are, therefore, less deformable than tendons, but they are similar in their resistance to tensile loads. Cartilage is mainly subjected to compressive loads and acts as a shock absorber in the articulations for distributing loads among bones. The resistance of cartilage to compressive loads, similar to the resistance in tendons and ligaments to tensile loads, derives principally from collagen. However, in articular cartilage the fibers are reinforced by cross-linkage with

glycosaminoglycans, which also attract water, forming a practically incompressible tissue that makes it well suited to withstand compressive loads.

Function of Muscles

Muscles are essential for movements of both the skeleton and inner organs. They are classified in two major groups, striated and smooth, according to the characteristics of their cells. Skeletal muscle (a type of striated muscle, associated with bones and cartilages; see appendix) can contract and maintain a variable or constant force over time. As Giovanni Borelli (noted in chap. 2) stated, muscles only perform work during contraction, never in extension, though muscle fibers can contract even if the whole muscle is experiencing extension, as during eccentric isotonic contraction. That muscle fibers work only by contraction implies that when a muscle contracts and performs a function, another muscle must perform the opposite function to return the system back to the original state. In the simplest cases, two muscles are involved, one contracting to provide the force for a particular movement and the other relaxing to allow the movement; the former is considered the prime mover or agonist and the latter the antagonist (fig. 3.7A). In accomplishing the opposite movement, the muscles switch roles as prime mover and antagonist.

Commonly, more than a pair of muscles is involved, and muscles that help the prime mover are termed synergists (fig. 3.7B). For instance, in elbow flexion the main muscles providing the force are the brachialis and biceps,

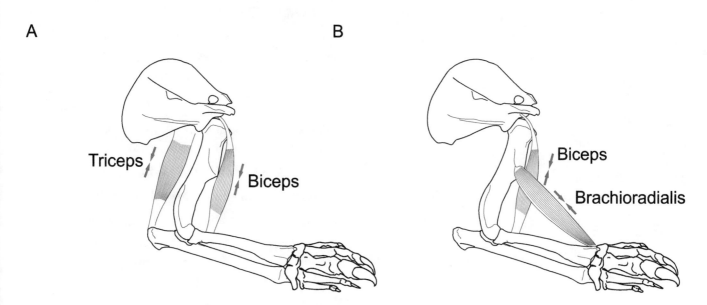

3.7. *A*, antagonistic muscles. Some muscles contract to execute a particular movement or function, and others execute the opposite action, returning the system to its initial position, or act at the same time to control movement. In the example, the biceps muscle is a flexor of the elbow of a lesser anteater, whereas the triceps extends it. *B*, synergistic muscles. Several muscles can cooperate in performing a function. In the example, the biceps and brachioradialis muscles cooperate in flexing the elbow of the lesser anteater.

with (e.g., in humans) the former being the prime mover and the latter a synergist. The triceps is the antagonist. In elbow extension, the roles of these muscles are reversed. However, antagonistic muscles may operate during a particular movement, acting in a limited way to help control movement. Indeed, the degree of activity of muscles involved is greater than as just described for simple flexion and extension. Consider, for example, the release of a weight, such as a barbell, from a flexed forearm position. In a controlled release, the forearms are gradually extended. As noted, the triceps are the main movers in such movements, but in a controlled release of the barbell, the biceps and brachialis are also engaged in helping to control the extension of the forearms. These muscles are contracting at the cellular level but lengthening overall, undergoing eccentric isotonic contraction. Another function of muscles is to act as "stabilizers." Many muscles of short-range motion collaborate in movement to ensure that the articulations are not dislocated or forced into harmful angles by the movement of main muscles.

In short, the forces exerted by muscles produce different effects according to the type of activity that the organism performs. Some terms are presented in the text box "Terms Describing Muscle Actions" and in figure 3.8. It should be noted, however, that some of these actions are not exclusive. For instance, the masseter muscle functions as a mandibular elevator but may also be regarded as an adductor, and it may even be described as a flexor of the craniomandibular articulation.

Muscular Force

Since the classic study by Gordon et al. (1966), we have known that the amount of force that skeletal muscle generates depends on its length, which is defined in terms of sarcomere (see appendix) length. As muscle-fiber length is positively related to muscle excursion, and excursions of the individual sarcomeres in series are additive (see Bodine et al. 1982; Winters et al. 2010), it can be assumed that the maximum force produced by a muscle is equal to the contraction force of one of its fibers multiplied by the total number of fibers. However, muscle architecture (determined by the number and arrangement of muscle fibers within the muscle) is the best predictor of a muscle's capacity to generate force and movement (Lieber and Ward 2011). The physiological cross-sectional area (PCSA; fig. 3.9) is the most recurrent parameter for estimating the magnitude of isometric force generation; it is defined based on muscle mass, pinnation angle, muscle density, and fiber length. PCSA is an approximation, and it assumes that all fibers are parallel to each other and to the axis of the insertion tendon and that all fibers are the same length. This is the case in a straplike muscle (e.g., thyrohyoid muscle, teres major muscle). Other muscles become

TERMS DESCRIBING MUSCLE ACTIONS

Flexors reduce the angle between articulated skeletal elements.

Extensors increase the angle between articulated skeletal elements.

Adductors move elements toward the midsagittal plane of the body or toward the axis of a limb or its more medial part (e.g., to close hands); the term *adductor* can also be used to describe a muscle that moves one part of the body toward another (e.g., the muscle that moves the lower jaw toward the upper jaw is called, except in mammals, the adductor mandibulae).

Abductors move an element away from the midsagittal plane of the body part or from the axis of a limb (e.g., to open hands).

Elevators raise an element in the dorsoventral direction; generally, the reference frame is the animal in horizontal position (e.g., the levator hyomandibulae).

Depressors lower an element in the dorsoventral direction; generally, the reference frame is the animal in horizontal position (e.g., the depressor mandibulae, but not in mammals, in which the muscle is called the digastric, although it still functions as a depressor; the difference in names, in this case, serves to emphasize the difference in construction and homology of the two muscles).

Protractors move an element away from its base.

Retractors move an element toward its base (e.g., tongue).

Rotators spin an element along its longest axis.

Supinators rotate the palm of the hand or sole of the foot upward.

Pronators rotate the palm of the hand or sole of the foot downward.

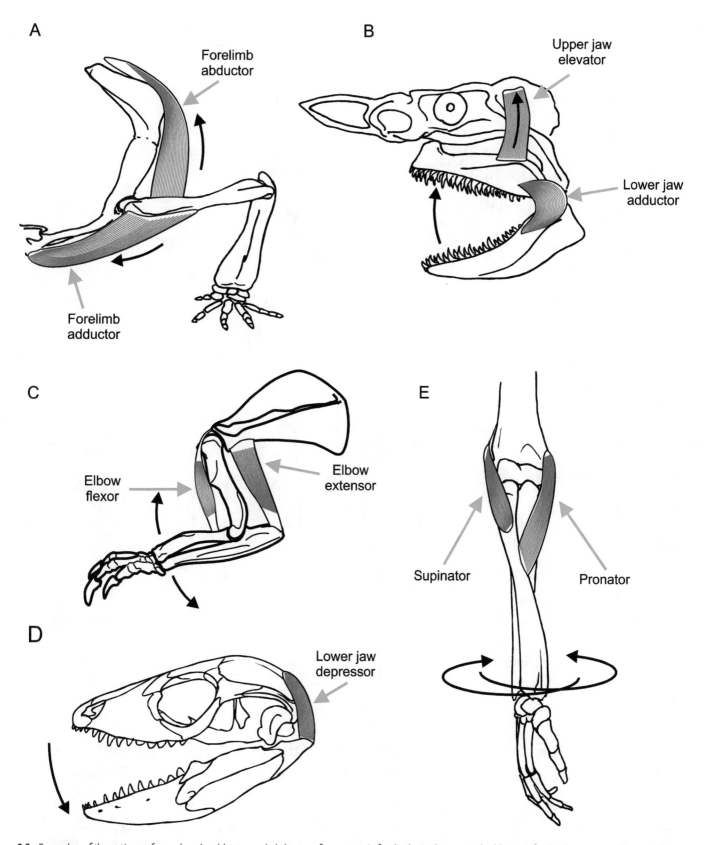

3.8. Examples of the actions of muscles. *A*, adductor and abductor of an anuran's forelimb. *B*, elevator and adductor of a shark's upper and lower jaws. *C*, extensor and flexor of an armadillo's elbow. *D*, depressor of a lizard's lower jaw. *E*, pronator and supinator of a human's forelimb.

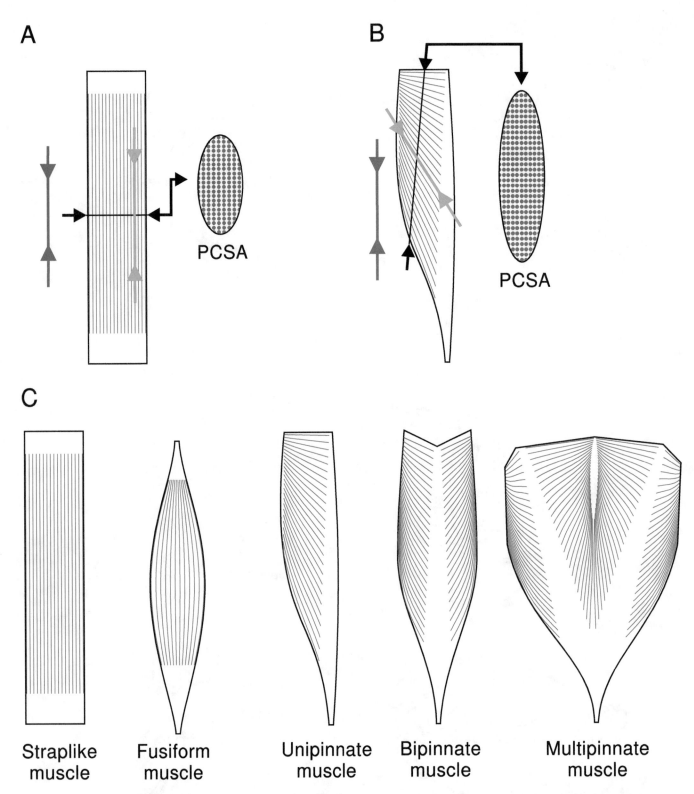

Straplike muscle   **Fusiform muscle**   **Unipinnate muscle**   **Bipinnate muscle**   **Multipinnate muscle**

3.9. Muscular force. The maximum force produced by a muscle depends on the number of fibers involved in its physiological cross-sectional area (PCSA). *A*, in strap-like or fusiform muscles (*C*, the two *left-side* images), fibers are parallel to each other and to the axis of the tendon of insertion and perpendicular to the PCSA, which corresponds to the cross section of the muscle transverse to its fibers. *B*, in pinnate muscles (*C*, the three *right-side* images), fibers insert obliquely onto one or more tendons that are placed to one side or internally in the muscle, allowing accommodation of a larger number of fibers; thus, PCSA is larger than in strap-like muscles. The *blue arrows* represent the direction of fiber shortening, and the *red arrows* represent the main direction of muscle shortening. In *A*, fiber direction is parallel to the shortening direction of the muscle, and in *B*, fiber direction is oblique to the shortening direction of the muscle; therefore, the contraction of pinnate muscles has a shorter range of motion.

thinner at one end for insertion on a bone (e.g., deltoid muscle) or at both ends, forming a fusiform muscle (e.g., biceps muscle). Thus, the force is concentrated in a smaller area.

In turn, the force produced by the muscle varies with the orientation of its fibers. One type of muscle is called pinnate muscle, where the fibers are short, arranged diagonally, and insert on a tendon along one side of the muscle (unipinnate) or in one or more tendons positioned in the center of the muscle (bipinnate or multipinnate). Pinnate muscles increase the contraction force by increasing the number of fibers per physiological unit area, but the obliquity of the fibers with respect to the direction of muscle movement reduces the length of contraction of the muscle as a whole. Examples are the extensor digitorum longus muscle (unipinnate), the rectus femoris muscle (bipinnate), and the deltoid muscle (multipinnate).

These variables in muscle architecture affect the degree, speed, and force of contraction. Each one has functional advantages and disadvantages. The degree of contraction of a muscle depends on fiber length. Straplike and fusiform muscles have longer fibers than pinnate muscles of equivalent mass, and they can therefore contract extensively (suited for moving light loads over long distances). Conversely, pinnate muscles develop more strength than fusiform or straplike muscles, but their range of contraction is shorter (suited for moving heavy loads over short distances). Because they have many short fibers, they can produce more force per muscle unit. The useful force (i.e., capable of generating mechanical work) produced by a pinnate muscle is not the force developed along the line of action of its muscular fibers, but rather the force resultant along the tendon axis. The other force resultant generates heat (and is unable to generate work), but the final useful force (work) far exceeds that developed by a muscle composed of parallel fibers.

4 Body Size

The size of organisms affects practically all aspects of life, imposing constraints on shape, physiology, behavior, and ecology. Body size, like these other aspects, is an attribute of an organism that may drive evolutionary change, and there is some debate over which predominates in cause and effect—that is, whether in most cases behavioral, ecological, and functional changes precede morphological change or vice versa (see Lister 2014 and references therein). Be that as it may, the last four decades of the twentieth century witnessed a significant output of scientific literature exploring the correlation of body size, and more specifically body mass (Campione and Evans 2020), with a great diversity of features in an animal's life (Peters 1983; Schmidt-Nielsen 1984; Damuth and MacFadden 1990; van Bergen and Phillips 2005) such as metabolism (Farlow 1976; McNab 1988; Hoppeler and Weibel 2005; Nagy 2005; White and Seymour 2005); locomotory habits (Alexander et al. 1979; Alexander 1985, 1989; Fariña et al. 1997; Biewener 2005); diet, population density, and home range (Damuth 1981a, 1981b, 1987, 1991, 1993; Lindstedt et al. 1986; Reiss 1988; Swihart et al. 1988; Nee et al. 1991); and extinction processes (Flessa et al. 1986; Lessa and Fariña 1996; Lessa et al. 1997; Barbero et al. 2020). But only more recently has the role of body size in the structure of trophic networks and other ecological networks begun to be explored systematically (e.g., Marquet et al. 2005; Woodward et al. 2005; Lyons and Smith 2013; Morgan Ernest 2013).

Large and small animals enjoy different advantages. The largest ones have fewer predators and have advantages in the competition between males as part of reproductive behavior; small animals can survive with less food in hard times (at least in an absolute sense, because a larger animal requires less energy per unit of body mass than a smaller animal), whereas the large ones are more sensitive to environmental changes that can produce fluctuations in available resources.

Regarding body size and patterns in the structure of the trophic network of communities, predators are generally about one to three orders of magnitude larger than their prey in terms of body mass, though there are some significant exceptions (e.g., host-parasite and some host-parasitoid systems, pack-hunting carnivores, and baleen whales). This general biological phenomenon illustrates the links between the trophic structure of entire communities and body size.

Body size also affects the temporal and spatial scale of trophic networks. Seasonal differences in diet, prompted by size changes during ontogeny, seem to account for many of the "intraguild feeding loops" seen in food networks, as between, for example, insects with aquatic larvae: large late-stage trichopteran larvae hunt megalopteran larvae, which, in turn, eat small early-stage trichopteran larvae. Indeed, the main determinant of a predator's trophic position within a food network is often its size (variable), rather than its taxonomic identity (fixed). Food networks are frequently considered as separate entities within clearly marked spatial boundaries (e.g., woodlands, savannas, deserts, lakes), but such boundaries may, in fact, be blurred due to body size. For instance, a large predator may forage over an area that includes several trophic networks.

Alterations in the size spectrum of a community, by events such as excessive fishing, species extinction, and invasion of new species, may dramatically alter community stability and the functioning of ecosystems to such a degree as to induce cascade extinctions, as is evident in the sharp drops in populations of the common seal (*Phoca vitulina*), the northern sea lion (*Callorhinus ursinus*), Steller's sea lion (*Eumetopias jubatus*), and the sea otter (*Enhydra lutris*) in the northern Pacific Ocean. These declines seem to have been the result of increased predation by killer whales (*Orcinus orca*), which preyed on large whales before industrial whaling drastically decreased their numbers. The invasion of a great predator in these coastal food networks seems to have caused a collapse in pinniped and otter populations on the coast. The sharp drops in the populations of the smaller predators resulted in an explosion in the number of sea urchins, which led to a greater grazing pressure on sea algae (Woodward et al. 2005).

Large animals undergo physical limitations that influence body shape, and, reciprocally, shape imposes restrictions on size. The larger the animal, the more its design has to be modified for counteracting gravity (i.e., for carrying its own weight). For terrestrial forms, the maximum size limit may be reached when the limbs supporting weight have to be so massive that locomotion would become impracticable. The same can be said for inertia; larger animals have to spend proportionally more energy to deal with inertia than do smaller ones. Body parts involved in defense or mating behaviors also can grow at different proportions from the rest of the body (i.e., allometrically; see below), such as the case of ram horns, showing that size and shape can be linked to biological function.

Studies of functional morphology and biomechanics frequently use measurements that must be correlated with size before functional interpretations can be made about extinct organisms. This raises several concepts that should be clarified before we proceed. Among the most important is the concept of size. In a trivial sense, any measurement can be used as a measure or assessment of size, such as length or height. However, these last two do not necessarily capture size in as biologically meaningful a sense as mass does. An example is easily furnished by considering which among a large adult African elephant, black rhinoceros, or giraffe is larger. Certainly, a giraffe is much taller than the other two, but what does this indicate with respect to the attributes noted above, such as metabolism? A giraffe is taller, but the elephant is longer. How would length be measured in a comparable way? Would we include length from head to tail? Should the elephant's trunk be included in overall length? In which position would the head of the giraffe be oriented, compared with the other two? Would we include the tail? These, of course, are rhetorical questions, posed to convey the idea that using straightforward linear measurements can be ambiguous in considering size and shape (see next section).

However, assessing size in terms of mass (hence body mass or body size) allows size to be compared more readily and thus provides a much more meaningful measure, because it gives us information on biological traits related to volume and its relations with other dimensions (areas and lengths), and indirectly to metabolism (which we will see is fundamental for understanding size in biology). Despite the great differences in height, giraffes and black rhinoceroses largely overlap in body size, as measured by mass, whereas elephants are considerably larger than both (Nowak 1999).

Obtaining a body mass for extant vertebrates is, in most cases, fairly straightforward. Mass, after all, can be measured directly using scales, but there are factors, such as *when* a mass is obtained, that should be taken into consideration, because mass may vary considerably depending

on time of the season (e.g., pre- and posthibernation), sex in cases of sexually dimorphic species, and even after copious feeding. It is also true that for some, such as large cetaceans and terrestrial mammals, weighing becomes a challenge, and dismemberment is required. Even so, the possibility of obtaining a direct measure of mass for extant species remains.

However, this is not the case for extinct species, for which, obviously, direct body masses are not available, so that methods for estimating mass are required. One such method is making use of correlations between body mass and measurements, such as length or width, of skeletal elements. Whereas sound correlations have been established between some skeletal elements and body mass for some groups of vertebrates, they are not always equally applicable to other groups, particularly to extinct forms that do not have extant analogous representatives (see below). And, as usual for fossil taxa, we must deal with skeletal remains that are often fragmented or incomplete and the fact that some taxa are represented by better-preserved specimens than others. In any event, it is important to grasp the relationships among linear measurements, areas, and volumes and to consider how they relate to body mass; these concepts are considered below.

Proportions: Geometric Relationships between Length, Area, and Volume

Around 300 BC, the Greek mathematician Euclid (325–265 BC; fig. 4.1) wrote *Elements*, which formally considered properties of points, straight lines, and regular geometric shapes, both planar and solid, such as polygons, circles and spheres, and triangles and cones, among others. In most morphometric studies, researchers employ linear measurements (e.g., width, length, and height) or distances between anatomical points (or landmarks); all of them are lengths. As noted in chapter 2, length is a fundamental physical quantity; that is, it cannot be defined in terms of other quantities and, therefore, is a magnitude of dimension 1. Area is a measurement of the surface of a two-dimensional shape, and is obtained by multiplying lengths. Area is a magnitude of dimension 2. Surface area is the measurement of the surface of a three-dimensional object. Volume is an extension measurement of the space occupied by an object. It results from multiplying lengths in their three dimensions and, therefore, is a measurement of dimension 3.

Relationships among such variables are described mathematically by equations termed *power equations*, which explain the kind of scaling relationships and have the general form

$$y = ax^b,$$

where the exponent b indicates the dimensional relationships between x and y; in other words, where one variable (y, the dependent variable) changes according to another variable (x, the independent variable) multiplied by a coefficient a (i.e., conversion factor, also called the scaling coefficient in physiology, that converts the values and units, but can be related to shape as well; see below), and raised to a power b (also called the scaling exponent in physiology). As an example, consider a cube with length of its sides equal to 1 cm (fig. 4.2). For such a cube, it is fairly easy to work out in your head that the area of one of its faces is 1 cm², its surface area is 6 cm² (because a cube has six faces), and its volume is 1 cm³, but we can apply the power formula to illustrate its use.

For the area of one of its sides (conversion factor equals 1):

Surface = a(length)²
Surface = 1(1 cm)²
Surface = 1 cm²

For its surface area (because a cube has 6 faces, the conversion factor equals 6):

Surface = a(length)²
Surface = 6(1 cm)²
Surface = 6 cm².

For its volume:

Volume = a(length)³
Volume = 1(1 cm)³
Volume= 1 cm³.

Joining together 27 of these cubes produces a larger cube with edges 3 cm in length; an area, for each face, of 9 cm²; a surface area of 54 cm²; and a volume of 27 cm³. It is straightforward to verify each of these values by using the power equations, as was done above, but what is more interesting, at least with respect to organisms, is the way that changes in size affect these values. Comparing the values of a cube of 1 cm length and another of 3 cm length reveals a fairly straightforward relationship: a threefold increase in length increases the surface of one face by a factor of this increase (i.e., 3) squared (i.e., 3² = 9), the surface area by six times the increase squared (6 × 3² = 54), and the volume by the increase cubed (3³ = 27). A further notable aspect is to consider how the surface area compared with volume changes between smaller and larger cubes. The

4.1. Euclid of Alexandria (325–265 BC), a mathematician and geometer who developed his studies in the Alexandria School in Egypt, is known in the West as the "Father of Geometry." His most influential work, *Elements*, is a treatise comprising 13 volumes in which definitions, postulates (axioms), propositions, and mathematical demonstrations (theorems) are described simply and logically. Painting by Jusepe de Ribera (Spanish/Italian, 1591–1652); Getty Museum Collection.

smaller cube, with a surface area of 6 cm², has a surface area to volume ratio of 6/1 or 6, whereas the larger cube, with a surface area of 54 cm², has a surface area to volume ratio of 54/27 or 2. That is, the smaller cubes have a larger relative surface area. This is true for any set of forms that display geometric similitude—forms that differ in size but have the same shape. For instance, for a hypothetical human giant whose height is three times greater than that of a typical human being, the cross-sectional area of his leg bones (and of body surface area as well) will be nine times larger than ours, but his volume will be 27 times larger and surface area to volume ratio will be one-third that of ours.

Mass is a scalar measurement that refers to the amount of matter that constitutes a body, and it is expressed in kilograms (kg). The concept of mass is closely related to those of capacity and volume, referring to how much matter an object contains and the space it occupies. This

4.2. Geometric relationships among length, area, and volume. In a cube with edges 1 cm in length (*green* cube), each side has an area of 1 cm², the surface area is 6 cm², and the cube occupies a volume of 1 cm³. Tripling the length of each edge is equivalent to gathering 27 cubes of the same size (*golden* cubes), which constitute a larger one (*red* cube) with edges 3 cm in length, a total area of 54 cm² and a volume of 27 cm³. That is, tripling length will increase the area by nine times and the volume by 27 times.

relationship is given by equivalencies among these three measurements and water. A cubic decimeter of water (volume) is contained in a liter (capacity) and constituted by a kilogram of matter (mass). That is why we say that mass is a scalar measurement of dimension three. Thus, volume and mass are effectively the same thing for water; for vertebrates, this relationship holds nearly as well because the global density of the entire body (averaging different tissues with different densities) is very close to that of water. We will see the importance of this later.

We also established above that volume scales with the cube of the length and that mass increases by the same factor as volume. Thus, if the mass of a small cube, with length 1 cm, is 1 g, then the mass of a larger cube, with length 3 cm, is 27 g. This only applies, however, if the cubes (or any set of objects being compared) have the same density. The relationship does not hold if we compare a solid ball with a soccer ball, in which most of the mass is contained in the covering of the ball (i.e., highly heterogeneous density). So, the inner structure and density of objects are important.

Linear measurements are frequently and appropriately used as a proxy for size in paleobiological studies; however, as illustrated by the example of the African elephant, black rhinoceros, and giraffe, their application could lead to ambiguous interpretations if the context is not correct. In the end, it is a matter of shape. Scaling within phylogenetic

groupings is often more straightforward, because shape is typically maintained; among felids or canids, for example, sizes may change within each clade, but the shapes of different felids or canids stay much the same. In any case, the same dimensional relationships would still hold for differently shaped animals (e.g., African elephant, black rhinoceros, and giraffe); that is, the exponent, *b*, in describing relationships among length, surface area, and volume. However, the coefficient, *a*, would be different. Meeh (1879) expressed the body surface area to volume formula as the ratio between surface and volume to the power of 2/3, which generates an *a* value when *y* is a surface and *x* is a volume, depending on shape. In the power equation, while *b* = 0.67 (or 2/3), *a* is 4.84 for a sphere and 6 for a cube; for animals, *a* is a species-specific value termed the *Meeh constant* that is about 10 for "compact-shaped" mammals and birds and >10 for elongated species (e.g., *Mustela* spp.; Withers et al. 2016; see also the table in Lusk 1928, 123). In our example, we would expect a similar value of *b* for relationships between lengths and volume (i.e., an exponent of 3) and different values of *a* depending on the animals' body shape as determined by the Meeh constant between surface and volume.

Although in paleobiological studies linear measurements are frequently correctly used as a proxy for size, as we have seen in the example of the African elephant, black rhinoceros, and giraffe, their application could produce ambiguous interpretations if the context is not correct. For certain biological traits, such as weight bearing, basal metabolic rate, and heat production/loss, it is more convenient to deal with an individual's body mass rather than its volume, area, or lengths.

Allometry

The term *allometry* was first used by Julian Huxley (1887–1925; grandson of "Darwin's Bulldog" Thomas Henry Huxley, brother of the famous writer Aldous Huxley and the Nobel Prize in Medicine winner Andrew Huxley) and Georges Teissier in 1936 (Huxley and Teissier 1936; Gayon 2000; Shingleton 2010) for referring to changes in relative proportions of body parts correlated with changes in overall size (fig. 4.3). Thus, the aim of allometric studies is to assess the variation of morphometric variables or other features of organisms associated with size variation (Huxley 1932; Cock 1966; Gould 1966; Alexander 1985). These relationships are also formalized mathematically by means of power equations of the type $y = ax^b$ (see above).

Applying the logarithmic transformation converts the power equation into a linear equation:

$$\log (y) = \log (a) + b \log (x)$$

where y is a biological attribute (e.g., morphometric, physiological, ecological), $\log(a)$ is the ordinate to origin or normalization constant, b is the slope or allometric coefficient, and x is body size.

When a power equation is log-transformed, converting it to a linear equation, the power coefficient (b) and the conversion factor (a) change so that $\log(a)$ becomes the ordinate to origin and b equals the slope of the linear equation, thus representing the dimensionality of the ratio of y and x. For example, when the independent variable x is a mass or volume, and y is a linear measurement, b will be a multiple of 1/3, whereas if y is a surface, b will be a multiple of 2/3. Of course, if the lengths of two skeletal elements are compared, isometry would be indicated by an exponent of 1; if the length of one bone doubles, then so does the length of the other. In other words, theoretical values of isometry are expected to be the same as the dimensionality ratio of y and x.

When the change in a biological attribute follows a geometric similitude with respect to volume (mass)—that is, proportions are maintained in spite of changes

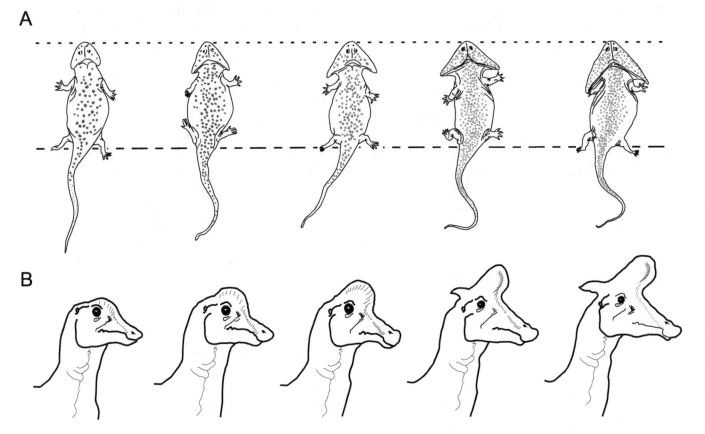

4.3. Allometry. Changes of relative proportions of body parts related to the increase in the total size in two ontogenetic sequences. *A*, the basal tetrapod *Diplocaulus* (Lepospondyli), scaled by head + trunk length; redrawn from Rinehart and Lucas (2001). *B*, the ornithischian dinosaur *Lambeosaurus lambei* (Hadrosauridae), from a young stage with skull length 40% of an adult size (*left end*) scaled to 100% of adult size (80 cm; *right end*); redrawn from Evans (2010).

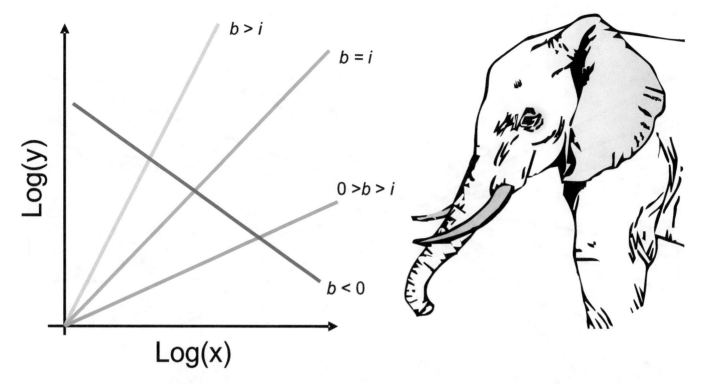

4.4. Relationships among variables. The *red straight line*, with negative slope (*b* < 0), corresponds to an enantiometric relationship (i.e., a decrease in the dependent variable *y* with the increase of the independent variable *x*). The *green straight line* represents an isometric relationship, with positive slope (*b* = *i*), where an increase in *x* is followed by the same increase in *y* (i.e., they maintain geometric similitude). According to the dimensionality of variables, *i* = 1 (two lengths), *i* = 2/3 (volume for *x*, area for *y*), *i* = 2 (length for *x*, area for *y*), *i* = 3 (length for *x*, volume for *y*). *Straight yellow* and *blue lines* represent positive and negative allometric relationships, respectively. An example from elephant ontogeny illustrates these principles. Ears maintain geometric similitude with body size, but the eye has a negative allometric relationship and the tusks a positive one; that is, proportionally adult elephants have ears of proportionately the same size as those of younger elephants, but their eyes are proportionately smaller and their tusks proportionately larger.

in size—we say that it follows an isometric relationship. For linear features such as lengths, widths, or depths, for example, the allometric coefficient *b* is a multiple of 1/3.

Deviations from this isometric relationship, with an allometric coefficient *b* higher or lower than theoretically expected, indicate that the involved variables follow an allometric relationship: allometry is the nonisometric change in an attribute (e.g., shape) that occurs with change in size and can be positive or negative, depending on whether the variable increases more (positive) or less (negative) quickly than the reference part in a smaller organism. As a result, larger organisms have for such features a higher or lower value than that expected for their body sizes (fig. 4.4). When a variable decreases with the increase of another, it is referred to as enantiometry. A striking example is the shrinking of larval structures during metamorphosis, such as the gills and tail of anuran tadpoles; another, more subtle, example is the relative cranial reduction during growth in primates (Corner and Richtsmeier 1991).

Reciprocally, allometric equations have been used for estimating body size (expressed as mass) from linear measurements. This application is very useful and is commonly used in paleobiology.

In figure 4.5, allometric changes of several biological attributes of eutherian mammals are reflected in relation to body size expressed as body mass on a logarithmic axis. Thus, body surface is proportional to body mass to the power of 2/3 (note the equivalence with volume, mentioned above) and skeletal mass to the power of 1.09, as occurs in cases of geometric similitude. Some attribute values, such as speed and home range, also increase with body size, though less quickly, as indicated by the lower slope, whereas others, such as population density and metabolic rate, decrease, as reflected by a negative slope.

Elastic Similarity

McMahon (1973) expressed the concept that the elastic deformation of an animal under its own weight should be similar for any body size. In theory, an organism cannot be larger than the size at which the applied stress surpasses the yield stress of its biomaterials. However, McMahon claimed that the elastic modulus should also be taken into account, given that biological structures should withstand bending stresses as well as compressive static forces. Considering the elastic modulus renders

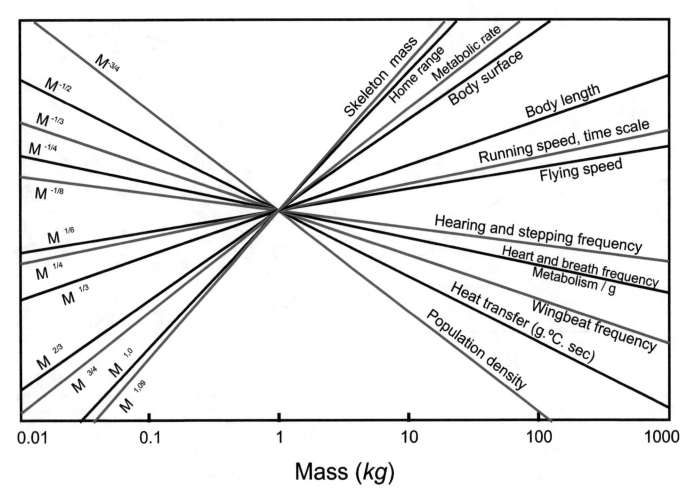

4.5 Generalizations of allometric changes for diverse biological attributes regarding body size, mainly in eutherian mammals. When using a double logarithmic scale, exponential curves are transformed into straight lines, where the exponents to which masses are raised (*left side*) correspond to their slope. For instance, body surface areas of animals, like the areas of geometric figures, are proportional to mass and volume raised to a 2/3 power. In a graph of double logarithmic scale, the exponent 2/3 is the slope of the straight line; that is, when the mass grows by 10³ units, the area does so by10²; or, in other words, when the mass increases by the third power (i.e., 3) the area does so by the second power (i.e., 2). Modified from Calder (1996).

maximal lengths proportional to 2/3 power of diameters, rather than a power of 3/4 as predicted only by compression on cross-sectional areas. In practical terms, this implies that all organisms would be designed for having a similar likelihood of elastic failure of the structures supporting them. That is, if this similitude existed, mice and elephants would have the same likelihood of suffering fractures in their limb bones. According to a recent review by Bertram (2016c), although McMahon's ideas initially seemed to receive support from a range of data that suggested elastic similarity may be an important feature of mammalian scaling, there is no structural reason why supporting bone deflections should remain proportional to limb bone length. For example, even if resistance to deformation is the main function of the long bones (Currey 1984), one might suppose that a deflection limit relative to bone width might be even more functionally relevant (Bertram 1988). Elastic similarity proportions will result in increased stress in bending (Alexander 1977), albeit to

a lesser degree than geometric similarity. Ultimately, this will lead to diminished support capability, just as geometrically similar proportions should. Actually, McMahon's concept seems to work quite well within phylogenetically (and to some extent size) restricted groups, such as bovids and primates, but not at broader taxonomic scales. Bertram (2016c) claimed that, taken as a whole, the accumulated data on structural shape changes over the size range of terrestrial mammals may indicate a complex of scaling relationships, each pertaining to a different size or phylogenetic group. While these aspects are interesting and important, their further consideration is beyond the scope of this book.

Size and Metabolism

The allometric relationship between basal metabolic rate and body mass has been one of the most studied of biological phenomena. In earlier investigations, it was thought that the effect of size on basal metabolism could be determined by the rate of heat loss through the surface area of the body, and thus proportional to mass raised to the power of 2/3. According to Brody (1945), the first scientists who noted that the effect of size on metabolism could reflect simple geometric scaling were Sarrus and Ramaeux in 1838. Max Rubner's studies on dogs in 1883 demonstrated that the relative metabolic rate of a given animal decreased with the same rate as its surface-to-volume ratio. This ratio was known as Rubner's surface law of metabolism. However, since the formalization of the allometric ratio by the physiologist Max Kleiber (1932) and several subsequent studies, it was found that the interspecific animal variation of the metabolic rate at rest is proportional to 3/4 of body mass (a little higher than 2/3; but see White and Seymour 2003, 2005 for a critical review of the data on which metabolic rates have been compared). For each fourfold increase in mass, only an increment of approximately three times the oxygen consumption is produced. It was subsequently also observed that the coefficient b, rather than being a multiple of 1/3, was actually a multiple of 1/4 (e.g., cardiac and respiratory rate; Peters 1983; Calder 1996; Brown and West 2000).

West et al. (1997, 1999a, 1999b) proposed that the exponent multiples of 1/4 found in many biological attributes assessed with regard to body mass stem from the fractal designs of structures and system functions (e.g., exchange surfaces of the respiratory system), and therefore empirically obtained exponential relationships mathematically describe hierarchical organization and fractal design (dynamics and self-similarity) over a wide range of scales (Brown et al. 2002).

Regardless of the mechanistic origin of allometric coefficients, the ratio of metabolic rate to body size to the power 3/4 is too distant from the ratio 2/3. If the metabolic rate is greater than 3/4, larger animals surpassing the optimal mass would certainly overheat, whereas small mammals would be more susceptible to cold temperatures than they really are. If the metabolic rate were lower than 2/3, the opposite would happen.

Size, Metabolism, and Ecology

In figure 4.6, several known examples are shown on the role of body size and metabolism in the structure of trophic and other ecological networks. Ingestion rates of consumers (grams of ingested prey items/time) and secondary production (grams of produced biomass/time) vary allometrically with an exponent close to 3/4, so that the amount of food required per individual consumer is proportional to its metabolic rate and so that a constant

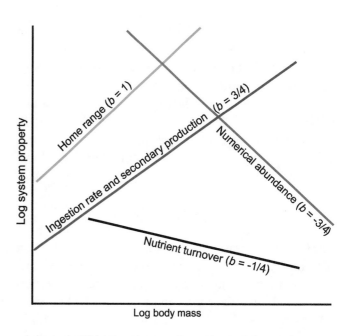

4.6. Examples of the most common allometric relationships between body mass and variables of ecosystem properties. The axes represent logarithms of variables and b indicates the value of the slope. Redrawn from Woodward et al. (2005).

fraction of metabolism tends to be allocated to biomass production. The home range of animal species is expected to scale to mass, that is M^1, a relationship that has been supported by field observations. Numeric abundance of species through trophic levels is expected to scale as M^{-1}, whereas field observations indicate it actually scales as $M^{-3/4}$ within individual trophic levels using a common energy source. The average rate of replacement of common elements (e.g., carbon or oxygen) and several important nutrients (e.g., phosphorus) scales as $M^{-1/4}$.

Over the past several decades, the metabolic rate of organisms has been considered as the main biological parameter that governs most of the patterns observable in ecology. This extension of Kleiber's law is called the metabolic theory of ecology (MTE; Brown et al. 2004; Savage et al. 2004) and is based on an interpretation of relationships among body size, body temperature, and metabolic rate. The main implication of MTE is that body size and temperature influences on metabolic rate define the main limitations to ecological processes. If this applies from the individual to the ecosystem levels, then attributes of life history, population dynamics, and ecosystem processes could be explained by the relationship among metabolic rate, body size, and body temperature. According to the MTE, metabolism restricts growth, reproduction, and death patterns at the organismal level: small organisms with fast metabolism have high feeding rates, and this produces fast growth, accelerates senescence, and promotes an early death, which is why natural selection promotes early reproduction. This, in turn, explains the diversity of reproductive strategies (*r* and *k*) as a consequence of the metabolic restrictions of organisms at the level of populations and communities. At the ecosystem level, the MTE explains the relationship between temperature and biomass: production rate of average biomass is higher in small organisms than in larger ones, which is, in turn, affected by temperature due to the variations it produces on metabolic rate.

Body-Size Estimate

The first step in estimating body size is the quantification of form, which requires that form be converted into a matrix of numeric attributes by using measurable variables. To this end, several devices can be used, such as calipers, 3-D digitizer arms and scanners, digital photos, and software, depending on the requirements and aims of a given study. A particularly cogent and comprehensive review and comparison of methodologies for estimating body mass is presented in Campione and Evans (2020). In chapter 2, we described the most common approaches to quantification of form: linear and geometric morphometrics. Historically, utilization of the former has been considerably more common in most investigations, including those on body-mass estimates (Campione and Evans 2020), whereas, as noted below, the application of analytical tools specific to and data generated from geometric morphometrics are a more recent development. We have already pointed out the importance of size and its influence on nearly all biological features. Studies on functional morphology and biomechanics often use measurements that should be tested for allometry in extant organisms before functional interpretations based on them can be applied to extinct organisms. Size, therefore, is one of the first attributes that should be considered in attempts to understand the biological traits of species of the past.

As noted earlier, body mass (*M*) is a scalar quantity that refers to the amount of matter expressed in kg. By contrast, weight (**W**) is a vector quantity that reflects the gravitational force (**g**) acting on the body and is expressed in newtons (N; SI units) or, its equivalent non-SI units, kilogram-force, kg-F (force exerted on 1 kg of mass by the Earth's average gravitational force of 9.81 m/s²; Newton's law of motion). So, on the Earth's surface an object of 1 kg of mass weighs 9.81 N. The relationship between them is given by the equation $M = \mathbf{W}/\mathbf{g}$. Although physicists often castigate biologists on the bad habit of using mass and weight interchangeably, in practice, given that the value 9.81 N equals 1 kg m/s² (kg-F), on the Earth's surface, weight and mass, expressed in their respective non-SI

(kg-F) and SI (kg) units, are virtually identical. However, as we stated, they are conceptually and quantitatively distinct, and care must be taken when used in equations that involve a combination of force and mass. Here, we will mainly refer to size as mass values (scalar quantity) throughout the text.

The most common approach for predicting body mass is based on simple or multiple linear regressions of different parts of the body (Scott 1983; Anderson et al. 1985; Damuth 1990; Gingerich 1990; Janis 1990; van Valkenburgh 1990; Myers 2001; Egi et al. 2004; Farlow et al. 2005; De Esteban-Trivigno et al. 2008; Toledo et al. 2014). Body mass may also be estimated based on whole articulated skeletons by combining the principle of geometric similitude with scale models. Such models may be fashioned, for example, by an artist or by means of a virtual or digital model produced with computer software programs (e.g., Bargo et al. 2000; Vizcaíno et al. 2011a). The recent improvements in the application of 3-D imaging techniques to paleontology (e.g., surface laser scanning, computed tomography, and photogrammetry) have allowed for the rapid digitization of fossil specimens (Sutton et al. 2014; Brassey et al. 2015; Brassey 2017; Otero et al. 2020). These volumetric body-mass estimation methods are therefore becoming increasingly popular. Another method involves geometric morphometrics, where the size marker of a landmark configuration is the centroid size, which has been demonstrated as following the same behavior as body mass (Hood 2000).

Physical Scale Models

Physical scale models have been used to estimate body mass in several extinct vertebrates such as dinosaurs (Alexander 1985, 1989), glyptodonts (Fariña 1995; Vizcaíno et al. 2011a), and ground sloths (Bargo et al. 2000). Such models are constructed to convenient scales based on mounted fossil skeletons and reconstructions of animal musculature and form (fig. 4.7). The volume of models may then be obtained by using Archimedes's principle: an object submerged in a fluid experiences a vertical and upward thrust equal to the weight of the fluid displaced by the object (Alexander 1983; see chap. 5, "Mechanics of Fluids").

If we consider an object of weight **W** (recall that $M = \mathbf{W}/\mathbf{g}$) and density ρ that is submerged in a fluid of density ρ', where **g** is acceleration due to gravity, the volume (V_o) of the object may be calculated as

$$V_o = \mathbf{W}/\rho\, \mathbf{g}$$

Hence, the volume (V_d) of the displaced fluid is

$$V_d = \mathbf{W}\rho'/\rho$$

The downward force acting on it is $\mathbf{W}'$ and given by the equation

$$\mathbf{W}' = \mathbf{W} - \mathbf{W}\rho'/\rho \qquad (1)$$

Most extant mammals have nearly the same density as water (1,000 kg/m³); this is assumed, very reasonably, to be true of extinct mammals, and they too are assigned this ρ value. For the calculation, scale models are submerged in water, and thus ρ' is also equal to 1,000 kg/m³. The volume of displaced water is then weighed and the mass of the model obtained from equation (1) is multiplied by the cube (that is, the third power) of the linear proportions with respect to those of the actual size of the animal.

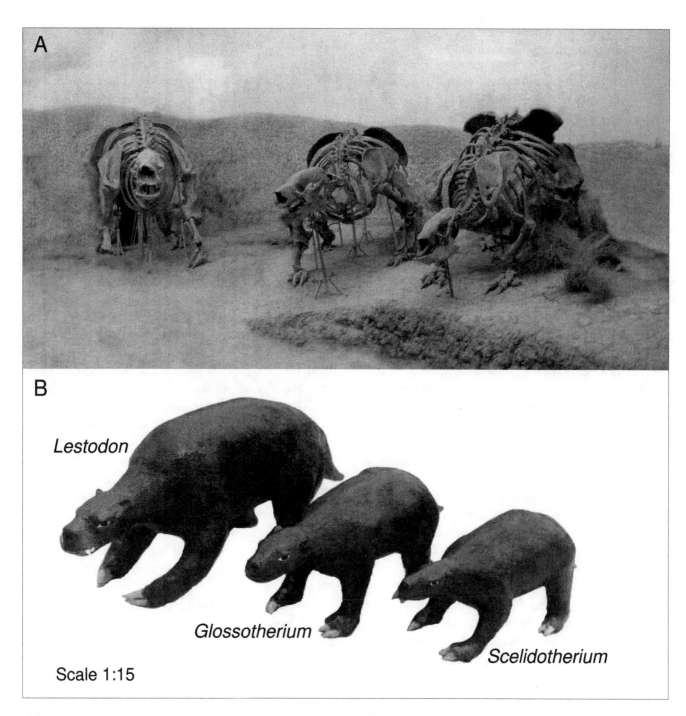

4.7. Physical scale models. *A*, skeletons of three Pleistocene ground sloths exhibited in the Museo de La Plata, Argentina. The body mass of *Lestodon* (to the *left*) may be close to 4 tonnes, of *Glossotherium* (*center*) 1.5 tonnes, and of *Scelidotherium* (*right*) 850 kg. *B*, physical scale models made based on the skeletons illustrated in *A* and used by Bargo et al. (2000) for estimating body mass.

In the example of body-mass estimation in terrestrial sloths, Bargo et al. (2000) used 1/40 scale models, and the mass of each model was multiplied by 40^3 (i.e., 64,000). Using a precision scale, the error obtained in estimating mass was lower than 0.5 grams. Consequently, the final error when multiplying by the scale should be lower than 32 kg. Given that the ground sloths analyzed by these authors had body masses on the order of hundreds and even thousands of kilograms, the error source was considered irrelevant.

Mathematical or Geometrical Models

Bramwell and Whitfield (1974) developed a method to estimate body mass based on reconstructions built in the form of scaled engineering drawings. These authors used simple geometric bodies (cones, cylinders, and spheres) to represent the head, limbs, neck, and trunk; these were subdivided into a series of slices from which total body volume was calculated as the sum of the volumes of these segments. According to Brassey (2017), this approach subsequently formed the basis of the more sophisticated 3-D mathematical slicing technique of Henderson (1999). This requires dorsal and lateral 2-D reconstructions of the extinct organism of interest, comprising a fleshed-out, soft-tissue contour outlining an articulated skeleton. Straight lines are drawn across 2-D profiles in a computer-aided design (CAD) package, and their intersections with the edge of the body contour exported as coordinate data. The intercept data are then used to define the major and minor radii of a series of elliptical slices along the body, with each pair of slices defining a volumetric "slab" with parallel/subparallel ends (fig. 4.8). Assuming a body density of 1,000 kg/m³, estimation of mass and center of mass are obtained quickly and easily and can be applied to any organism with bilateral symmetry.

For estimating the mass of ground sloths, Bargo et al. (2000) proceeded by estimating the transverse and dorsoventral diameters of the body at regular intervals (20 cm) along the skeleton, from the snout to the tip of the tail. Each pair of measurements was interpreted as the longest and shortest diameters of an ellipse, using PC-MATLAB (1990). Adjacent margins of the ellipses were joined to produce a continuous body contour, and the limbs were treated and measured as cylinders. Based on this computerized geometrical model, mass is estimated.

As mentioned above, the application of 3-D imaging techniques, such as surface laser scanning and photogrammetry, to paleontology has become popular for the digitization of fossils and estimating body mass using geometric methods. An example of the application of photogrammetry on free-living right whales represents an interesting analogy for the accuracy of its application to large extinct vertebrates. Christiansen et al. (2019) used a combination of aerial photography and historical catch records to estimate body mass of right whales (*Eubalaena*). Several parameters (body length, width, and dorsoventral height) of the southern right whale (*Eubalaena australis*) were measured by aerial photogrammetry, and the data were then used to estimate body volume by modeling the whales as a series of infinitely small ellipses. Further, the whales' body girth was calculated at three positions (across the pectoral fins, umbilicus, and anus), allowing the development of a linear model to predict body volume from the body girth and length data. The model was then used to estimate the body volume of several North Pacific right whales (*Eubalaena japonica*) that had been lethally caught and for which body mass was measured, yielding a conversion factor between the model-predicted mass and that measured. The conversion factor was then applied to predict the body mass of the free-living whales. Christiansen et al. (2019) noted that their model had a mean error of 1.6% (*s.d.* = 0.012) as opposed to 9.5% (*s.d.* = 7.68) for less sophisticated body length-to-mass model and, thus, that it is considerably more accurate in predicted body mass.

Allometric Equations

Estimating body mass from allometric equations is the method used most often (e.g., Damuth and McFadden 1990; Fariña et al. 1998; Millien and Bovy 2010). It is based on using a group of extant species, for which body mass and the length between homologous anatomical points of bony elements are known, as a frame of reference; as explained below, the choice of which group to use as a reference is not a trivial matter. Regression equations by ordinary least squares (bi- or multivariate) are applied using mass as the dependent variable and the measurement(s) as independent variable(s). Regression models by least squares are the most suitable when the aim is to estimate the value that the dependent variable would take for a given value of the independent variable (Warton et al. 2006; fig. 4.9).

Several authors, such as Damuth and MacFadden (1990), have postulated that more accurate mass estimates are obtained from postcranial elements (e.g., humerus and femur), because they support an animal's body. This is especially important for extinct species that have no extant representatives or that exceed the size range of extant species. Therefore, postcranial elements have been used for estimating body size (Alexander et al. 1979; Scott 1990; Farlow et al. 2005; De Esteban-Trivigno et al. 2008; Toledo

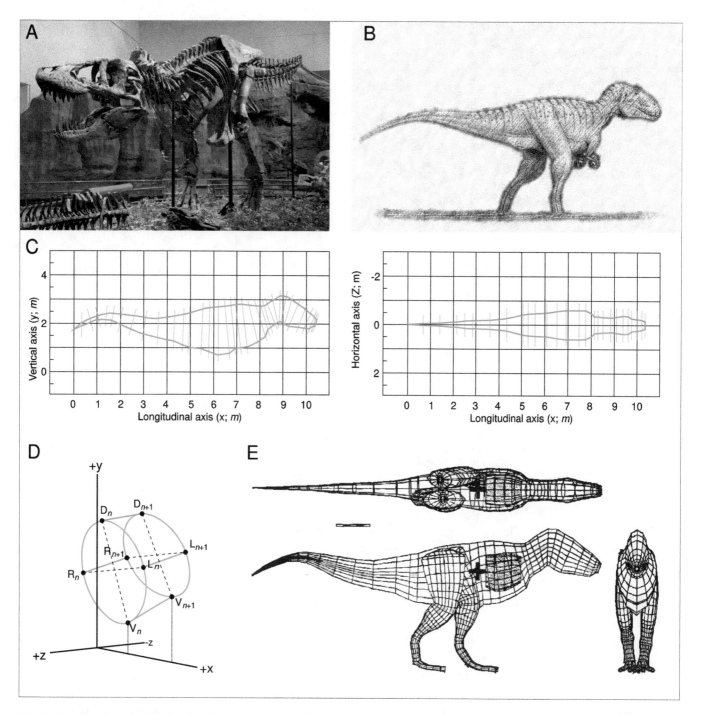

4.8. Body-mass and center-of-mass estimation through mathematical or geometrical models. *A, Tyrannosaurus rex* (holotype) exhibited in the Carnegie Museum of Natural History, Pittsburgh, PA (photo by Scott R. Anselmo under Creative Commons 3.0 license). *B*, life reconstruction of *Tyrannosaurus rex* (redrawn from Paul 1988). *C*, lateral and dorsal profiles of the tyrannosaur based on the life reconstruction, with cross sections (*dotted blue lines*) superimposed on the contour (*green*); combining *x* and *y* points with those of *x* and *Z* of the cross sections, ellipses in 3-D are obtained (modified from Henderson 1999). *D*, hypothetical configuration of a body segment, using the 11th segment as an example (modified by Henderson 1999); the extremes are defined by cross sections *n* and *n* + 1. *D, V, R*, and *L* are dorsal, ventral, right, and left points, respectively (*colors* as for *C*). *E*, dorsal, lateral, and anterior views of the wireframe model of the tyrannosaur used for calculating mass and center of mass (modified from Henderson 1999); the volume of the pulmonary cavities is also represented.

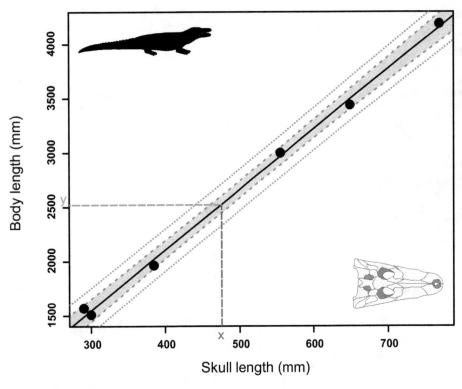

Regression line equation
y = 5.572x-125.35

Cricosaurus araucanensis
x = 475.2 mm
y = 2522   mm

Dakosaurus andiniensis
x = 780 mm
y = 4221 mm

4.9. Body-length estimates of metriorhynchid crocodiles of the Neuquén Basin (Argentina) based on skull length, obtained graphically from regression lines or analytically (modified from Herrera 2012). *Black solid line*, regression line; *gray shading between dotted gray lines*, confidence interval of 95%; *region between dotted blue lines*, prediction interval. For *Cricosaurus araucanensis*, graphic extrapolation from the skull length x = 475.2 mm (*dashed red line*) results in a body length of y = 2,522 mm (*dashed green line*). The same value can also be calculated from the regression equation. Note: the author demonstrated that the relationship between the logarithm of both variables (lengths) does not differ from 1 (same dimensionality), and, therefore, constructed the equation based on original measurements, thus avoiding bias introduced by logarithmic retransformation (see text).

et al. 2014). However, the fossil record is often fragmented and/or incomplete; further, as cranial and dental remains (particularly with regard to mammals) are more frequently recovered, given their heightened interest in systematics, they are also used for estimating mass (e.g., Gingerich et al. 1982; Myers 2001; Mendoza et al. 2006).

As suggested above, the interest in and investigations on body size witnessed an important boom in the 1980s and led to publication of the volume on body size in extinct mammals, edited by Damuth and MacFadden (1990), which includes numerous bivariate equations for estimating body mass in specific taxonomic groups, based both on postcranial elements and on the skull and teeth. Among the chapters in this volume, those by Damuth (1990), Janis (1990), Scott (1990), and MacFadden and Hulbert (1990) for ungulates, by van Valkenburgh (1990) for carnivores, and by Jungers (1990) and Ruff (1990) for primates are noteworthy. It became evident from these pioneering works that highly reliable equations (i.e., accurately predicting body mass) cannot be derived from all skeletal elements. In addition, Smith (1984; see also De Esteban-Trivigno et al. 2008) had noted that it is not sufficient to rely on equations with a high determination coefficient

(R^2), but that the percent prediction error (%PE), which provides an indication of the average error of the estimate, should also be considered.

Allometric equations have been shown to be useful and valid (despite the violation of certain statistical assumptions, such as absence of independence in the data; see below); the implications of using these allometric equations are best illustrated by an example. Cassini et al. (2012b) employed extant ungulates as appropriate analogs for notoungulates, a clade of extinct South American ungulates. The choice of which equation to employ is therefore not negligible, because the equation will provide a body mass that is the average of the group on which the equation was determined; if we plug into the equation based on bovids a particular value of *x*, let us say skull length, for the notoungulate *Adinotherium*, the body mass (*y*) generated for *Adinotherium* is the average body mass for a bovid with that particular skull length. In practice, then, we seek analogs that have skeletal dimensions similar to those of the extinct organism for which body mass is to be estimated; that is, such elements reflect similarity in overall size and shape. It would not be appropriate to use an equation based on primates to estimate body mass for

Adinotherium; were we to do so, we would be assigning a body mass to *Adinotherium* as though it were comparable in body form and size to a primate. If these precautions are taken into consideration, allometric equations offer a mathematically simple and robust model for estimating body size.

The procedure for obtaining mass estimates by allometric equations is the same for any skeletal element. First, equations that account for the existing empirical relationship between one or more variables with body mass must be generated, preferably through linear models. For this purpose, it is essential to have available a good sample of extant specimens of known mass that are phylogenetically related to or, at least, good analogs of the extinct taxa for which mass is to be estimated. Perhaps the most straightforward way to proceed is by making use of the osteological collections of natural history museums and taking the required measurements. However, because the specimens in such collections are only rarely associated with body-mass information, it is usually also necessary to rely on values reported in bibliographic sources (e.g., Janis 1990; Scott 1990), a methodology frequently used for this kind of approach (as by, e.g., Mendoza and Palmqvist 2006, 2008; Vizcaíno et al. 2006a, 2012a, 2018; De Esteban-Trivigno et al. 2008; Toledo et al. 2014). It should be kept in mind that some taxa in the reference sample could be sexually dimorphic for body size. In such cases, it is necessary to consider mass for each sex, according to the availability of information recorded in institutions and the literature; in the absence of such information, the information available is taken as representing average values for a species (Janis 1990).

We have not yet dealt with the difficulty of the lack of phylogenetic independence of our data. Several authors (e.g., Felsenstein 1985; Harvey and Pagel 1991; Diniz-Filho et al. 1998) have voiced their objections to this method of generating equations because it does not consider the influence of phylogenetic legacy in the relationship between osteological measurements and body size. Because organisms are linked by hierarchical phylogenetic relationships, their dimensions cannot be treated as independent measurements taken from a normal distribution, as linear regression assumes (Felsenstein 1985). Indeed, it is parsimonious to expect that closely related organisms will share similarity in shape and size, given that they share a common ancestor (see chap. 1). Hence, part of the variation can be explained by phylogeny, and the tendency for related species to resemble each other more than they resemble species drawn at random from the tree has been referred to as *phylogenetic signal* (Blomberg and Garland 2002) or *phylogenetic effect* (Derrickson and Ricklefs 1988). There are several proposed methods that consider the influence of phylogenetic legacy, such as phylogenetic eigenvector regression analysis (PVR; see Diniz-Filho et al. 1998). However, as explained in chapters 1 and 2, if the group under study is extinct (i.e., it has not left extant representatives) or the phylogenetic relationship between the extinct and extant groups is ambiguous (e.g., as is the case for South American native ungulates or extinct xenarthrans), these methods are not applicable.

Another aspect to consider is the taxonomic composition of the reference samples. As it is often the case that different groups do not have the same representation in number of species, regression analyses would be biased

A. Weighting example from Mendoza *et al.* (2006)

$$X_i = \frac{n_{(species)}/n_{(families)}}{n_{(species\ in\ the\ i\text{-}nth\ family)}}$$

B. Weighting example from De Esteban-Trivigno *et al.* (2008)

Formulas for genera

$$n_{(theoretical)} = \frac{n_{(species)}}{n_{(genera)}} \qquad X_i = \frac{n_{(theoretical)}}{n_{(species\ of\ the\ genus)}}$$

Formulas for families

$$n_{(theoretical)} = \frac{n_{(genera)}}{n_{(families)}} \qquad X_i = \frac{n_{(theoretical)}}{n_{(species\ in\ the\ family)}}$$

4.10. *A,* weighting formula from Mendoza et al. (2006). Value *Xi* is calculated for each family and is assigned as a new variable (weighting variable) to each species belonging to that family. Value $n_{(species)}$ and value $n_{(family)}$ correspond, respectively, to the total number of species and families represented in the sample. *B,* weighting formula from De Esteban-Trivigno et al. (2008). A weighting subvalue *Xi* is calculated at different taxonomic hierarchies (e.g., genus, family, order) from which a theoretical *n* that represents the average value is calculated. Value $n_{(species)}$ (*green*) and value $n_{(genera)}$ (*blue*) correspond, respectively, to the total number of species and genera represented in a given family. For *Xi* $_{(families)}$, the value $n_{(genera)}$ (*orange*) and value $n_{(families)}$ (*purple*) correspond to the number of genera and families represented in a determined order in the sample. Then, these *Xi* values are multiplied together to obtain a weighting variable that is used for considering the bias produced by taxa overrepresentation in the considered taxonomic hierarchies.

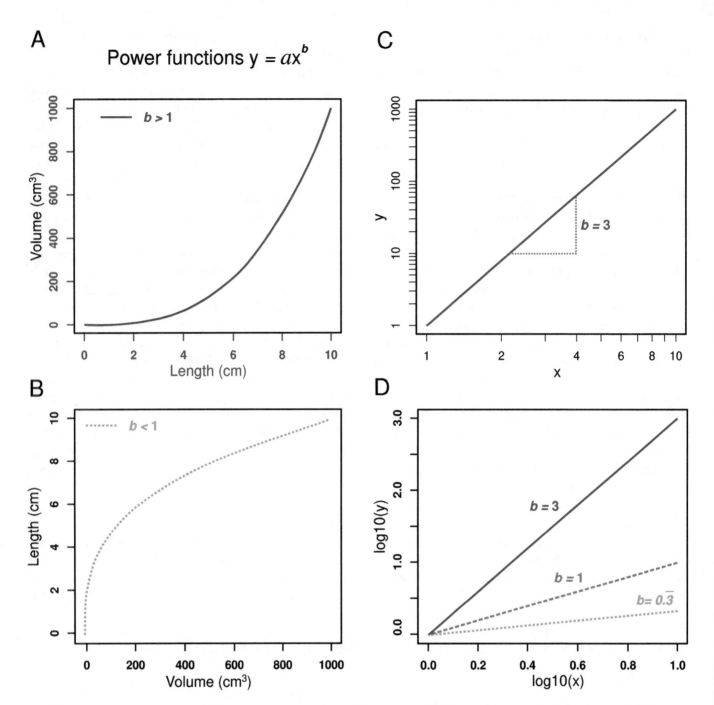

4.11. Power relationship and double logarithmic relationships. *A* and *B*, graphic representation of a power equation of type $y = ax^b$. In *A* (*solid blue line*), the exponent *b* is higher than 1, resulting from a variable *x* that increases linearly (length) and another variable *y* that does so to the third power (volume); that is $b = 3/1 = 3$. In *B* (*dotted green line*), the exponent *b* is less than 1, resulting from a variable *x* that increases to the third power (volume) and another variable *y* that does so linearly (length); that is, $b = 1/3 = 0.3$. *C*, linear representation of the equation $y = ax^3$ (*solid blue line*) at double logarithmic scale. In this case, the growth rate of variable *y* with respect to variable *x* equals the exponent *b* ($b = 3$). *D*, graph at normal scale resulting from applying the logarithm to both variables. Three examples of the values that *b* can take are represented: in *solid blue line* when $b > 1$ ($b = 3$), in *dashed gray line* when $b = 1$, and in *dotted green line* when $b < 1$ ($b = 1/3$). The three cases correspond to isometric slopes expected for the relationship between a linear variable and a cubic variable, two linear variables, and a cubic variable and another linear variable, respectively.

toward the better-represented groups. To avoid such biases, it is desirable to include as diverse a sample as possible and to weight the contribution of underrepresented taxa compared with those that are better represented. Such weighting functions serve to equalize the contribution of each taxon in the regression equation. To this end, several

approximations, of which we describe two examples, have been proposed.

Mendoza et al. (2006) proposed a straightforward method, in which the value (X_i) of the weighting factor for each species of a particular taxon—family, in this case—must equal the average number of species per

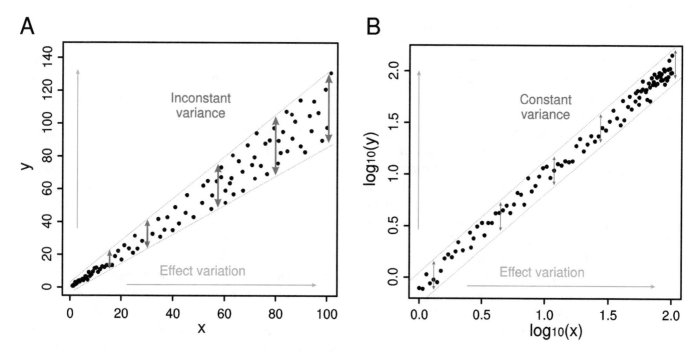

4.12. Heteroscedasticity and homoscedasticity. *A*, heteroscedasticity, data dispersion (variance) increases with the increase of the values of variables (*x* and *y*). *B*, homoscedasticity, when transforming data to logarithm (10-based in this case, but any base can be used), a homogenization of the data dispersion is obtained (constant variance).

family (number of species/number of families) divided by the number of species belonging to that family. Thus, the final contribution of each family to the analysis is the same (fig. 4.10*A*).

De Esteban-Trivigno et al. (2008) presented a somewhat more sophisticated method: for a family that has a genus with four species and another only with two, the total number of species (six) divided by the number of genera in the family (two) yields the theoretical value of an equal number of species for each genus (three). Finally, the actual number of species in each genus divided by the theoretical value (three, for this example) yields the value (X_i) for each species in the variable used for weighting (fig. 4.10*B*).

As mass is a quantity that increases by the third power, whereas lengths between anatomical points (landmarks) do so linearly, variables should be log-transformed (base 10 or natural logarithm) converting a power equation ($y = ax^b$; see above),

$$M = al^b$$

where *M* is body mass and *l* is any length (such as width, height; i.e., the measurement between two anatomical points), into a linear equation,

$$\log(M) = \log(a) + b \log(l)$$

where $\log(a)$ is the ordinate to origin and *b* the slope (fig. 4.11). This transformation also corrects for data heteroscedasticity (i.e., differences in data dispersion, or variance, among data of different orders of magnitude; fig. 4.12*A*), rendering it homoscedastic (i.e., homogeneity in the dispersion around the arithmetic mean in linear regressions; fig. 4.12*B*). As can be observed from the equation, arriving at an estimate of mass requires that the value of a particular measurement taken on a fossil skeletal element be plugged into the equation, from which mass may then be obtained by calculating its antilogarithm. Several authors (e.g., Smith 1993) have pointed out that applying the antilogarithm for obtaining the mass value in kilograms leads to a distortion of the obtained value due to the difference between the normal distribution of the logarithmic residuals and the transformed values. That is, when the logarithms of values form a normal distribution (i.e., symmetrical), the original (antilog) values are log normally distributed (i.e., asymmetrical with a pronounced tail to the right). Cassini et al. (2012b) proposed, as a means of avoiding such distortions, the use of nonparametric regressions per quantiles, which do not have dimensionality problems caused by logarithmic transformations and with which we will deal further below.

For the case of parametric regressions, several indices or correction factors must be obtained for each equation, which then have to be multiplied by the retransformed mass value. The most frequently used is that of Snowdon

(1991), who proposed the ratio estimator (RE) to correct an obtained value. For simply restoring the dimension, the prediction value should be multiplied by the RE corresponding to each equation. Values such as R^2, RE, and %PE will yield the validity and usefulness of the obtained equations for making estimates.

In summary, allometric equations constitute one of the oldest and currently most widely used methods for estimating body masses. The choice of the reference group (or of previously published equations) will depend on the variables that can be measured on the fossil specimens, the availability of specimens housed in collections, and whether there are phylogenetically related extant forms or extant putative ecological analogs that may be relied on. It is possible to begin with simple models involving only one variable and progress to multivariate models, or to summarize the information in a unique variable (centroid size) from morphogeometric data and refine prediction models as much as possible.

Center-of-Mass Estimation

If we consider an object at rest, such as an animal standing, we may view its mass as concentrated at a unique point, known as its center of mass. This is the point, to put it simply, where an animal is balanced. As an animal moves and changes the relative position of its parts, its center of mass changes from one instant to the next.

However, if we consider the animal moving as a whole, rather than in terms of the separate movement of its parts, we may still recognize a unique point where its center of mass is located. The position of an animal's center of mass allows us to know how the mass is distributed, which has important implications for both static equilibrium (e.g., standing on its legs) or dynamic equilibrium (walking, running, or flying) situations.

In solving or considering mechanical problems, it is useful from the start that an animal's center of mass be located. This may be accomplished by simple means. Consider that the animal is suspended from a wire. There are two forces acting on it: an upward force, due to the tension of the wire, and a downward force, due to gravity, that corresponds to the animal's weight. Equilibrium of the animal's body is achieved only when the two forces, which of course are vectors, act in precisely opposite directions. The upward force acts along the length of the wire, and weight acts at the center of mass and in line with the center of the Earth. Consequently, the suspended body will be at rest when the center of mass is aligned with the wire.

For determining the center of mass, scale models can be used as described by Alexander (1983, 1985). First, a model is suspended from a point along the sagittal plane of its head, as shown in figure 4.13, or tail; the center of mass would fall along line AB. Then, it is suspended from the middle part of its back; the center of mass would now fall along line CD. The body's actual center of mass lies

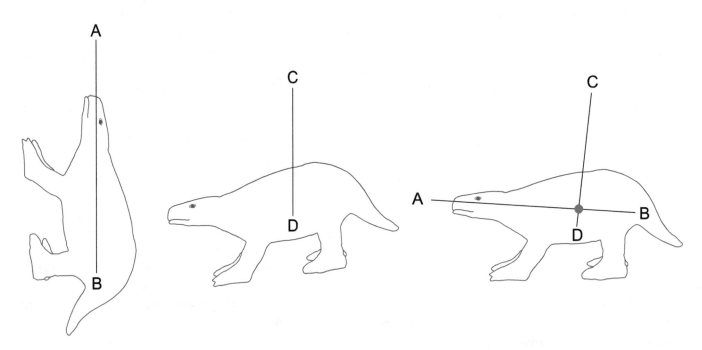

4.13. Center-of-mass estimates using physical scale models of a mylodontid ground sloth, as described by Alexander (1983, 1985). First, a model is suspended from a point along the sagittal plane of its head or tail; its center of mass is indicated as falling along *line AB*. Then, it is suspended from the middle part of its back; its center of mass is indicated as now falling along *line CD*. The model's actual center of mass lies at the intersection of *lines AB* and *CD* (*red circle*).

at the intersection of lines AB and CD. Alexander (1983, 1985) applied a correction to take into account the air-filled volume of the lungs because they comprise a region of lower density than the rest of the body that has a small, but measurable, effect on the position of the center of mass. Assuming that bone is distributed uniformly in the body and lung volume comprises 10% of body volume, the position of the center of mass is moved backward by 10% of body length. The center of mass can nowadays be calculated from tridimensional models generated by engineering and design software (e.g., Autocad, Rhino3D, 3D Studio).

In the static position represented by models, the center of mass of mylodontid ground sloths is located in the abdominal region, just anterior to the hind limbs (fig. 4.13). This means that when such an animal stood on four legs, 60 to 70% of its weight was supported by its hind limbs and 30 to 40% by its forelimbs. The same mass distribution was obtained for glyptodonts (Fariña 1995) and assumed for the giant sloth *Megatherium* (Casinos 1996). The mass distribution observed in these giant xenarthrans is exactly the opposite of what occurs in large extant quadrupedal mammals, such as elephants and ungulates (40% supported by the hind limbs and 60 % by the forelimbs; Alexander 1985; Alexander and Pond 1992).

Examples of Using Different Methods to Estimate Body Size of South American Endemic Fossil Mammals

Body-Size Estimates in Xenarthrans

Fariña et al. (1998) estimated body mass of 13 species of megamammals of the Lujanian Age (Late Pleistocene–Early Holocene), including seven xenarthrans, one notoungulate, one litoptern, one perissodactyl, one proboscidean, and two carnivorans. These taxa are known from very complete remains, both cranial and postcranial, offering the possibility of evaluating the reliability of the estimates obtained from the different elements. In the case of xenarthrans, represented by the glyptodonts and ground sloths in their study, the diversity of Lujanian forms is considerably greater than this group's present diversity (see Vizcaíno and Bargo 2014; Vizcaíno et al. 2018), so that phylogenetically based equations are unavailable, owing either to the size range or lack of morphological similitude with extant xenarthrans. Bivariate equations taken from the literature and defined for skull and tooth (Janis 1990) and postcranial measurements (Anderson et al. 1985; Scott 1990) were used (table 4.1). The authors concluded that tooth measurements were not suitable for estimating body mass of extinct xenarthrans, due to the fact that tooth homologies to other mammals are not clear and because many xenarthrans have a reduced and homodont dentition, so that body mass is underestimated. Some limb measurements, by contrast, yielded ludicrous overestimates that had to be discarded. For instance, the estimate obtained from the transverse diameter of the femur, which in the giant ground sloth *Megatherium* is unusually wide, yielded a value close to 98 tonnes for this sloth, the skeleton of which suggests a size similar to that of an African elephant (around five tonnes). In short, the obtained body mass values seemed to be highly dependent on the chosen measurement and morphological distances between the "problem" taxa and those used as a reference frame, and it was concluded that the arithmetic mean of all the equations used yielded reliable estimators.

De Esteban-Trivigno et al. (2008) developed a series of multivariate equations for body-mass estimation, with the specific aim of applying them to extinct xenarthrans. For this purpose, they obtained multivariate equations for a vast sample of extant quadrupedal mammals. To

Table 4.1. Examples of body size estimates in some Pleistocene xenarthrans, based on bivariate equations defined for skull, tooth, and postcranial measurements of extant mammals, excluding Xenarthra, taken from Fariña et al. (1998)

	Equation	*Megatherium americanum*	*Glyptodon reticulatus*
Humerus length (H1)	Log mass = 3.4026 * log H1 − 2.3707	6,279 kg	658 kg
Femur length (F1)	Log mass = 3.4855 * log F1 − 2.9112	2,169 kg	558 kg
Femur transverse diameter (F6)	Log mass = 2.821 * log F6 + 0.9062	97,417 kg	6,982 kg
Third lower molar length (TLML)	Log mass = log TLML * 3.183 + 0.801	1,095 kg	—
Seventh lower molar length (7LML)	Log mass = log 7LML * 3.201 + 1.13	—	253 kg
Total calculated equations		44	43
Maximum value		97,417 kg	7,005 kg
Minimum value		524 kg	31 kg
Arithmetic mean		6,073 kg	862.3 kg

Note: Some of them produce notable over- and underestimations. Equations with humerus measurements (H1) and femur (F1 and F6) according to Scott (1990); equations with tooth measurements according to Janis (1990).

Table 4.2. Equations developed and applied by De Esteban-Trivigno et al. (2008) for estimating masses of extinct species of xenarthrans

Predictive Function	Adjusted R^2	RE	Mean %PE (LOOCV)
.457U2 + .836R2 + .969F4 1 1.04H1 – .515U1 – 6.259	.995	1.0374	13.0
1.037T3 + .688H7 + 1.138F5 – 5.938	.990	1.0539	15.5
1.473F4 + 1.098S1 + .250H5 – 6.836	.993	1.0354	14.4
1.320R2 + .521F7 + 1.632H1 + .634R1 – 7.458	.993	1.0298	16.1
.988S1 – .883R1 + 1.062F1 + 1.232H2 + .495H4 – 8.472	.993	1.0042	14.7

Note: Adjusted R^2, determination coefficient adjusted by the number of explanatory terms in the model; RE, "ratio estimator", the correction coefficient of the logarithmic transformation; Mean %PE, mean percentage of prediction error of the estimate obtained using the leave-one-out cross validation (LOOCV) method. Equations are of form ln (mass in kg) = $\Sigma_i\, b_i$ (ln X_i) + a_i, where ln is the logarithm to base e (natural), Σ is the sum, X_i represents the *i-th* morphological variable (in mm), b_i the slope, and a_i is the constant term (intercept).

avoid taxonomic bias, they used the greatest diversity possible in the reference sample (xenarthrans, artiodactyls, rodents, and carnivorans, among others) and carried out sample weighting so that each taxon contributed equally to the equation, using the procedure described in the section "Allometric Equations" and figure 4.10*B*. These equations involve traditional linear variables that are easily obtainable (e.g., total lengths, diaphysis widths) and each one combines measurements of different postcranial elements. Apart from reporting the usual statistics (R^2, *p*), they also provided %PE, allowing assessment of the robustness and reliability of the prediction equation. In so doing, they found that equations with combined measurements of the scapula, humerus, ulna, and femur were the most reliable (table 4.2) and obtained mass values for a Miocene glyptodont from Patagonia (*Propalaehoplophorus australis*, 81 kg) and for some Pleistocene giant sloths (*Scelidotherium*, 580 kg, and *Lestodon*, 3,600 kg). However, these equations require the measurements of all the elements involved in the equation. As it is not usual to have such complete specimens available in paleontology, the application of these equations is necessarily very limited.

Toledo et al. (2014) obtained multivariate equations using a taxonomic sample of terrestrial mammals with different grades of arboreality or digging habits (xenarthrans, pangolins, carnivorans, and primates). The aim was to obtain equations for estimating the body mass of Miocene sloths from the Santa Cruz Formation of Patagonia (see chap. 9) for which arboreal and semiarboreal habits were hypothesized (chap. 6). As most specimens do not have all postcranial elements, the authors obtained equations for each postcranial element separately (scapula, humerus, radius, ulna, pelvis, femur, tibia, fibula, astragalus, and calcaneum; table 4.3). The authors performed the taxonomic weighting proposed by Mendoza et al. (2006). Before obtaining the equations, they tested the influence of phylogeny on body mass by means of orthonormal transformation analysis of the variance. For attaining the equations, they performed a stepwise analysis of the model using Akaike Information Criterion (AIC) to obtain regression equations for which the global values of redundancy were the lowest possible. A posteriori, variance inflation factor (VIF), estimators of predictive power, %PE, and quotients of logarithmic retransformation can be used. Among the

Table 4.3. Statistics for each predictive equation

Element	R^2	R^2 adjusted	F(df)	p-F	RE	%PE
Scapula	.916	.908	125.048 (2, 23)	4.39E-13	1.018	13.963
Humerus	.970	.963	130.357 (5, 20)	1.49E-14	.979	9.263
Ulna	.880	.875	175.608 (1, 24)	1.56E-12	1.213	21.167
Radius	.862	.849	71.548 (2, 23)	1.34E-10	1.031	41.363
Pelvis	.836	.821	58.448 (2, 23)	9.62E-10	1.301	18.663
Femur	.968	.955	77.674 (7, 18)	3.76E-12	1.055	7.833
Tibia-fibula	.934	.917	56.428 (5, 20)	4.17E-11	1.126	13.598
Astragalus	.862	.850	72.080 (2, 23)	1.24E-10	1.034	18.419
Calcaneus	.802	.794	97.417 (1, 24)	6.34E-10	1.270	47.838

Note: R^2, squared determination coefficient; adjusted R^2, determination coefficient adjusted by the number of explanatory terms in the model; F, value of Fisher test for null dependence and associated freedom degrees (*df*); p-F, probability of *F*-test; RE, "ratio estimator" is the correction coefficient of the logarithmic transformation; %PE, percentage of prediction error of the estimate (from De Esteban-Trivigno et al. 2008).

Table 4.4. Averaged body masses of Miocene sloths (Xenarthra, Folivora)

Family	Genus	Mean body mass (kg)
Megalonychidae	*Eucholoeops*	59.52
Megatherioidea	*Hapalops*	43.266
Megatherioidea	*Analcimorphus*	64.395
Megatherioidea	*Schismotherium*	37.992
Megatheriidae	*Prepotherium*	107.793
Mylodontidae	*Analcitherium*	88.226
Mylodontidae	*Nematherium*	89.329

Note: Data estimated from multivariate equations obtained for postcranial elements, according to Toledo et al. (2014).

postcranial elements, equations for the humerus, femur, and tibia-fibula had the lowest %PE and yielded estimates with less deviation (table 4.3). Table 4.4 shows the averaged results of the body-size estimates for the sloth genera of the Santa Cruz Formation.

Vizcaíno et al. (2018) compared body-size estimates generated by linear and multiple regressions based only on extant xenarthrans with those based on a wider sample of extant mammals. They found that equations based only on extant xenarthrans yielded remarkably low underestimates when compared with other methods, probably due to morphological differences between extant and extinct xenarthrans, supporting the idea that anagenesis could introduce bias when the reference frame is constrained by the phylogeny. Recently, Dantas (2022) proposed a standardized major axis regression to estimate the body mass of Pleistocene xenarthrans, an approach that has been questioned by Hubbe and Machado (2022), who

considered that a combination of an insufficient sample and a sample biased to smaller sizes makes the equation derived by Dantas (2022) unreliable for estimating xenarthran body masses.

Body-Size Estimates in Native Ungulates

The most common remains of the fossil native ungulate fauna are isolated teeth or partial mandibles. It is thus not surprising that paleontologists use dental variables and the respective equations provided by Damuth and MacFadden (1990), as in the work of, for instance, Fariña et al. (1998), Croft (2000), Villafañe et al. (2006), and Reguero et al. (2010). Fariña et al. (1998) estimated the body mass of *Toxodon* and *Macrauchenia* (Notoungulata and Litopterna, respectively), the most recent representatives of South American native ungulates, which lack living descendants. They used previously published equations from measurements of the skulls, teeth, and limbs of artiodactyl and perissodactyl ungulates (Anderson et al. 1985; Janis 1990; Scott 1990). In cases where more complete remains are available (because multivariate analyses require multiple measurements to be taken on the same specimen), authors such as Cassini et al. (2012a) employed several bivariate and multivariate equations from Janis (1990) and Mendoza et al. (2006), respectively (table 4.5). They followed the proposal by Christiansen and Harris (2005) for obtaining a mean weighted by %PE of each equation so that the values obtained from equations with high predictive errors could contribute less to the final mass value

Table 4.5. Examples of body masses obtained for some Miocene native ungulates (Santa Cruz Formation)

Family	Species	Janis (1990)			Mendoza et al. (2006)			Mean Body mass (kg)
		Mean	s.d.	n	Mean	s.d.	n	
Astrapotheriidae	*Astrapotherium magnum*	933.74	357.10	4	908.90	223.07	4	921.32
Macraucheniidae	*Theosodon gracilis*	112.44	60.09	2	130.66	83.20	2	121.55
Proterotheriidae	*Diadiaphorus majusculus*	72.03	14.85	9	92.08	22.28	7	82.05
	Tetramerorhinus cingulatum	33.09	4.87	3	50.34		1	41.71
	Thoatherium minusculum	20.55	4.75	6	27.84	1.09	2	24.20
Homalodotheriidae	*Homalodotherium* sp.	405.08		1				405.08
Toxodontidae	*Adinotherium ovinum*	100.29	9.18	8				100.29
	Nesodon imbricatus	644.85	203.26	12	630.17	112.09	10	637.51
Interatheriidae	*Interatherium robustum*	1.80	0.45	13	2.96	0.58	12	2.38
	Protypotherium australis	5.12	0.45	4	10.35	1.24	3	7.73
Hegetotheriidae	*Hegetotherium mirabile*	7.20	2.66	5	8.21	1.78	2	7.71
	Pachyrukhos moyani	1.62	0.16	4	2.64	0.23	2	2.13

Note: Data based on equations by Janis (1990) and Mendoza et al. (2006). Modified from Cassini et al. (2012a). *s.d.*, standard deviation.

reported for each specimen. Besides this, the difficulties do not differ from those described for xenarthrans.

Body-Size Estimation from Centroid Size: Native Ungulates and Sparassodont Metatherians

This methodology is also based on allometric equations, but landmarks (in either two or three dimensions) and tools typical of geometric morphometrics are used rather than multivariate equations from several measurements. In this method, the form of the object under study is constituted by size and shape (fig. 4.14).

Unlike the methods of linear morphometrics, where size is defined according to the type of analysis, in geometric morphometrics the size marker of a landmark configuration is the centroid size (*cs*; Kendall 1984; Bookstein 1986; Goodall 1991; Dryden and Mardia 1998). A landmark configuration is a matrix of *p* rows (number of landmarks) and *k* columns (number of dimensions; 2 or 3). A function at *x* is denoted by mathematicians as *f(x)*, such that the centroid size may be considered as a function of size that meets the following property:

$$f(a\ lc) = a\ f(lc)$$

where *a* is a positive scalar quantity that can be extracted from the landmark configuration matrix (*lc*). The *cs* of a landmark configuration is defined as the square root of the squared sum of the distances of each landmark to the centroid of the configuration (calculated as the column mean), expressed by

$$cs = \sqrt{\sum_{i=1}^{p} \sum_{j=1}^{k} (lc_{ij} - \overline{lc_j})^2}$$

where *lc* is the configuration matrix of landmarks for each specimen, of *p* rows (number of landmarks) and *k* columns (number of dimensions).

The *cs* is used in geometric morphometrics, as it is not correlated with any shape variable when there is no allometry (Bookstein 1986; Cressie 1986; Kendall 1986). Hood (2000) assessed the potential use of *cs* in studies of size sex dimorphism in the muskrat *Ondatra zibethicus* (Rodentia, Muridae) and of eight bat species (Chiroptera, Pteropodidae). This author concluded that the *cs* is a geometric quantity that follows the same behavior as body mass and is, therefore, an excellent indicator of body mass (fig. 4.15).

In the case of the ungulates from the Santa Cruz Formation, for which there are no extant representatives,

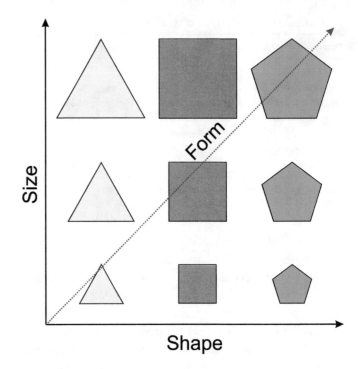

Above, 4.14. Changes in form as a result of changes in size and shape.

Facing top, 4.15. Graphs of body mass and centroid size illustrating size sexual dimorphism in pteropodid bats. For each species, female data are on the *left* (*red dots*) and male data on the *right* (*blue dots*). The *stars* indicate that sexual dimorphism was statistically significant in ANOVA tests (value of $p < 0.0001$), and *ns* indicates that it was nonsignificant. *A*, body mass. *B*, centroid size obtained from the landmark configurations (modified from Hood 2000).

Facing bottom, 4.16. Body-mass estimates obtained through centroid size of a landmark configuration (modified from Cassini et al. 2012b). *A*, regression of the mandibular centroid size and species dispersion. Symbols: *crosses*, Hyracoidea; *squares*, Perissodactyla (*black*, Equidae; *gray*, Tapiridae; *white*, Rhinocerotidae); *white triangles*, Hippopotamidae; *black triangles*, Suina; *black inverted triangles*, Tragulidae; *white inverted triangles*, Antilocapridae; *black rhombuses*, Camelidae; *circles*, Bovidae; *gray circles*, Cervidae; *solid straight line* of regression by ordinary least squares (OLS); *dashed line*, nonparametric regression of lower quantile; *dotted line*, nonparametric regression of upper quantile. *B*, intercepts and slopes (values on the y axis) for OLS regression represented by a *horizontal solid black line*, with *dashed red lines* indicating the confidence interval of 95%. The quantile is indicated on the *x* axis, with *points* indicating the value for the quantile intercept and slope and the *orange shaded zone* the confidence interval for such parameters.

Cassini et al. (2012b) used a reference sample constituted by extant representatives of all families (and subfamilies) of the orders Hyracoidea (hyraxes), Artiodactyla (e.g., pigs, camelids, antelopes) and Perissodactyla (horses, tapirs, and rhinoceros), with a total of 155 species with published mass values. Regressions were obtained by ordinary least squares (OLS) and quantile regression, with the upper (UpperQ) and lower (LowerQ) quantiles calculated using decimal logarithm of centroid size of a 3-D landmark configuration (36 landmarks for the skull and 14 for the

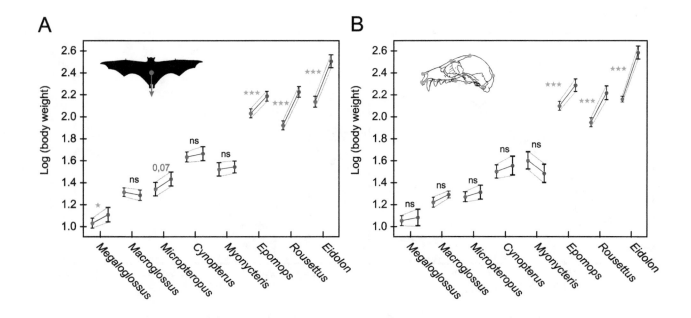

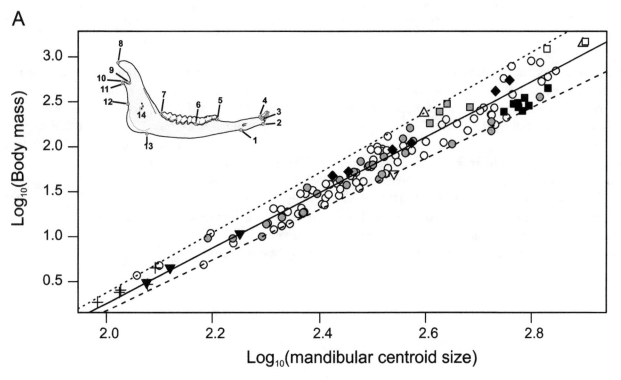

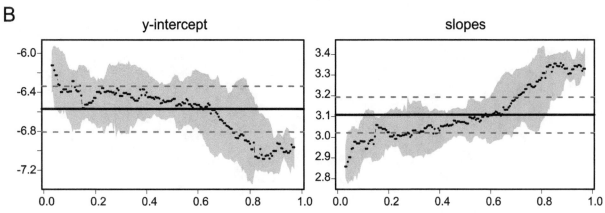

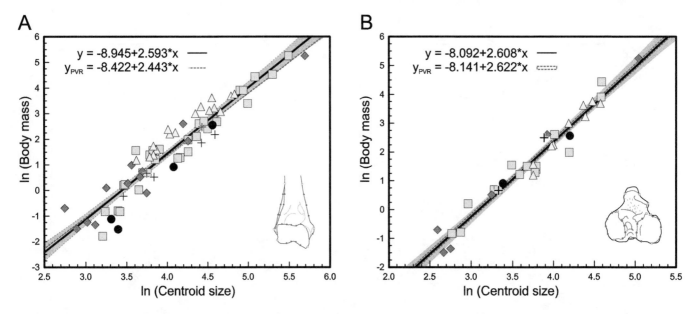

4.17. Biplots and regression lines previously (*straight black line*) and subsequently (*dashed red line*) obtained based on the phylogenetic adjustment between centroid size and body mass of 71 species of marsupials and placental predators (one family of Didelphimorphia, three of Dasyuromorpha, and nine of Carnivora), transformed by the natural logarithm for landmark configurations and the corresponding equations. *A*, humerus, and *B*, tibia. The symbols indicate groups of locomotory habits: *cross*, semifossorial; *yellow triangles*, cursorial; *brown diamonds*, terrestrial; *green squares*, climbing; *black circles*, arboreal. For the humerus, both *straight lines* differ, whereas for the tibia they are practically the same (modified from Ercoli and Prevosti 2011).

mandible) versus decimal logarithm of body mass (fig. 4.16). Quantile regression is a nonparametric method that does not require the assumptions of normality and homoscedasticity of the OLS regression.

These authors compared %PE of the obtained equations considering sex dimorphism, measurement per species, and the two weighting examples mentioned above, and they obtained equations with %PE close to 6%, which are close to those obtained for the postcranium. They concluded that the equation obtained by the upper quantile yielded an estimator that could be interpreted as the theoretical maximum for a given centroid size, within the context of a particular reference sample.

Ercoli and Prevosti (2011) also used centroid size of long-bone epiphyses (humerus, ulna, and tibia), in two dimensions, for estimating mass of sparassodont metatherians. In this case, information on phylogenetic relationships was available; hence, the authors used regression analysis of phylogenetic eigenvectors (PVR) to correct the obtained predictive equations by accommodating the phylogenetic signal. Thus, they managed to improve the %PE of equations to around 17% (fig. 4.17).

Substrate Preference and Use: Locomotion in Fluids

5

The form of a vertebrate's skeleton and its associated musculature are tightly knit to the environment that it inhabits, and thus they reflect the type of substrate in which the vertebrate can live, the activities it can carry out, the uses to which the substrate can be put, and the manner through which these actions may be accomplished. Study of the biological attributes that elucidate the relationship between a vertebrate and the substrate requires that these attributes be clearly defined.

One approach is to analyze substrate preference by characterizing a vertebrate according to the type of substrate in which it lives and performs its activities. An organism may be considered *aquatic* if it carries out its activities in the water, *terrestrial* if it does so in or on the ground, and *arboreal* if it is mainly active in trees. In the aquatic environment, the substrate on which forces are exerted in swimming is the water itself. Similarly, in a terrestrial environment the substrate on which forces are exerted in running could be any number of surfaces, such as the ground or a tree limb, and in the aerial environment, the air is the medium involved in flight. A more finely grained focus on these broad environmental categories reveals, of course, that they are not homogeneous. For example, the aquatic realm may involve a broad spectrum of habitats (riverine, deltaic, estuarine, or marine, among others) that include differences in bottom conditions, vegetation, hydrostatic pressure, presence of formations such as reefs, subaqueous topography (crests, trenches, plateaus), and differences in behavior of the water (bottom, pelagic, and rip currents). Each of these will have an impact on activities other than locomotion. Terrestrial environments include a tremendous variety of substrates, some as obvious as the ground, rocks, and trees, and others that are less often considered, such as ice, air, and air-water interface. Vertebrates commonly alternate between different substrates in performing activities such as feeding, reproducing, seeking refuge, or resting. The lesser anteater (*Tamandua*) may feed on the ground or in trees, but it prefers to rest in trees; sea lions feed in the water, but rest and reproduce on land; and many anseriforms (birds such as ducks, geese, and swans) feed and mate in water, rest and care for their offspring on land, and travel through the air, often flying long distances to migrate (fig. 5.1).

For some vertebrates, combined categories of the "semiarboreal" or "semiaquatic" type are usually used. Alternatively, Miljutin (2009) proposed a dual categorization for mammals that is independent of the mode of locomotion and based instead on substrate utilization strategies, such as the type of substrate used for rest and foraging. For example, the underground-terrestrial category is applied to marmots (*Marmota*), which rest (and hibernate during winter months) in burrows in the substrate and forage on the ground. This classification can be made much more complex by combining substrate utilization and feeding strategies, which are largely independent of one another; this combination provides a matrix of strategies that may serve as an ecological classification. Such classifications can indicate the position of species within the multidimensional continuum of form and ecology. For instance, using broad categories, within the arboreal and semiarboreal class, the silky anteater (*Cyclopes*) classifies as an arboreal/semiarboreal animalivore and the sloth (*Bradypus*) as an arboreal/semiarboreal herbivore (see Miljutin 2009, table 5).

Vertebrates may also be categorized according to substrate use; that is, according to the different ways in which a vertebrate interacts with one or more types of substrate, in some cases actively modifying it. For example, many animals dig for food or build temporary or permanent shelters in which they protect themselves and their offspring from environmental factors (Hildebrand and Goslow 2001). Such categorization, in turn, can be independent of substrate type: tetrapods include animals that dig on the ground and those that dig on tree trunks, and similarly animals that run on the ground as well as others that run along tree limbs. We point this out here to emphasize the conceptual difference between substrate preference (e.g., aquatic, arboreal) and substrate use (e.g., runner, digger).

The main uses for substrates, with several examples provided in parentheses, are:

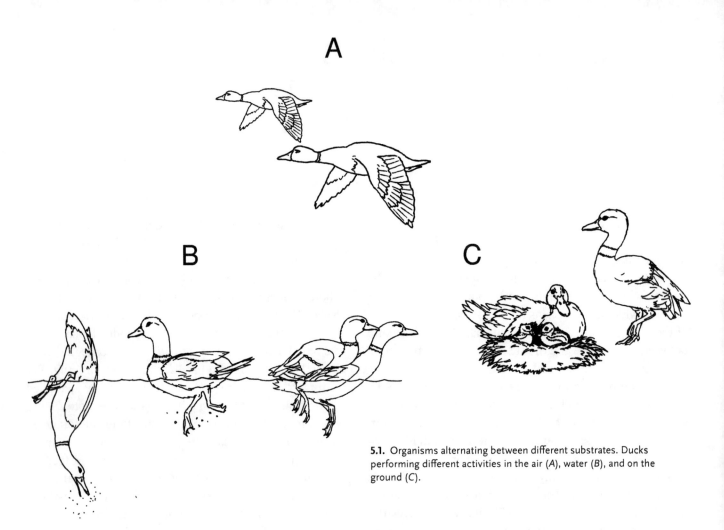

5.1. Organisms alternating between different substrates. Ducks performing different activities in the air (*A*), water (*B*), and on the ground (*C*).

1. *Locomotion*: the substrate is used for spatial displacement, with the vertebrate generally making use of static and dynamic friction forces generated between its body and the substrate. Such displacement can fulfill numerous biological roles (search for food and shelter, escape, and interaction with conspecifics).

2. *Shelter*: an animal uses the substrate to protect itself from environmental factors, including weather and other organisms (predators, parasites, competitors, and conspecifics).

3. *Obtaining food*: the organism uses the substrate to obtain food. Many animals move on the substrate that contains the food (grazing herbivorous mammals and some diving birds) or obtain food by ingesting the substrate itself and then straining out the food items. This latter strategy is observed in several aquatic vertebrates that feed by straining plankton or other suspended organic matter from the water (basking sharks, baleen whales) or contained in the mud (shad, *Prochilodus*).

These uses are not mutually exclusive, and many organisms combine several of these activities to varying degrees. Numerous invertebrates (e.g., earthworms) move (locomotion) through ingestion of the substrate (food), thereby creating burrows in which they live (shelter). In most vertebrates, though, such overlap is not as extensive; examples are ram feeders, as demonstrated by the whale shark (*Rhincodon typus*), which ingests phytoplankton (food) while moving (locomotion) through the sea.

Such overlaps make it difficult to categorize locomotion in a straightforward manner for morphofunctional and ecomorphological studies. Moreover, it is common that in these studies, the use of locomotion categories is mixed with the most comprehensive concepts of preference and substrate use described here. Thus, it is common to find the terms *arboreal* and *climber* used as synonyms and *digger* as a category of locomotion. Another difficulty faced by these approaches is the diversity of biological roles in which limbs, the main locomotory organs in most tetrapods, participate. These include obtaining food (search, selection, fragmentation), escape from predators, construction of housing sites (caves, nests), interaction with

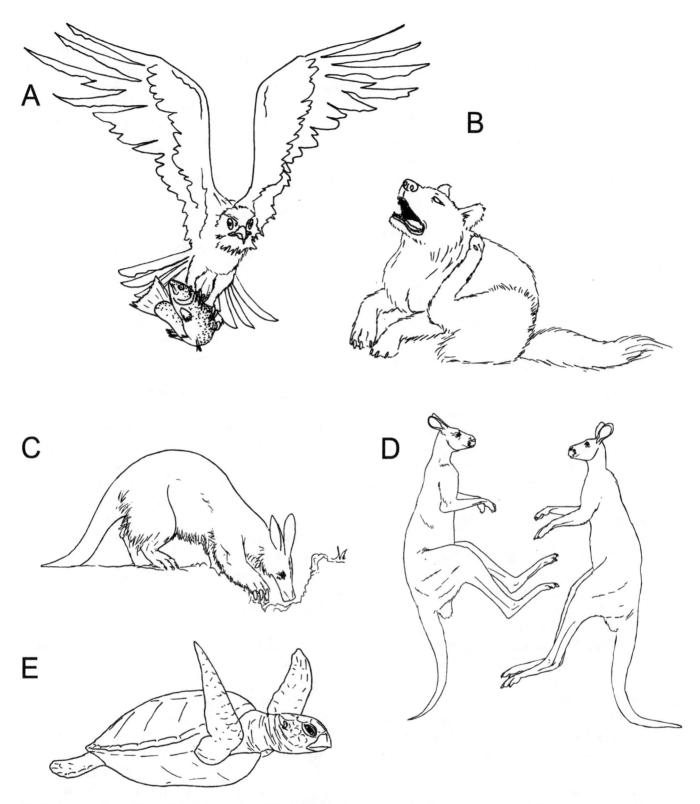

5.2. Limbs can perform different biological roles. *A*, aerial locomotion and prey capture. *B*, self-grooming. *C*, removal of substrate in search of food. *D*, interaction with conspecifics. *E*, aquatic locomotion.

conspecifics (intercourse, aggression, grooming, transport of young) or actions performed on the individual itself (idiomotions), such as self-grooming (fig. 5.2). For example, in the southern long-nosed armadillo (*Dasypus hybridus*) the first two roles overlap with the third: when digging

they build their burrow and look for food at the same time, and the same burrows serve as a refuge to rest and escape from predators. In any case, classifications have to be properly contextualized according to the goals pursued by the research; for example, the lesser anteater (*Tamandua tetradactyla*) can be considered as a digger in one study and as a climber in another.

Among the researchers who considered the problem of locomotory classifications, particularly as applied to mammals, was Charles Oxnard. In *The Order of Man*, Oxnard (1984) proposed that the complexity of animal behavior is so extensive as to make it impossible to describe locomotory patterns by assigning a taxon to categories based on only one descriptor (such as *jumper, runner,* or *climber*). This is so because even when the spectrum of an organism's behaviors is known in detail from field studies, its complexity is such that it defies attempts to define the organism's locomotion by means of a single label. On the other hand, as animals' behaviors include complex repertoires of movements, it is tempting to increase the detail of the classification. However, extremely fine-grained classifications can lead to the creation of as many categories as animals analyzed, whereas the point of classification should be the establishment of categories that allow the recognition of shared attributes or patterns among organisms in one group as compared with organisms in other groups. As we learn more about the habits of organisms, taking a too fine-grained approach defeats the purpose of classifying or, at least, leads to a breakdown in the usefulness of categories, because we would ultimately be led to consider each organism as unique (i.e., in its own group), thus obscuring the recognition of shared attributes. Moreover, locomotory classifications, although based on behavioral realities, sometimes are derived partly from prior knowledge of morphology, resulting in a circularity that further devalues their usefulness for evaluating extinct and extant forms. Recognizing this, Oxnard (1984, 102) proposed a different methodological approach to deal with these difficulties (see chap. 2) in defining "averaged biomechanical situations." This approach considers the combination of the influences of functions on a given structure, taking into account not only the resultant of all forces involved in producing functions, but also considerations of time, such as duration, frequency, and occurrence during life history of the functions. Some activities, such as sprinting, can be highly intensive but relatively infrequent, whereas others, such as walking, are less intensive but much more frequent. In classifying an animal that exhibits both of these activities, which of them should take precedence? This average biomechanical situation must be studied and defined prior to the search for a biological significance (biological role). A similar approach is the use of "functional sequences" as, for example, by Elissamburu and Vizcaíno (2004).

It follows, then, that in the analysis of the relationship between morphology and function of, mainly, the postcranial skeleton, the identification of the types and uses of substrates precedes that of locomotory types. The most common approaches in morphofunctional and ecomorphological analyses of substrate use are those geared toward characterization of how animals displace themselves in or on a substrate—that is, locomotion—and we will therefore focus on locomotory types. Although the objective may be the same, the problems faced by vertebrates that move in the water or the air differ markedly from those that do so over hard substrates. To facilitate their consideration, we will, following an overview of locomotory types developed among vertebrates, treat locomotion in fluids in this chapter and terrestrial locomotion in the next chapter.

Diversity in Locomotory Types among Vertebrates

Although the oldest known vertebrates date to more than 500 million years ago (such as the jawless *Haikouichthys* from the Early Cambrian of Chengjiang, China; Shu et al. 1999, 2003), they began to become important components in the biodiversity of marine ecosystems just over 470 million years ago during the Early Ordovician. Radinsky (1987) showed considerable prescience in proposing a hypothetical ancestral vertebrate, based on living basal chordates, that includes features later discovered in *Haikouichthys*. The functions of several ancestral vertebrate features were considered in greater detail by Alexander (1990), Liem et al. (2001), and Pough and Janis (2019). This ancestor was probably a soft-bodied, vaguely wormlike creature a few centimeters long that moved by lateral undulations of its body though the muddy or sandy bottoms of shallow seas and fed by filter or suspension feeding—that is, by filtering out small organic food items suspended in the water (fig. 5.3). Most of its body wall posterior to the pharynx was composed of metameric muscular blocks, or myomeres. Alternating contractions of myomeres on opposite sides of the body produced the undulating propulsive movements for locomotion. As in living basal chordates, the presence of an elongated rodlike notochord prevented the body from telescoping, or shortening, during contraction of the myomeres; the elastic nature of this structure functioned as an antagonist to the myomeres, straightening the body when the musculature relaxed. Muscular contractions, coordinated by the nervous system, produced posteriorly progressing waves along the trunk that resulted in lateral undulating movements, pushing backward on the water and, in accordance with Newton's law of action and reaction, generated a thrust that propelled the animal forward. A dorsal finlike fold that extended posteriorly and

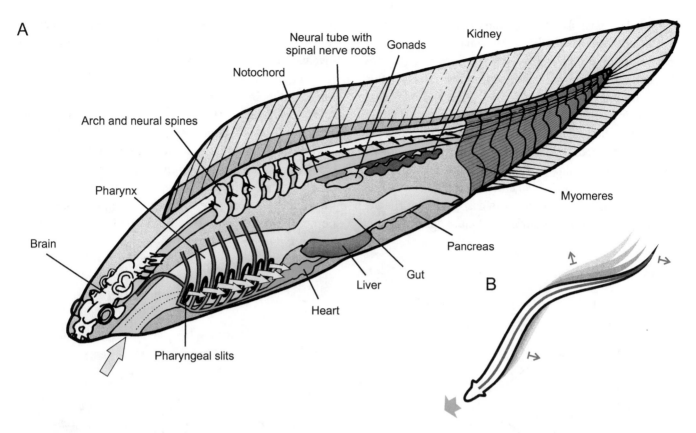

A

- Neural tube with spinal nerve roots
- Gonads
- Kidney
- Notochord
- Arch and neural spines
- Pharynx
- Brain
- Pharyngeal slits
- Heart
- Liver
- Gut
- Pancreas
- Myomeres

B

5.3. *A*, general anatomy of an idealized ancestral vertebrate. Note the pharyngeal branchial system and metameric arrangement of musculature; *green arrows* indicate path of water. *B*, efficient undulatory swimming due to the presence of notochord: *blue arrow*, direction of movement; *red arrows*, force exerted by body wall pressing on water.

ventrally onto the postanal tail increased the surface area that pushed against the water, thereby increasing propulsive capacity while also generating stability by preventing roll along the longitudinal axis of the body.

The most notable subsequent step in the evolution of vertebrates was the appearance of jaws. Roles in ventilation and biting have been hypothesized as functional explanations for the appearance of jaws, allowing more efficient respiration and access to previously unexplored food sources (Pough and Janis 2019). However, in gnathostomes, or jawed vertebrates, the action of jaws is not limited to the capture and processing of food items (see chap. 7). Jaws that can be actively closed have the faculty of grasping, allowing the manipulation of objects, and thus they constitute a multifunctional tool that may be employed in several biological roles: to dig, move stones, build nests, grasp a mate during courtship and reproduction, and hold or move offspring, among many other potential uses. In the aquatic environment, the efficacy of jaws in carrying out such functions depends on applying the jaws onto objects (although many fish also use suction to drag objects into their mouth), and this requires precise control of movements to position the body in three-dimensional space. In other words, the jaws, and the body connected to

them, must be guided toward an objective. For example, when a prey item tries to escape, rapid changes in the direction of travel are required to keep chasing it. Thus, the appearance of jaws in gnathostomes was coupled with substantial changes in locomotion that were associated with the emergence of paired fins supported by internal skeletal elements or girdles.

Cartilaginous (Chondrichthyes) and bony (Osteichthyes) fishes share a body plan, with two sets of paired (pectoral and pelvic) fins, one or two unpaired dorsal fins, one anal fin, and a caudal fin (fig. 5.4). A *heterocercal* caudal fin—that is, asymmetric, with uneven upper and lower lobes—is most common. In these vertebrates, the vertebral column bends into the larger upper lobe; technically, this type of heterocercal fin is termed *epicercal* (some vertebrates, such as anaspid jawless fishes, ichthyosaurs and mesosaurs, have a heterocercal tail in which the vertebral column bends ventrally; this type of heterocercal tail is termed *hypocercal*). Externally, however, the caudal fin can adopt a wide variety of profiles.

Within this general body pattern, in living actinopterygians (these osteichthyans comprise the vast majority of living bony fishes) there is an extraordinary diversity in size, body shape, and modifications of the locomotory system. The axial skeleton and the axial musculature are reinforced, the tail is modified and usually externally symmetric (or homocercal; see below), the fins are highly flexible, and the position of the paired fins is variable. Further, the scales are very thin, lightening them and increasing their flexibility, which enhances locomotory performance. Lungs, which are present basally among osteichthyans, have become modified into a swim bladder (see below).

The functional significance of an ossified versus a cartilaginous skeleton is not yet entirely clear. It is tempting to postulate that in bony fish the ossification of the vertebrae, with their long neural and hemal spines and the presence of dorsal and ventral ribs, resulted in a skeleton reinforced for the insertion of the powerful musculature involved in generating forces to produce thrust from the tail and caudal fin. However, this explanation is not entirely convincing, as sharks are also very powerful swimmers, yet have a cartilaginous internal skeleton. Other hypotheses involve bone acting as a reservoir for phosphorus and being advantageous due to its greater resistance to blood acidification (Wagner and Aspenberg 2011).

The caudal fin of actinopterygians was modified through posterior shortening and abrupt dorsal deflection of the vertebral column from which the mineralized

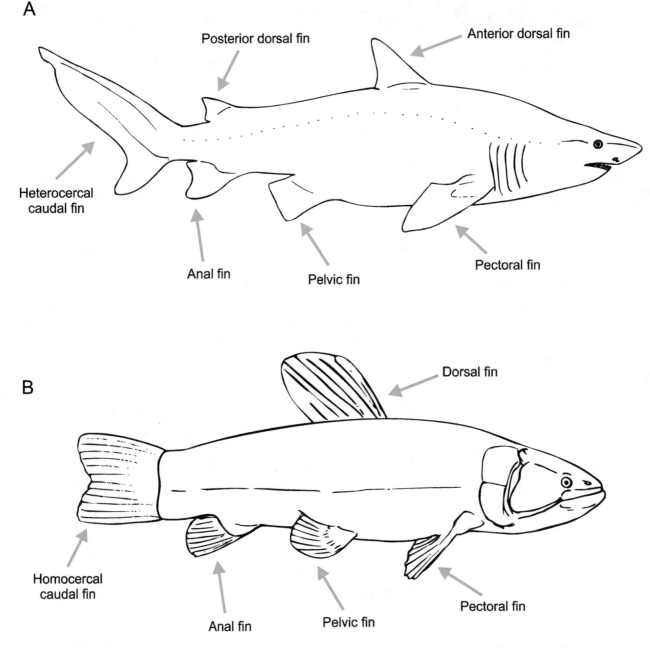

5.4. Fin types in fishes. *A*, elasmobranch chondrichthyan (shark), able to maintain high speeds. *B*, teleost osteichthyan (wolf fish), able to generate explosive acceleration.

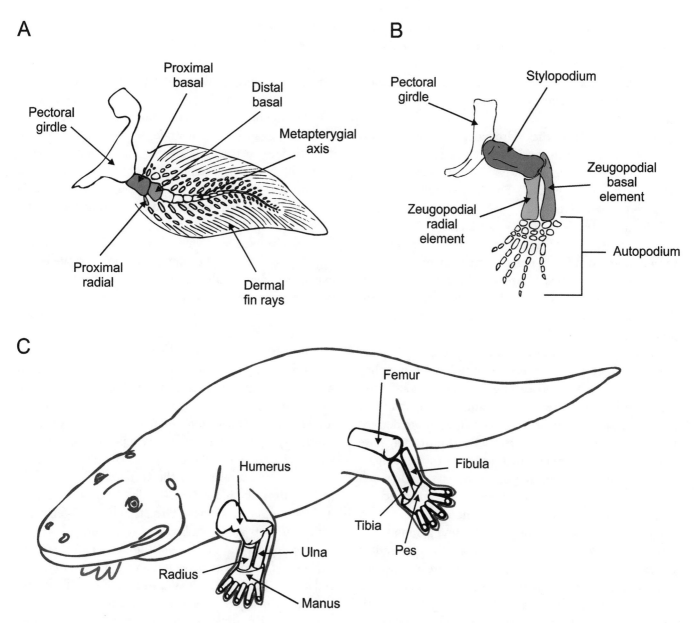

A

- Pectoral girdle
- Proximal basal
- Distal basal
- Metapterygial axis
- Proximal radial
- Dermal fin rays

B

- Pectoral girdle
- Stylopodium
- Zeugopodial basal element
- Zeugopodial radial element
- Autopodium

C

- Femur
- Humerus
- Fibula
- Tibia
- Pes
- Radius
- Ulna
- Manus

5.5. Comparison of the basic common anatomical structure of a fin of basal sarcopterygian fish (*A*) and an idealized chiridium limb (*B*), showing the underlying homologies in the evolution of a limb capable of allowing motion on the terrestrial substrate. *C*, idealized tetrapod showing the configuration common to both limbs.

rays (*lepidotrichia*) are arranged so that the caudal fin is externally symmetric. Due to this apparent symmetry, it is called *homocercal*. This arrangement has been interpreted as allowing fine control over the intensity and direction of thrust through subtle modifications in shape of the caudal fin (Pough and Janis 2019). The rays (also termed lepidotrichia) of the remaining fins were reduced in number with respect to their analogs, the *ceratotrichia* in cartilaginous fish, and were associated with a muscular system that provides great control over the shape of the fins during swimming.

The pectoral fins moved dorsally, or higher up to about the midheight of the body, to lie behind the operculum,

and contributed to counteracting the forward reaction generated by the exit of the water from the pharynx, thus helping to halt the animal's forward progress. However, the braking action is accompanied by elevation of the anterior part of the animal, which, in turn, is compensated by repositioning the pelvic fins near the former lower position of the pectoral fins. This change in the position of the paired fins is thought to have been associated with the modification of the lungs into a swim bladder, in which gas accumulation and elimination is accomplished, in more derived actinopterygians, by way of a network of vessels. The swim bladder thus functions as a hydrostatic organ affecting overall body density and thus buoyancy, allowing actinopterygians to adjust their internal air pressure to external water pressure (i.e., more or less the way a buoyancy compensator can be used by a diver to adjust air volume to control buoyancy) to remain static at a certain depth, reducing the energy expenditure otherwise necessary to do so.

Further details on the hydrodynamics and other aspects of locomotion in fluids, in both aquatic and aerial media, will be discussed later in this chapter (see the section "Locomotion in Fluids"). For now, we note that pisciform vertebrates developed a great diversity of forms that reflect different types of locomotory habits.

The physical properties of aquatic and terrestrial environments are so different that the transition of vertebrates from the former to the latter required considerable modifications in almost all systems. This transition gave rise to a clade of limbed vertebrates, Tetrapodomorpha, that has become highly diverse. The earliest members of this clade, such as *Acanthostega* and *Ichthyostega*, had evolved by the Late Devonian and were limbed (see below), but they were not terrestrial, or at least not habitually so; although they may have had occasion to scamper onto land, they remained aquatic vertebrates (Clack 2012). More derived forms became habitually terrestrial and ultimately gave rise to Tetrapoda, the clade that includes extant members of terrestrial vertebrates and their extinct kin. However, tetrapods are not completely independent of water because it remains an essential element of life processes. Many extant amphibians (anurans, urodeles, and gymnophionans) have an aquatic larval stage and are thus dependent on water for reproduction. Many of them are also aquatic during adulthood. Amniota, the clade of tetrapods that overcame this strict dependence by evolving the amniotic or cleidoic egg, includes mainly terrestrial vertebrates, but several current and extinct lineages reverted to an aquatic existence, in some cases to the degree of not being able to survive outside that environment. Radinsky (1987) clearly identified the problems to be solved in the transition from the aquatic to the terrestrial environment and the solutions developed in the evolution of tetrapods (fig. 5.5).

The most important distinction in the physical properties of aquatic and terrestrial environments is the difference in the densities of water and air. While that of water is 1 g/cm³, air is less dense by three orders of magnitude: 0.001 g/cm³. In the aquatic environment, weight support (recall that weight is mass due to the acceleration of gravity) is a relatively minor consideration. In vertebrates, the density of muscle, which constitutes most of the body mass, is 1.05 g/cm³ and that of bone is about 3 g/cm³. The body, therefore, is not much denser than the water that surrounds it and is sustained largely by the water's

upward buoyant force (i.e., Archimedes's principle; see the section "Locomotion in Fluids"). Such support is lacking in the terrestrial environment. Modifications to the vertebral column, girdles, and ribs act in part to prevent the body from collapsing on the lungs and the rest of the viscera. Another important factor in terrestrial locomotion is friction against the substrate. Whereas air, due to its density, does not produce determining frictional forces (at least within known locomotion speeds in land animals), friction between the body and the ground is considerable. This can be both an advantage and a disadvantage; friction between the ground and the limbs or other parts of the body can be used to produce thrust but dragging or sliding the limbs and/or the body along the substrate is very expensive energetically. Avoiding this expense requires not only that the trunk be lifted from the substrate during locomotion, but also that the limbs be raised, advanced, and again engaged to produce thrust.

The vertebral column, then, became differentiated into several regions, including cervical, thoracic, sacral, and caudal regions; in fish, by contrast, only trunk and caudal regions are recognized (in osteichthyans, the pectoral girdle is connected to the neurocranium by skeletal elements). In addition, new joints appeared between vertebrae, and these skeletal elements articulated in such a way as to resist collapse of the vertebral column by the action of gravity and reduce the degree of torsion along its longitudinal axis during terrestrial locomotion. These articular facets are located on processes, termed *pre-* and *postzygapophyses*, at the base of the neural arches. Their function is not only to prevent the vertebral column from collapsing, but also to allow it to move laterally without dislocation of adjacent vertebrae. The strong, but relatively simple, trunk muscles of fishes were extensively modified, forming a less bulky but very complex musculature associated with the vertebrae, girdles, ribs, and limbs. Further, highly developed ligaments extend between the ventral and dorsal parts of vertebral centra and between the free ends of the neural processes. The structural arrangement of these components implies that the vertebral column of tetrapods can be considered more of an "arch" than a "bridge," with the neural spines and ribs acting as lever arms of the intervertebral joints.

The paired appendages of tetrapods are termed *chiridia* (*chiridium* in the singular) and are composed of three main segments arranged in proximal to distal order. The segment closest (i.e., most proximal) to the body, the stylopodium (or propodium), consists of a single skeletal element in the arm (or brachium) and in the thigh, the humerus and the femur, respectively. The middle segment, the zeugopodium (or epipodium), has two skeletal elements, the ulna and radius in the forearm (or ante-brachium) and the tibia and fibula in the leg (or crus). In the most distal segment, or autopodium (manus or hand, and pes or foot), numerous elements are present. An autopodium is composed of three proximal to distal segments. The most proximal of the manus is the carpus, whereas the tarsus is that of the pes. The digits articulate with the carpus and tarsus. These include the metapodium (the middle segment of the autopodium), composed of metacarpals in the manus and metatarsals in the pes. Following these is the acropodium, the most distal segment of the manus and pes, which includes the phalanges that make up the fingers and toes (see appendix). According to some interpretations, the digits developed as neomorphs on the postaxial side of the metapterygial axis (primary support axis) of the fin of sarcopterygian fish (Hinchliffe 1994), although debate continues on the matter (see Laurin 2011). The paired appendages of tetrapods, unlike those of fish (and tetrapods that have secondarily lost or reduced their appendages; see chap. 6), must support the weight of the body in addition to participating in locomotion; and, indeed, they provide nearly all of the propulsive force. Movement in the terrestrial environment thus required considerable modification of the appendicular apparatus, including both the skeletal system and the muscular system, which became extremely complex and large. Part of this musculature's function is to compensate for moments generated by an animal's weight at the pivot points of the joints, to prevent the limbs from collapsing, although some joints may also be held stable by ligaments.

In tetrapods, the girdles were expanded, especially ventrally, to support the weight of the body and serve as the origin for much of the appendicular musculature, which itself increased in volume and complexity. In basal forms, the stylopodium was oriented more or less laterally to the long axis of the body, and contraction of the musculature attaching ventrally on the girdles drew it toward the midline of the body (adduction), thus raising the trunk from the substrate. Other, more dorsal, muscles pulled the limb forward or backward during locomotion. Unlike the condition in bony fish, the pectoral girdle lost its connection with the skull, allowing the skull to move independently of the body by way of a newly distinct region of the vertebral column, the neck or cervical region. This also prevented direct transmission to the head of vibrations produced by interactions of the anterior limbs with the substrate. By contrast, the pelvic girdle gained a sturdy contact with the rest of the body through articulation with the vertebral column, defining a sacral region that transmitted the propulsive thrust of the posterior limbs to the body.

Locomotion in Fluids

As already noted, vertebrates originated and underwent much of their evolution in water, a fluid medium, but many of them, current and past, live or lived, moving and feeding, in two characteristic fluids of this planet: air and water.

Let us consider fluids more carefully. In nature, matter occurs in different states of aggregation. Traditionally, three fundamental states—solid, liquid, and gaseous—were recognized, although now a fourth, plasma, is also documented. All except solids are considered fluids, but for the purposes of this book, it is sufficient to consider only the liquid and gaseous states. A fluid is an amorphous, continuous medium in which the molecules that form the substance are held together by weak attractive forces and are unable to withstand shear forces. Fluids can thus change form, as they lack restorative forces that allow them to regain their original form. Whereas gases are considered compressible and do not have constant volume, liquids are incompressible and have nearly constant volume.

The study of locomotion in fluid media such as water and air includes both commonalities and differences that depend on the physical properties that characterize them (i.e., density, compressibility, viscosity, and surface tension). The laws and principles of physics that explain the forces involved in the interactions between bodies and fluids are part of the branch of the mechanics of continuous media known as fluid mechanics (see chap. 2).

Fluid Mechanics

This branch of physics is a subdiscipline of continuum mechanics that assumes that fluids comply with the general laws of physics, such as the conservation of mass, energy, and momentum, and the first and second laws of thermodynamics. The general equations that describe the movement and force balance of fluids are generally referred to as the Navier-Stokes equations.

In physics textbooks, hydrodynamics and aerodynamics are defined as the part of mechanics that studies the movement of fluids. Many of the concepts involved in this field of study refer to the movement of fluids in tubes according to physical laws such as Torricelli's theorem, the Poiseuille law, and the Flow's or volumetric flow's law, among others. Here, and in the subsequent sections on hydro- and aerodynamics, we will focus on the concepts of fluid mechanics that treat the interactions between the fluid and the contour of the object on which it acts—in our case, the body and appendages of vertebrates. Some

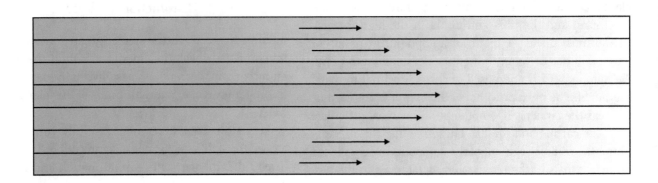

5.6. Laminar (A) and turbulent (B) flows. Note the parallel arrangement of the layers in laminar flow, in contrast to the disorderly arrangement in turbulent flow, which may give rise to vortices. In either case, velocity is higher in the cross-sectional center than near the walls (compare *arrow lengths*), represented by the *upper* and *lower boundary* of each illustration.

of these concepts are common to both fluids (liquids and gases) and are discussed below.

Bernoulli's theorem, derived from the law of conservation of energy (i.e., the relationship between pressure and a fluid's potential energy), states that the internal pressure of a fluid decreases as its velocity increases (and its internal pressure increases as its velocity decreases). In the same way, when a fluid increases its velocity, its pressure decreases. These factors have profound implications for both swimming and flying. It follows that in a moving fluid, the sum of pressure and velocity at any point remains constant. Its mathematical expression, in simplified form, is given by the equation

$$P + \mathbf{v} = k$$

where P is pressure, $\mathbf{v}$ is velocity, and k is a constant determined by calculation. Other variables, such as fluid density and gravitational acceleration, are also considered in the complete formula, but the simplified equation is sufficient for grasping the concept.

The well-known Archimedes's principle postulates that a given object, totally or partially submerged in a resting fluid, receives an upward buoyant force that is equal to the weight of the volume of fluid displaced by the object. This fluid volume is equivalent to the entire volume of the object if it is fully submerged or to the portion of the object that is submerged below the surface of the fluid. The force acting upward on an object may be described as a buoyant (or hydrostatic) force, concentrated at the center of mass of the object and measured in newtons. It is given by the following mathematical expression:

$$\mathbf{F}_b = m\mathbf{g} = \rho V\mathbf{g}$$

where $\mathbf{F}_b$ is the force, m the mass of the object, ρ the density of the fluid, V the volume displaced, and $\mathbf{g}$ the acceleration of gravity.

FLOW REGIME

For our purposes here, only two fluid regimes need be considered: laminar flow and turbulent flow (fig. 5.6). In laminar flow, the fluid moves in an orderly and stratified manner in parallel sheets, which do not mix, along a smooth path, also called a streamline. In turbulent flow, the movement of the fluid is chaotic or messy, producing small vortices or whirlpools at nonperiodic (i.e., irregular) intervals.

The Reynolds number (*Re*) is a dimensionless quantity that characterizes the type of fluid regime or movement. It is defined conceptually as the ratio between the forces of inertia and viscosity. In general, when at low *Re* values, the flow is laminar, and at high values the flow is turbulent, but its use is broader. The experiment conducted by Reynolds involved the movement of fluids through pipes, so that in most physics textbooks its mathematical expression is based on the diameter of a pipe:

$$Re = vD\rho/\mu$$

where ρ is the density and μ (the Greek letter mu) is the viscosity of the fluid, relative to the force of viscosity, as μ/ρ is the kinematic viscosity. The inertial forces are represented by the average speed, *v*, of the fluid and the diameter, *D*, of the pipe. Here, we will focus on a mathematical formula that is not so much concerned with fluids circulating through pipes as with fluids passing around submerged objects and their supporting surfaces, technically termed *foils*, where what matters is the nature of the flow surrounding the object. In the equation for the Reynolds number for the chord (*Rec*) for lift surfaces,

$$Rec = vc\rho/\mu$$

where **v** is flight speed, *D* is replaced by the wing chord (*c*; distance between the leading, or attack, edge and the trailing edge of a wing), and μ/ρ represents the kinematic viscosity of the air (1.460×10^{-5} m²/s).

The boundary layer is the fluid layer in immediate proximity to a contact surface, where the effects of viscosity are significant. In this layer, the velocity of the fluid with respect to the surface ranges from zero in the vicinity of the surface to the speed of the undisturbed current, which occurs where the layer ends.

In fluid dynamics, the *drag* or resistance is a mechanical force, $\mathbf{F_d}$, generated by the contact and interaction of the surface of an object and a fluid, also known as friction. This force opposes the movement of the object in a fluid and depends on the properties of both the object and the fluid. It is defined mathematically as

$$\mathbf{F_d} = 1/2 \, \rho \, \mathbf{v}^2 \, A \, Cd$$

where *Cd* is the drag coefficient, *A* the reference cross-sectional area (perpendicular to the direction of the flow), ρ the density, and *v* the relative flow speed of the object with respect to the fluid. However, this force is very difficult to calculate, because it depends on, among other things, the viscosity of the fluid, the body surface area, the texture of the surface, the relative flow speed of the fluid on the surface, and whether the fluid is laminar or turbulent. Therefore, it is generally determined experimentally.

The region of disturbed flow of a fluid immediately behind a moving or stationary object, caused by the flow of

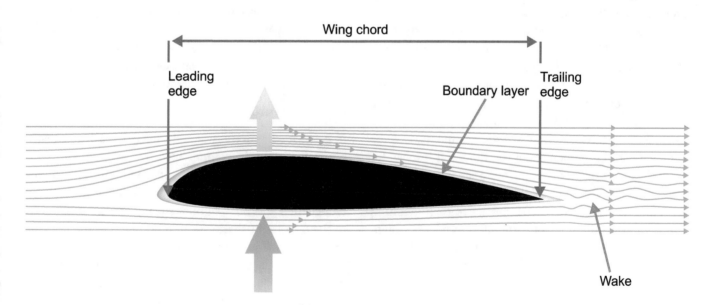

5.7. Forces acting on a fluid-dynamic profile that moves with respect to a fluid. Note that the fluid layers moving over the upper surface of the profile travel a longer distance per time unit and, therefore, move at a higher velocity than the layers moving below the lower surface, as represented by the *small blue arrows*. This difference of velocity manifests as a difference of pressure, which is higher below than above the profile, as represented by the *thick blue arrow*. The boundary layer, chord, and wake (formed by the turbulence of the exit flow) are also indicated.

the surrounding fluid, is called *wake* (fig. 5.7). It is usually turbulent. Related to this concept is the Froude number (*Fr*). Like the Reynolds number, it is a dimensionless ratio between the force of inertia and the force of gravity. It is used to determine the resistance of a partially or totally submerged object in a fluid and allows comparison between objects of different sizes.

Requirements for Locomotion in Fluids

In moving through a fluid, an animal faces several challenges, such as maintaining its vertical position, reducing resistance as it advances through the fluid, propelling itself through a continuous, amorphous environment, and maintaining the orientation and direction of its body while controlling stability. With regard to vertical position and in accordance with Archimedes's principle, because animals generally have a density just greater than that of water, the upward force received from the water is sufficient to provide buoyancy, requiring only a small expenditure of energy to generate the additional force necessary to maintain the position. However, in air, the density of which is much lower than that of animals, the buoyant force is insufficient, and lift must be produced to oppose the weight. We have noted that when a body moves through a fluid, it experiences a resistant force in the direction opposite to that of the body's movement. In fluid dynamics, this force is called drag or fluid resistance, and is proportional to speed in laminar flow or speed squared in turbulent flow. In laminar flow, it depends on both density and viscosity, and thus solutions to such drag forces will differ in water and air. In aerodynamics, drag is the result of two components, the friction between the surface of the body and the fluid (dependent on viscosity) and the pressure distributions over the body surface. In hydrodynamics, drag is dependent on whether the object moves completely submerged or near the surface; in the latter case, wave resistance should be taken in account (see wake, above). The solutions to controlling rotation around axes of the body depend on the density of the fluid and involve, for example, unpaired median fins in the aqueous media and stabilizing surfaces such as the tail and wing tips in the air. These concepts, the particularities of each medium, and the design solutions of vertebrates to move through these media are discussed further in the following two sections.

Locomotion in Water

As noted earlier, the first vertebrates lived and thus moved in water, the medium that requires the lowest energy expenditure in supporting body mass. Vertebrates that live in water are called *aquatic*. However, as already discussed, this term is used in the context of substrate preference rather than of the use and mode of movement in or on it. In this environment many animals, such as sponges, are sessile—that is, they remain anchored in the sea—and others, mainly small buoyant animals collectively referred to as plankton, typically take advantage of external forces (mass movements of the fluid, such as currents) for displacement. In general, however, aquatic vertebrates (and many nonvertebrates as well) are swimmers, mobile in the sense that they move actively (that is, they swim), although they too can take advantage of mass movements.

A first approach in classifying locomotion in aquatic vertebrates is to differentiate between primary swimmers (e.g., fish and the larvae of many lissamphibians) and secondary swimmers (e.g., ichthyosaurs, crocodiles, turtles, cetaceans, and pinnipeds) based on whether they are descended from aquatic or terrestrial ancestors, respectively (fig. 5.8). Various groups of secondary swimmers have convergently evolved features that facilitate movement in water. For example, some have developed elongated, hydrodynamic bodies, finlike paired appendages, and, indeed, a secondary caudal fin, which may be oriented vertically (ichthyosaurs) or horizontally (cetaceans). In cetaceans, this fin is supported by (in addition to the caudal vertebrae) a network of densely packed collagen fibers, rather than by integumentary rays as occurs in fish. Because body weight is largely supported by the upward push of the water, the aquatic environment does not impose many of the limitations to size that exist in the terrestrial environment, and it is among such vertebrates that extreme sizes are reached. Indeed, they include the largest vertebrates—the rorqual cetaceans (e.g., the blue whale)—that have ever existed. In the following sections, several useful concepts in the study of aquatic locomotion are reviewed.

HYDRODYNAMICS

In dense media such as water, the force of gravity is mostly, but not completely, counteracted by the water's upward buoyant force. Different groups of aquatic animals employ various mechanisms to reduce the effect of gravity. Some reduce their specific weight by storing lipids; for example, in the liver, as in sharks, or in other parts of the body, as in cetaceans (such as the blubber just under the skin of most cetaceans). Bony fish instead regulate their buoyancy with a swim bladder, and other vertebrates simply correct their location in the water column by movements of the body and fins.

Gravitational force is summarized as a single downward force acting on an animal's center of mass; on the

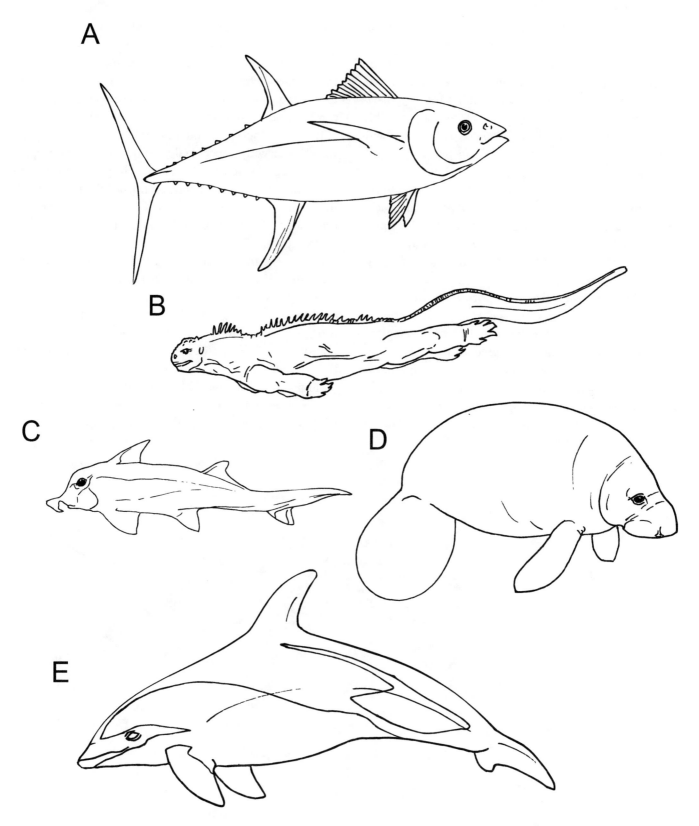

5.8. Primary (*A* and *C*) and secondary (*B*, *D*, and *E*) swimmers. *A*, tuna fish. *B*, sea iguana. *C*, rabbitfish (holocephalian). *D*, manatee. *E*, porpoise.

other hand, thrust generated by the density of the water is summarized as an upward force acting on the center of flotation. If an animal is of uniform density, the location of these centers coincides; if not, the flotation center lies a certain distance from the center of mass, producing a torque, proportional to the distance, that will tend to rotate the animal (fig. 5.9). This tendency is counteracted by the fins, planes that apply pressure on the water. As water is practically incompressible, the force applied by the fins generates a reaction in the opposite direction. In turn, if the propulsive force generated by the fins passes through the center of mass, where the force of gravity pulls down, displacement occurs in that direction. If the propulsive force does not pass through the center of mass, another torque is produced, which can also be counteracted with foils or an asymmetrical caudal fin that generates a propulsive force at an appropriate angle. Vertical unpaired (median in position) fins counteract roll (rotation around the anteroposterior axis of the body) and yaw (swinging side to side around the dorsoventral axis). Paired fins, which project horizontally near the anterior end of the body, control pitch (vertical rotation or tilt) and, arranged at an angle to water flow, can produce lift.

As already noted, resistance to forward movement depends on the type of flow regime, which depends on the density and viscosity of the medium. In each flow regime, a particular type of resistance predominates; frictional drag is greater in turbulent flow, whereas in laminar flow, form drag is greater. Body shape also has an influence; for animals of equal volume moving at similar speeds, the more fusiform (or elongated and tapered) the body, the greater the frictional resistance and the lower the form drag (such a body shape may be described as hydrodynamic). On the other hand, a shorter and flattened body minimizes frictional resistance but experiences increased form drag. In addition, differences between fluid layers can be influential; when animals swim near the surface of the water, the waves and different velocities of surface-water layers with respect to deeper layers increase drag.

To initiate movement, thrust must be greater than total resistance. The latter depends on the density of the fluid, the relative velocity of the animal with respect to the fluid, the drag coefficient (which depends on the shape of the animal and the Reynolds number), and the features of the surface of the animal contacting the fluid. Once in motion, a significantly higher thrust value is not necessary, as it is generally assumed that swimming at a constant speed requires only that thrust equal total resistance. Thus, swimming velocity is proportional to the resistance and to the power applied, and power depends on the metabolism of the animal.

Fluids are also affected by the force of gravity; that is, they have weight. As explained in chapters 2 and 4, weight is related to mass and, therefore, to density. In air, the

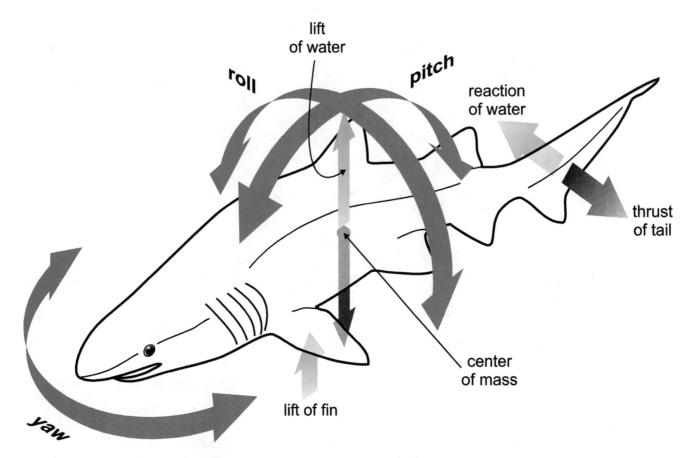

5.9. Control of body position by means of paired and unpaired median fins in fish. The caudal fin produces thrust, and the pectoral fins provide lift; the median fins (dorsal and anal) restrict roll and yaw, and the paired fins restrict pitch (*red arrows*).

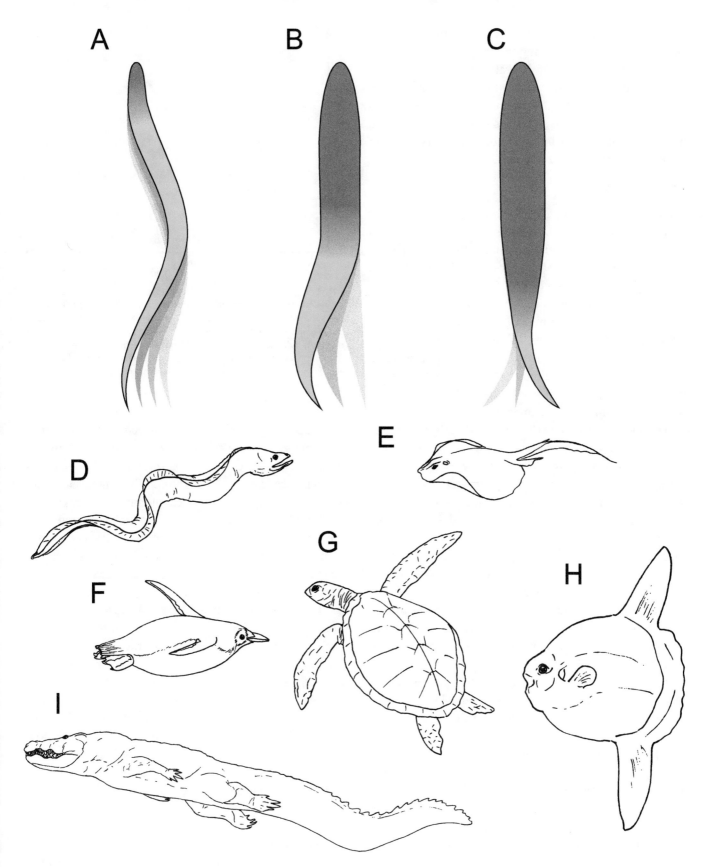

5.10. Undulatory swimming: wave propagation along the body (*A* and *B*). Note that when decreasing the length of the involved body proportion (*blue*), the kind of movement changes to oscillating (*C*). Undulatory swimmers (*D*, moray eel; *E*, dasyatid stingray; *I*, crocodile) and oscillatory swimmers (*F*, penguin; *G*, green turtle; *H*, sunfish).

weight of the air (the atmospheric pressure) exerted on surfaces is not a determining factor, but in aquatic environments, the pressure (hydrostatic pressure) exerted by the water is, due to water's higher density, a very important factor in locomotion. For example, friction increases with increased pressure, as occurs when diving at depth, because pressure increases 1 atmosphere every 10 m, in addition to that of the air pressure.

TYPES OF SWIMMING

Propulsion in water can be produced by undulatory or oscillatory movements. Undulatory swimmers produce a propulsive force through lateral (that is, side-to-side) wave-like movements of the body that extend from the trunk to the tail and the caudal fin (fig. 5.10). Wave propagation along the body increases in magnitude toward the tail, where it reaches a maximum (Hildebrand and Goslow 2001). Undulatory swimmers are usually classified according to the proportion of the trunk involved in producing undulating movements, although there are many animals in which undulation is restricted to propulsive fins (e.g., the giant rowing fish and some rays). When the proportion of the body that moves is very limited, the movement ceases to be wavelike and becomes oscillatory. Oscillatory swimmers generate propulsive force by oscillating their fins. Most primary swimmers (aquatic vertebrates with a long lineage of aquatic ancestry) are undulatory, and most secondary swimmers (aquatic vertebrates descended from terrestrial ancestors) are oscillatory, although there are many exceptions in either category. For example, among undulatory swimmers we may find hagfishes, tadpoles, eels, and crocodiles. Among the oscillatory swimmers, we may note chimaeras (e.g., the ratfish *Chimaera monstruosa*), sunfish (*Mola mola*), some rays, adult anurans, water and diving birds, and cetaceans (see Smits 2019 for further detail; fig. 5.10). In addition, some vertebrates can take advantage of the expulsion of water from the gill slits to generate propulsion, a type of locomotion known as pulsatile, which Smits (2019) noted is most common among invertebrates.

In addition, fins may act as foils and generate lift when moving through the water and are thus considered hydrofoils. Some animals, such as skates and rays, sea turtles, penguins and other birds, and pinnipeds, make extensive use of this property to remain vertically stationary in the water column or move around (underwater flight).

POWER AND ENERGY COST

The energy cost of locomotion in water depends strongly on the speed of movement; energy cost increases approximately with the square of speed (Hildebrand and Goslow

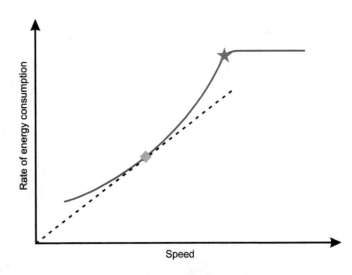

5.11. Energetic cost of swimming. The energetic cost in water increases approximately with the square of swimming speed. The *curved line* represents the maximum range of speeds. The *dashed line* extends from the origin to a point where it is tangential to the curved line. This point (*green rhombus*) indicates the velocity at which the transport cost is minimal (migration speed). The *red star* represents the maximum speed with aerobic metabolism. Modified from Hildebrand and Goslow (2001).

2001; fig. 5.11). Another factor involved is the duration of movement, or more specifically the capacity of an animal to develop the power to sustain a given speed for a given period of time; this is commonly termed *endurance*. The type of metabolism comes into play here; in general, sustained speed swimming involves aerobic metabolism, and explosive swimming (swimming at full speed for very short periods of time) involves anaerobic metabolism. Vertebrates with higher metabolic rates are capable of longer efforts in aerobic regimens (e.g., pinnipeds, dolphins, and penguins), although muscular and physiological adaptations allow several ectothermic vertebrates to achieve high speeds in sustained swimming (e.g., tuna and related fish).

A high metabolic rate, however, can be problematic for aquatic birds and mammals. Water is a medium of high thermal conductivity, and in the vast majority of environments its temperature is much lower than that of the animal's body, so that the animal experiences a continuous heat loss that must be countered or reduced in various ways, such as insulating layers of subcutaneous fat and countercurrent heat exchange.

FIN SHAPE

In fish, the primary propulsive organ is the caudal fin. Flexion of the posterior end of the notochord or vertebral column results in a heterocercal, and thus a more dorsoventrally extensive, caudal fin (dorsal flexion for epicercal and ventral flexion for hypocercal caudal fins). The increased development of one or the other lobe increases the area of the caudal fin and improves the generation of thrust during propulsion, thus favoring rapid acceleration.

Not all aquatic animals move at the same velocity. As we have explained, among animals capable of reaching high velocities, there are those capable of sustained swimming and others capable of explosive swimming. The optimal body design to minimize resistance during sustained locomotion propelled by the caudal fin is a hydrodynamic, fusiform, and compressed form (fig. 5.4). This morphology is appropriate to cover long distances in open environments, and these animals are known as "cruise swimmers" (e.g., many sharks, tunas, and dolphins). Those specialized in explosive swimming have short and wide caudal peduncles and low-profile caudal fins (fig. 5.4). At low speeds, such as those required in complex environments (e.g., highly vegetated reefs or lagoons), propulsion by means of median and paired fins generates lower energy expenditures than propulsion by means of a caudal fin, in addition to improving control in all three dimensions, especially when the fins act independently.

DIVING

For various reasons, many aquatic vertebrates need to venture to considerable depths and are termed *divers*. As noted above, due to the density of water, the hydrostatic pressure increases at approximately the rate of 1 atmosphere per 10 m of depth, so at a depth of 100 m the pressure exerted by the water on the surface of the diving animal will be greater than 10 atmospheres. Most diving vertebrates have adaptations in design and physiology that resist these pressures. This need is especially acute in secondary swimmers to prevent the collapse of some internal structures at

such increased pressures (e.g., the middle ear, blood vessels, and heart). Other problems inherent in diving have to do with the decrease in temperature and the amount of ambient light available for vision, but these are beyond the scope of this book.

APPLICATIONS IN PALEOBIOLOGY

The aquatic environment imposes severe restrictions, so that it is relatively easy to identify efficient general designs for locomotion in this environment. The recognition of these designs in extinct organisms allows reasonably accurate inferences on their swimming abilities. However, some of the most interesting parameters in swimming, such as maximum speeds, cannot be easily measured or calculated even for extant animals (Motani 2002) due to the complexities derived from current understanding of fluid kinematics and to difficulties in conducting experiments on aquatic animals. In the following paragraphs, we provide several examples of studies of locomotion in fluids as applied to paleobiology.

Massare (1988) analyzed a large sample of Mesozoic marine reptiles, including ichthyosaurs, marine crocodiles, plesiosaurs (plesiosauroids and pliosauroids), and mosasaurs, with the aim of estimating their maximum swimming speed and formulating hypotheses on their prey-capture strategy. This author reconstructed body shape (fig. 5.12), estimated body size for each group, and then calculated the increase in maximum energy required to overcome resistance of the water at different speeds. Massare (1988) categorized ichthyosaurs as axially oscillating swimmers based on functional analogy with dolphins and compared plesiosaurs to sea lions in the combined

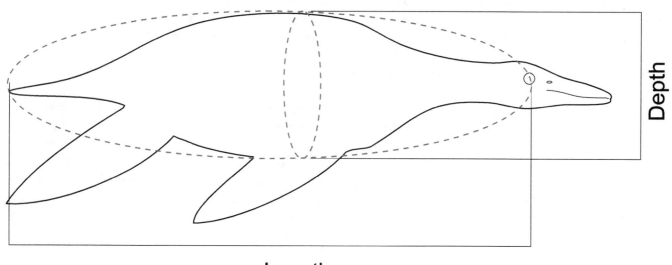

5.12. Approximation of plesiosaur body form modeled as a prolate spheroid. Redrawn from Massare (1988).

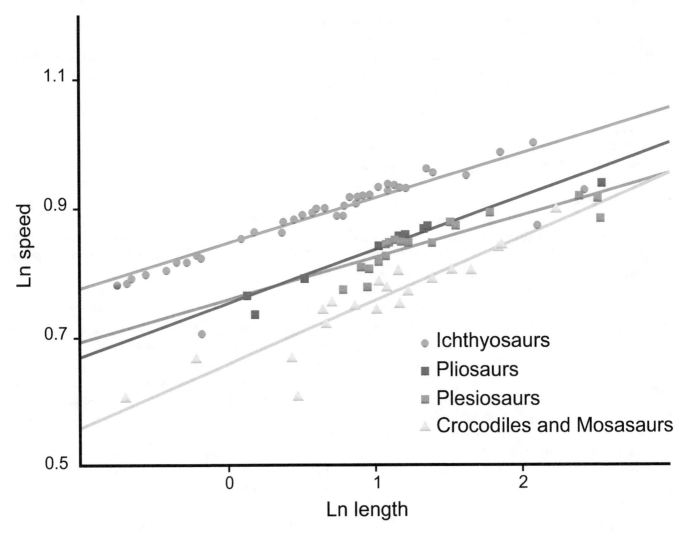

5.13. Body length versus speed (both natural log-transformed) scatterplot of aquatic reptiles with different body forms. Note that the less elongated forms (ichthyosaurs) are the fastest. Modified from Massare (1988).

use of paddling and underwater flight, in contrast to more traditional views that compared them to turtles and penguins. Extinct marine crocodiles, by analogy with living crocodiles, were categorized as axial ripple swimmers, a mode of propulsion also inferred for mosasaurs, owing to their similarity to crocodiles. The body surface (including both shape and area) was estimated for each of the animals under study, with the trunks modeled as prolate spheroids of differing degrees of elongation. Massare (1988) then calculated the minimum drag coefficient for the body shapes and thus obtained a maximum value of sustained swimming speed for each type of body shape (fig. 5.13). Finally, the proportion of metabolic energy available to produce thrust was estimated for each of the extinct Mesozoic marine reptiles, based on previously published assumptions on the basal metabolic rate, aerobic muscle efficiency, body mass, and average density of these reptiles. This author concluded that they would have had lower maximum speeds than living odontocete cetaceans. Among

the reptile groups, ichthyosaurs would have been fastest in sustained swimming (fig. 5.13), followed by both groups of plesiosaurs, and then marine crocodiles and mosasaurs. Differences in speed among plesiosaurs, crocodiles, and mosasaurs were less marked among taxa of larger body size.

Motani (2002) revised and expanded on Massare's (1988) methodology with the aim of achieving more accurate estimates with fewer assumptions. He proposed using values for certain parameters based on more appropriate analogs, such as for basal metabolic rate (distinguishing between values for endotherms and ectotherms), body size, surface area, and the allometric adjustment of the Reynolds number. One of the methodological problems this author encountered was that available values for metabolism were mainly those for terrestrial reptiles, so Motani included metabolic values for sea turtles in his studies of Mesozoic marine reptiles. To generate more accurate estimates for body volume and surface area, Motani

(2002) used the PaleoMass computer software developed by Motani (2001) and considered the implications of the several assumptions inherent for each set of equations. For example, given the greater body mass of the extinct Mesozoic marine reptiles, Motani (2002) proposed that the basal metabolic values of current reptiles may not be applicable to the former, as they may have enjoyed more stable and elevated body temperatures, which may have allowed them more energy available for swimming.

Locomotion in Air

Many small, arboreal vertebrates, including several lissamphibians and reptiles, for example, occasionally take advantage of air resistance, in combination with extensible body membranes functioning as parachute surfaces, to escape predation or reduce the consequences of an unexpected fall. Other vertebrates, including mammals from distinct lineages, are gliders and also make use of membranes, which, in addition to slowing down a fall, allow them to travel considerable horizontal distances. By contrast, those vertebrates that actively move through the air (i.e., are active fliers) must not only be able to generate and maintain lift, but also to gain height and propel themselves. These vertebrates, mainly birds and the extinct pterosaurs among reptiles and bats among mammals, have evolved convergently toward modifying the anterior appendages into wings, which not only provide the lift that

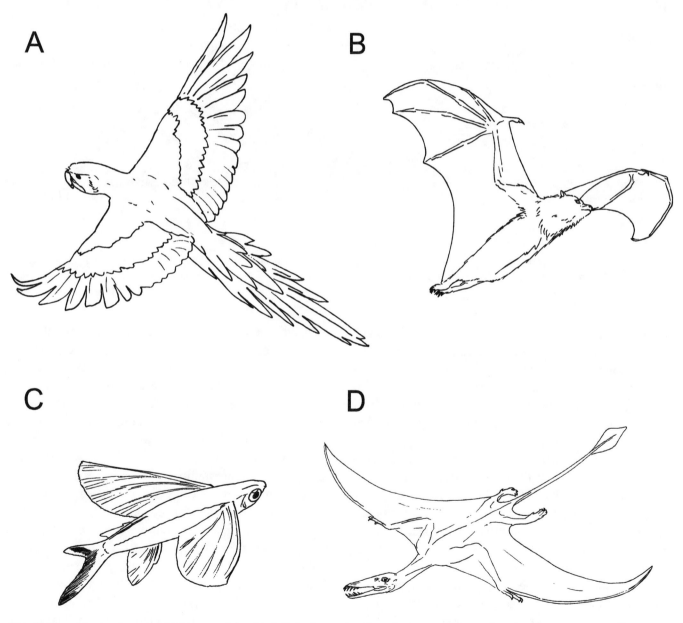

5.14. Flying vertebrates. *A*, modern bird (macaw, Psittacidae). *B*, frugivorous bat (flying fox, Pteropodidae). *C*, flying fish (Exocoetidae). *D*, life reconstruction of a rhamphorhynchid pterosaur.

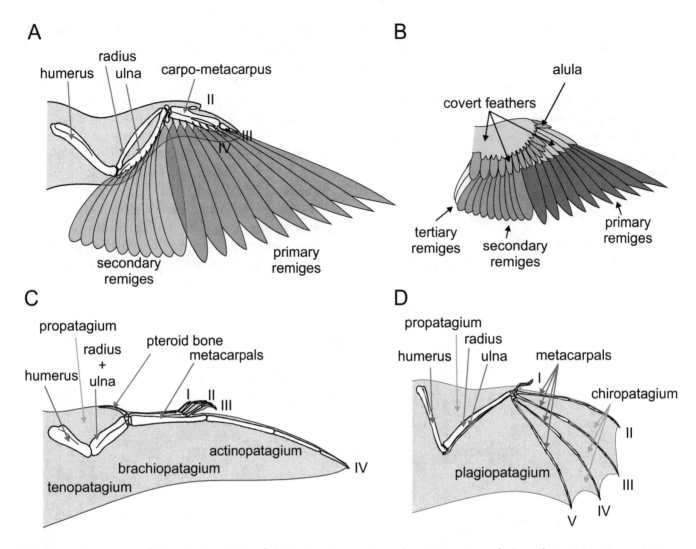

5.15. Comparative anatomy of wings. *A,* wing skeleton of a bird, with primary and secondary remiges. *B,* wing feathers of extant birds. *C,* wing skeleton and reconstructed membranes of a pterosaur (modified from Elgin et al. 2011). *D,* wing skeleton and membranes of an extant bat.

keeps the animal in the air by overcoming body weight, but also the propulsive force in the direction of travel. Although some bony fish (such as Exocoetidae) have modified their paired fins into true wings and can make fairly long flights through the air (fig. 5.14), the greatest diversity in form and mechanisms of flight among vertebrates has been achieved among pterosaurs, birds, and bats.

The members of these three groups of flying vertebrates can (or could, in the case of pterosaurs) alter the area, shape, and position of their wings with respect to the body during flight, abilities that greatly increase maneuverability, particularly in takeoff and landing. Due to the greater diversity among birds, we will focus on the wings of these vertebrates, but many of the concepts involved with flight and design aspects of, in particular, the wings, are common to those of other flying vertebrates.

The wings of a bird are located above the body's center of gravity and have a complex structure, especially the

autopod, where the distal carpals fuse with several metacarpals to form a carpometacarpus and only the phalanges of three digits are free. In contrast to pterosaurs and bats, in which the aerodynamic surfaces consist of membranes (*patagia*, plural of *patagium*) formed by the integument and stretched between elongated digits, in birds the lift surface is formed by feathers, which are integumentary derivatives. Feathers (fig. 5.15) enable a greater range of aerodynamic possibilities and cover nearly the entire wing. Membranes, by contrast, are limited to the front and back of the stylopodial-zeugopodial segments of the wing. The feathers of the avian wing differ morphologically and functionally from those that cover most of the rest of the body. The remiges (remex, in the singular) are large, long, and relatively rigid feathers arranged along the posterior or trailing edge of the wing. The primary remiges, commonly known more simply as primaries, are the most distal, or outermost, remiges, inserting on the carpo-metacarpus and phalanges. These feathers may be narrowed distally, so that when they spread apart, notches are produced between them, forming a slotted wing that reduces the drag (resistance) on the wings.

We will now consider several elementary concepts of aerodynamics that help illuminate the study of locomotion in air. Although they are discussed with reference to birds, they are applicable to all flying vertebrates.

AERODYNAMICS

In aerodynamic terms, the wings and tail constitute, as already noted, lift surfaces (fig. 5.16) that can be described mathematically by a series of parameters. The wingspan is the total distance across the outstretched wings, and the chord is a line extending between the front, or leading, edge of the wing and its rear, or trailing, edge; its length changes in tapered wings, as in those of birds and many airplanes. The shape of the wing's section is the profile. It is usually not constant over the wing's length, also decreasing with increased distance from the bird's body. As both the profile and chord length change, the profile is

characterized as based on a calculated mean chord. The thickness is the maximum distance between the upper and lower surfaces of the wing. In birds, the upper and lower wing surfaces are curved, though not necessarily to the same degree. Camber is used to describe this property of an airfoil profile—that is, it is asymmetric and the upper surface is usually more strongly curved than the lower surface (cases where the reverse happens do occur, as in some insects; see Alexander 2017). An important feature of aerodynamic performance is the degree of camber, the amount of difference in curvature between upper and lower airfoil surfaces. The mean camber is a line that extends from the leading to trailing edge and is, at each point, midway between the upper and lower surfaces of the wing. In a cambered (asymmetric) wing profile, the mean camber line is above the chord line. It follows, then, that a symmetric airfoil profile, in which the upper and lower surfaces have the same curvature and hence the mean camber coincides with the chord, is not cambered. However, such a wing can produce lift if the angle of attack is positive with respect to the airflow (see below).

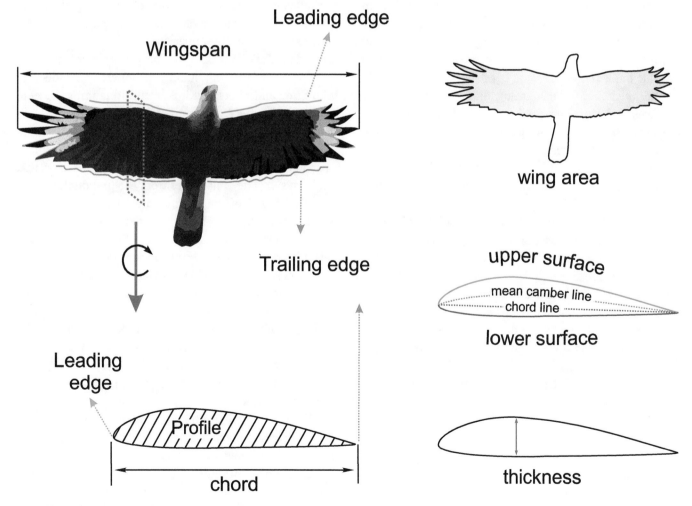

5.16. Wing attributes related to lift.

Forces transmitted by fluids act as pressures. Therefore, to generate lift, an airfoil must induce a pressure gradient in the surrounding air. This is achieved by a velocity gradient produced by the deflection of the relative flow of the air around the airfoil, either by means of a positive angle of attack, a cambered profile, or both. Air flowing below the airfoil is slowed down while air flowing above is sped up, especially at the trailing edge where the airflow is forced downward. According to Newton's law of motion, given that velocity is a vector, changing the direction and/or the magnitude of the airflow results in acceleration of the airflow. Thus, a force is generated if the airflow is induced to deflect, and this deflection changes the pressure distribution around the airfoil, establishing a pressure gradient with a region of lower pressure above and a region of higher pressure below the airfoil. These differences in pressure are consistent with Bernoulli's principle and produce a net upward force on the airflow, orthogonal to the direction of the airflow, which is the lift force. Note that it does not matter if the air is flowing around a static airfoil, an airfoil is moving in static air, or both are moving with respect to each other. Only the relative motion of the air and airfoil matters, and for this reason the airflow is often referred to as relative wind. It is also important to make the distinction between lift, as a force, and an increase in altitude; lift does not necessarily cause an object to gain altitude but is the force that keeps it aloft.

As noted above, constant flow of air around the wing surfaces is integral for producing lift. Bernoulli's principle, which was noted above as well as in the "Fluid Mechanics" section, helps explain the physical principles of lift. In considering the profile of the wing, it is evident that the leading edge is thicker than the trailing edge and that the upper surface is more strongly curved than the lower surface (fig. 5.16). When the airflow is deflected around the airfoil, air layers passing above it will speed up, while air layers passing below it will slow down. To maintain the value of the constant k (for more about Bernoulli's principle, see the "Fluid Mechanics" section), a particle that increases its speed must do so by decreasing its pressure, and vice versa. The pressure differential between the upper and lower wing surfaces produces, as mentioned in the previous paragraph, a force resultant that is perpendicular to the profile termed the *lift force*, defined mathematically as

$$\mathbf{L} = 1/2 \; \rho \mathbf{v}^2 A Cs$$

where ρ is the air density, $\mathbf{v}$ the velocity at a given point (the term $1/2 \; \rho \mathbf{v}^2$ is referred to as the dynamic pressure), A is the area of the lift surface, and Cs is the lift coefficient,

which depends on the type of wing profile and the angle of attack between the wing chord and the direction of airflow (fig. 5.17). The lift force is concentrated at a particular point, termed the *center of pressure*, along the chord length and is at right angles to the direction of airflow. The center of pressure is not stationary and depends on the angle of attack (McGowan 1994). Increasing the angle of attack has the same effect as increasing the camber of the airfoil, and thus also increases the difference in pressure above and below the wing and, in consequence, increases lift. However, this is true only to a certain point; an excessive angle of attack leads to a turbulent flow over the wing that dramatically reduces lift by separating the boundary layer from the wing surface. At a certain point, this results in a sudden decrease in lift, a phenomenon known as stall.

Antistalling devices, however, can be employed to help prevent stalling and are commonly used in aircraft. The basic idea behind such devices is to create a gap, termed a *slot*, toward the leading edge of the profile that extends from the lower surface upward and backward to the upper surface. The slot, which may be provided by a passage through the wing itself or by the addition of a narrow slat attached to the leading edge of the wing, channels a high-speed air current to a zone of separation of the boundary layer on the upper surface. This allows for increased angles of attack, which provide greater lift. In the usual course of flying, stalling is generally avoided, by both birds and pilots, but it is useful in landing; indeed, landing is, in effect, a controlled stall. Airplanes use stalling as a landing technique and have a variety of devices that enable this maneuver. Birds do so too, and they have a particular device, termed the *alula* (or bastard wing), that creates a slot. The alula arises from the first digit, located near the "bend" of the wing, and is usually made up of four alular quills (De Iuliis and Pulerà 2019). It acts as a hypersupporting element, essentially like a slat, that channels a high-speed air current to the upper surface. This allows increased lift at slower speeds and delays stalling. By delaying stalling of the wing, stalling speed and thus landing speed are reduced, making possible a more carefully controlled descent (McGowan 1994; fig. 5.17). The alula may be observed in action, as noted by McGowan (1994), by watching a bird such as a pigeon during landing; the alula may be seen to lift suddenly just before the bird touches down.

Movement of air around the wing produces an upward force, as already noted, but it also exerts drag, which acts parallel to the movement of the airflow and so imposes resistance on the propulsive force (momentum). Thus, the force acting on the wing is termed the *total reaction*, which may be resolved as two forces, lift (**L**) and drag (**D**), both

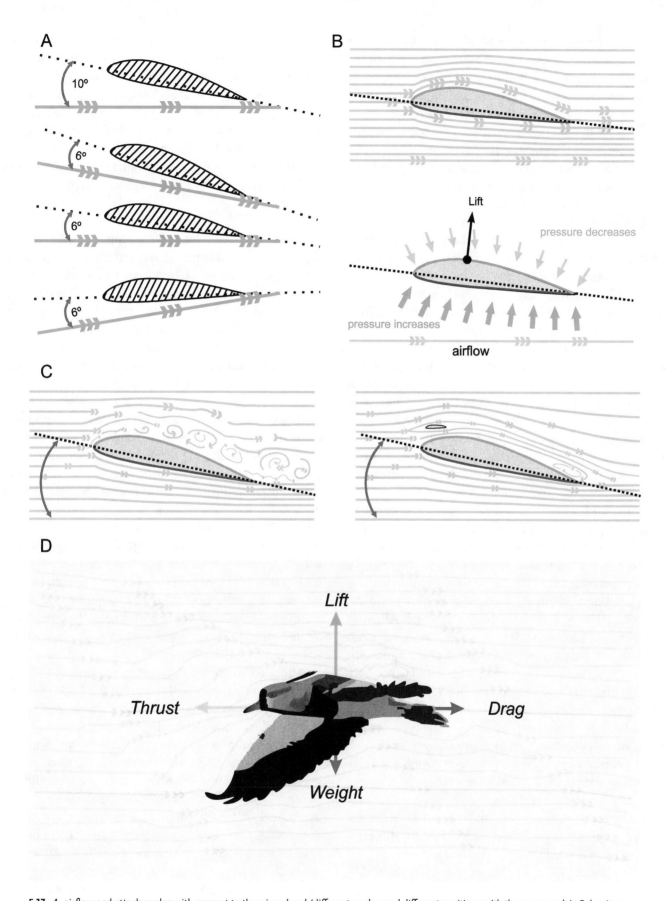

A

10°

6°

6°

6°

B

Lift

pressure decreases

pressure increases

airflow

C

D

Lift

Thrust

Drag

Weight

5.17. *A*, airflow and attack angles with respect to the wing chord (different angles and different positions with the same angle). *B*, laminar flow on the wing profile, higher speed on upper surface than on lower surface and pressure distribution with the resulting lift force. *C*, stall (*left-side* image) caused by turbulent flow, and alular effect (*right-side* image) for reestablishing laminar flow. *D*, pairs of lift forces, lift/weight and thrust/drag (depicted on a southern lapwing), that act on a flying vertebrate.

of which increase with the speed of airflow. Lift opposes weight, and drag opposes the propulsive force, so that the basic forces acting on a bird in flight come in opposing pairs, lift/weight, on the one hand, and momentum/drag (or resistance), on the other (fig. 5.17). Lift depends on the shape of the profile, the wing surface, the air density (which decreases with height), the relative wind speed, and the angle of attack of the chord. Based on these principles, various indices have been developed to characterize the performance of wing shape in different situations.

The relationship between the wingspan and the mean chord length is called the aspect ratio (AR), which gives an idea of the slenderness of a wing. It is easy to calculate for a rectangularly shaped wing and is expressed simply as AR = wingspan/chord length, because the chord length is the same everywhere along the length of the wing. For a tapered wing, the mean chord length must be used, which is also easy to calculate for a regularly tapered wing (whose edges are straight) by summing the root chord (its widest part, near the body) and its tip chord (the chord length at the tip of the wing) and dividing by two. For more complexly shaped wings, such as elliptical wings, AR is given by squaring the wingspan and dividing by the surface area of the wing. An easy way to think of the difference is that high AR wings are relatively long and slender, whereas low AR wings are short and stubby.

Wings with a higher aspect ratio have a higher lift-to-drag ratio (LD) than do wings with a lower aspect ratio, because they experience relatively less drag. LD is a measure of the efficiency of the wing, a concept that may be appreciated by considering the rate of descent in gliding, or the gliding angle (the minimum angle that a bird can fly while gliding). Compare a bird with a high AR, such as a gull, and another, such as a pigeon, with a lower AR; the gull will have a shallower angle of descent than the pigeon, meaning that the former will travel farther horizontally over the same vertical drop (McGowan 1994). One might then wonder, given this feature of high AR wings, why all wings are not long and slender. The reason is that there are also disadvantages to such wings, depending on the functions they are meant to perform. One such disadvantage is structural; a longer wing needs to be stronger, meaning more material is required for its construction, which involves increased costs and mass. Another is that longer wings, because of their higher moment of inertia, are less maneuverable.

Wing loading (WL) is the mass of the bird divided by the wing area. The lower the WL, the less power or effort is required to sustain flight. Most bats and passerine birds have low WL values. Overall, we may characterize long and slender wings as having high AR, LD, and WL values, a combination that is particularly useful in high-speed (i.e., in high relative wind), level flight; an example is the wandering albatross *Diomedea exulans*, for which the values are AR = 18:1, LD = 40:1, and WL = 1.5. By contrast, the short and wide wings of many passerines and birds of prey have low AR, LD, and WL values; for falcons and eagles, for example, WL values are approximately between 0.9 and 0.7, respectively (fig. 5.18). These kinds of wings are good for flying at low speeds and with great maneuverability.

TYPES OF FLIGHT

The propulsive force for powered flight is generated by the upward and downward movement, or flapping, of the wings. During flight, birds usually alternate between flapping and nonflapping phases (fig. 5.19). Two intermittent patterns have been described: flapping-gliding (with wings outstretched) and flapping-missile (with wings folded against the body). While the former appears to be more economical than continuous flapping at low speeds, the latter mode is more efficient at higher speeds.

During flapping, two patterns may be observed based on the type of vortex produced at the wing tip: ring vortex or continuous vortex (fig. 5.20). Wing-tip vortices are a component of drag termed *induced drag* (the other component is the drag not dependent on the lift, hence termed *parasitic drag*). When lift is generated only during the downstroke phase of wing movement, a ring vortex is produced, whereas when lift is also generated during the upstroke, a continuous vortex is produced. Regardless of wing shape, most birds and bats use a ring vortex stroke during slow flight. At faster speeds, species with low AR values (short, stubby wings) tend to continue with a ring vortex stroke, whereas those with high AR values (long, slender wings) tend to transition to a continuous vortex stroke (Scholey 1983; Rayner 1991).

Although a detailed explanation of wing movements during flapping is beyond the scope of this book, the functional differences in the two strokes with regard to lift may be understood in less technical terms. Essentially, the more proximal part of the wing (that segment including the secondary remiges, which insert on the ulna) undergoes little vertical movement and acts as if it were gliding, thus generating the lift that counteracts the force of gravity. By contrast, the lift resultant force at the distal segment of the wing (where the primary remiges are) also has an anterior component that produces propulsion. When the speed of flight through the air is constant, the forces acting at the distal and more proximal segments of the wing produce a set of summed vectors such that thrust exceeds total resistance, resulting in the animal moving forward respect to the air mass, and lift equals or exceeds body

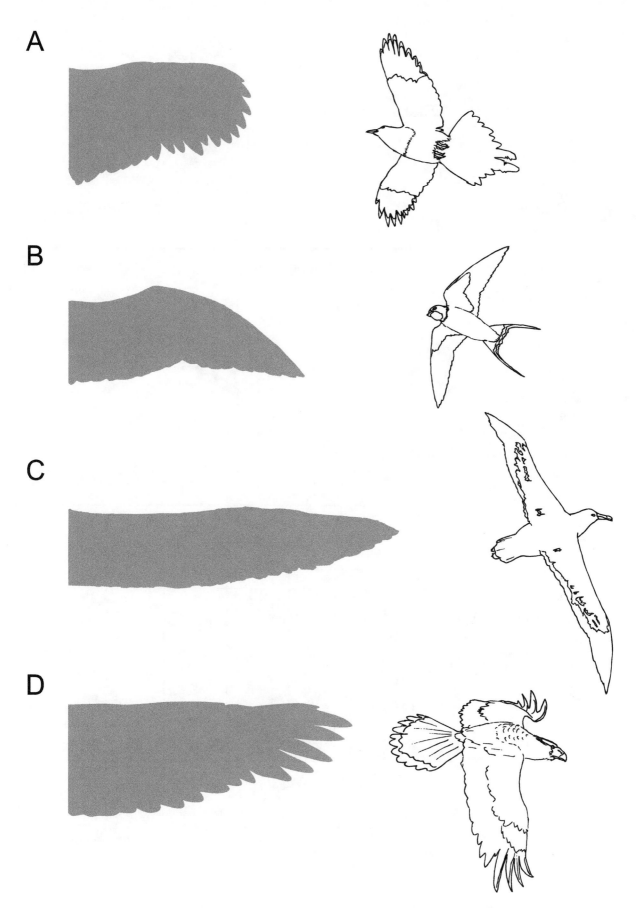

5.18. Types of wings in flying birds. *A*, elliptical wings of low wing load (magpie). *B*, fast wings of low wing load (swallow). *C*, wings of high aspect ratio, for dynamic soaring at high speed typical of seabirds (albatross). *D*, wings of low aspect ratio, for thermal soaring at low speed and high lift (caracara).

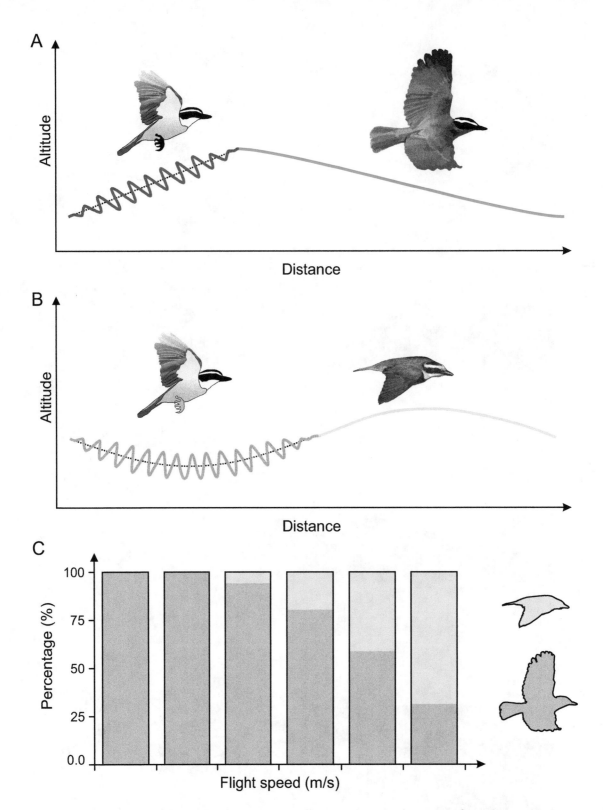

5.19. Alternation of powered flight with flapping and gliding or flapping and missile stages exemplified in the great kiskadee (*Pitangus sulphuratus*). *A*, trajectory described in flapping-gliding. *B*, trajectory described in flapping-missile. *C*, comparison between gliding and missile modes at different speeds (*blue*, gliding; *green*, missile). The gliding mode is the mode used most at low speeds and the missile mode at high speeds (modified from Tobalske 2007).

5.20. Flapping flight types based on the form of vortex produced. *A*, when lift is generated only during the downstroke phase, a ring vortex is produced. *B*, when lift is generated during both the downstroke and upstroke phases, a continuous vortex is produced. Modified from Tobalske and Dial (1996).

weight, resulting in the animal remaining horizontal or rising, respectively, relative to the air mass.

In soaring flight, weight is the force that generates propulsion and lift, and so is characteristic of large birds (fig. 5.21). During soaring, a bird always loses height with respect to the air mass; that is, it is falling with respect to the surrounding air. However, birds counteract this "falling" through updrafts (or more simply, rising air), which are used to maintain or increase altitude. Updrafts are generated by natural obstructions that deflect the wind upward, as in mountainous terrain, ravines, sea cliffs, and slopes, by meeting air masses, or by heating of the ground surface of flatter terrains, causing differential air temperatures that lead to upward columns of warm air called thermal currents. At the top of thermals, the air cools and falls again peripherally to the core column of warm air, so a bird must keep flying in a circular pattern to remain within the column of warm air (i.e., to avoid the downward flow of cooler air). Raptors are highly adept at

flying in thermals, and frigates (Suliformes) take advantage of thermals at sea in tropical areas under particular oceanic and atmospheric conditions that generate marked differences in temperature between adjacent areas of the sea surface.

Another strategy to gain height without flapping is to take advantage of wind gradients, as some seabirds do in dynamic soaring (fig. 5.21). The wandering albatross *Diomedea exulans* performs this maneuver by flying with the wind; that is, downwind. As it descends, it flies through successively slower layers of air and thus its relative airspeed increases; as a consequence, so does lift. Near the surface of the sea, the albatross turns back to fly against the wind at a very high relative speed. Even though it begins to lose ground speed, it maintains lift (because of its high relative speed) and rises, thus gaining altitude. In doing so, it converts much of its kinetic energy into potential energy, and then reverts to flying downwind. Thus, it is capable of flying for many hours and covering

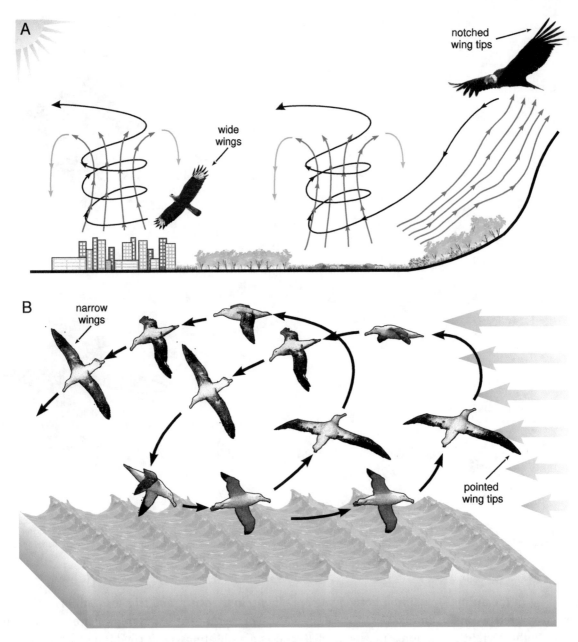

5.21. Soaring. *A*, static soaring birds that use ascending thermals, exemplified by the condor (mountains) and caracara (plains). *Red arrows*, ascending warm air currents (thermals); *blue arrows*, descending cool air currents. *B*, dynamic soaring birds that use wind gradient in marine environments, exemplified by the wandering albatross. *Black arrows*, bird's trajectory; *blue arrows*, wind direction and velocity.

many kilometers with little energy expenditure. Several flight-pattern variations involving wind gradients (or wind shear) have been described by, for example, Richardson (2011) and Richardson et al. (2018).

Last, there are birds capable of stationary flight, called hovering. Many seabirds do this by flying at a speed equal to that of the headwind, by either flapping their wings or gliding. Other birds, such as hawks, and bats engage in hovering over food sources in low wind-speed conditions by rapidly flapping their wings. Hummingbirds, hovering vertebrates par excellence, do so by maintaining their body

at an angle of about 45° from the horizontal and moving their wings in a horizontal figure-eight pattern, a behavior made possible by the particular anatomy of their shoulder joints, which allows them to generate lift during both the downstroke and the upstroke (fig. 5.22).

POWER AND THE ENERGETIC COSTS OF FLIGHT

The power (work per unit time) for wing movement is derived from the musculature. In birds, the pectoral muscle, lying on the ventral surface of the bird's broad sternum, produces the power to draw the wing downward (i.e., for

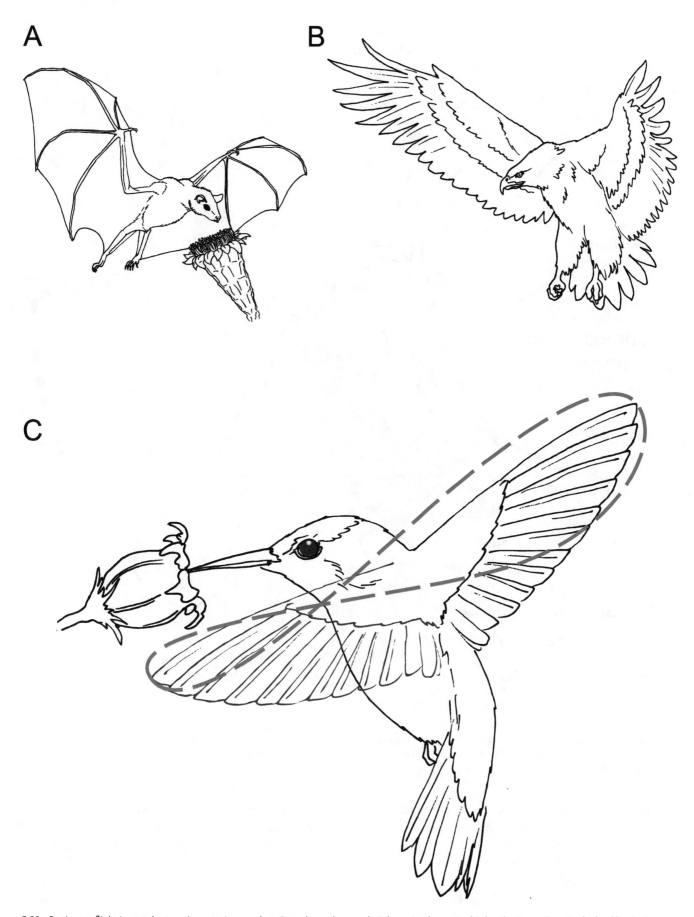

5.22. Stationary flight in vertebrates. *A*, nectarivorous bat. *B*, eagle on the prowl. *C*, hovering hummingbird gathering pollen. *Dashed red line* indicates the figure-eight pattern described by the trajectory of the wing tip.

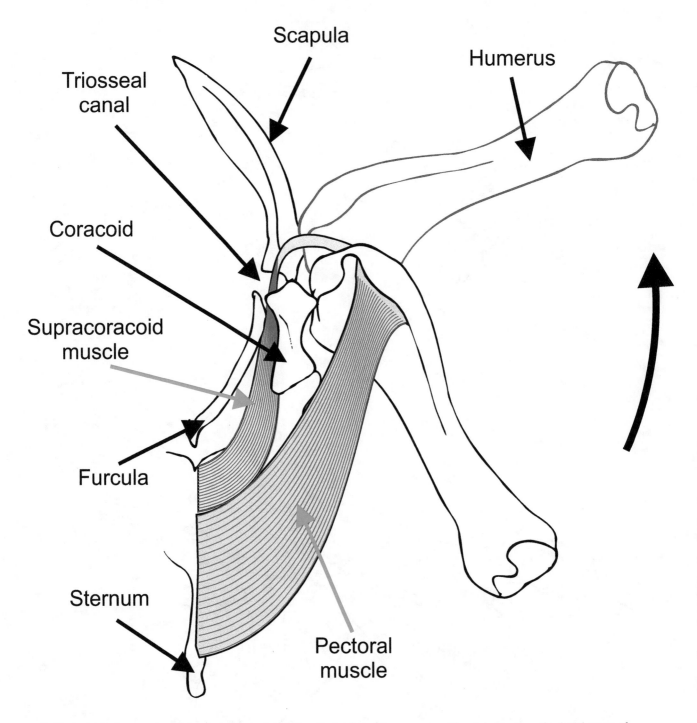

5.23. Pectoral wing muscles of birds (anterior view). Note how the tendon of the supracoracoid muscle inserts on the dorsal aspect of the humerus, passing first through the triosseal canal, and acts as a wing elevator and antagonist of the pectoral muscle. Modified from Liem et al. (2001).

the downstroke), whereas the supracoracoid muscle, lying deep to the pectoral muscle, is the main elevator of the wing, thus acting in the execution of a vigorous recovery stroke (the upstroke) that, combined with the movement of the primaries, also produces some thrust, thus preventing excessive deceleration (fig. 5.23). The supracoracoid muscle is more developed in birds capable of rapid takeoff or flapping hover. The arrangement and position of the supracoracoid is particular in that it lies below the element that it elevates, while at the same time enhancing stability for flight by maintaining the center of gravity below the wing. The supracoracoid executes its action by way of the pulley action of its long, stout tendon that angles toward the shoulder joint and, in most birds, passes through a passage, the triosseal canal, in the bony structure of the joint and onto the dorsal surface of the shoulder. It then extends laterally to insert on the dorsal surface of the humerus (De Iuliis and Pulerà 2019).

Birds have high metabolic rates, but rates are especially high in those that perform flapping flight, approximately seven to 15 times higher than the basal metabolic rate. Speed also contributes to the energy cost of flight, with lower and higher speeds being the most expensive. The lower speeds (e.g., as when taking off or landing) require more energy due to the increased turbulent flow of the air and, therefore, the induced drag. Smooth high-speed flight is expensive because it requires higher rates of flapping and muscle activity. In these circumstances, the muscles work in anaerobic metabolism, an energy regimen that cannot be sustained for extended periods of time. Takeoff is a critical moment of flight, and generating the necessary lift thrust involves different strategies depending on body size. Smaller birds take off by flapping their wings at very high frequencies, which is energetically expensive, whereas larger birds need to run (taxi) or jump from a height to gain lift that cannot be achieved by flapping alone. Hovering in low wind-speed conditions (as in hummingbirds and some birds of prey) can take approximately twice the power required to maintain a given speed. It is thus not surprising that hummingbirds commonly feed on nectar, an energy-dense food item.

WING SHAPE

Many birds have elliptical wings, characterized by low AR, marked camber, and primary remiges with notches that facilitate slow flight and great maneuverability (e.g., passerines). Elliptical wings are also characteristic of bats that fly through vegetation, for which high maneuverability is needed. Wings for fast flight are pointed, with moderate AR, little camber, and without grooves. They have a high WL, which facilitates penetration into the air mass, allowing, for example, advancing against the wind (as in migratory birds such as geese, and also in bats that live in open areas). Wings for dynamic soaring are long, narrow (high AR), with little camber, and without notches. They are typical of seabirds, such as albatrosses, that require a high vertical wind gradient, a condition produced by lower layers of air slowed by friction with the sea. Wings for thermal soaring have a moderate AR and are intermediate in shape between elliptical and long, tapered wings. They have low WL, pronounced camber, and marked notches, which allow considerable maneuverability, an ability that is important for staying within ascending air columns (figs. 5.18, 5.21). These types of wings are also observed

in some of the large fruit bats (Pteropodidae; see Norberg and Rayner 1987).

APPLICATIONS IN PALEOBIOLOGY

Birds and other vertebrates such as bats and pterosaurs have evolved a wide variety of forms and flight modes. As most of the lift surface of wings is composed of soft tissues, especially in pterosaurs and bats, its preservation tends to be quite exceptional (see later for several examples). Therefore, for extinct flying vertebrates, their body size, morphology of their appendicular elements, and their proportions provide the most typical data for inferring several other aspects of their aerodynamics. Norberg and Rayner (1987) identified the following three main variables as important for understanding the functional basis of ecomorphological correlations in bats: WL, AR, and tip-shape index. However, given that we know that both extant birds and bats can alter the shape and surface of their wings during flight, there are some limitations in applying the knowledge discussed here to extinct flying vertebrates.

The protocol for paleobiological studies established at the beginning of this book proposes that estimation of body mass be a first step. Any of the approaches described in chapter 4 can be applied to this end. In the following pages, we will note, among other things, how different estimates of body size and parameters such as metabolic rate can lead to quite different hypotheses on the flight capabilities and strategies in extinct vertebrates.

Vizcaíno and Fariña (1999, 2000) reexamined the evidence and hypotheses on the flight capabilities of *Argentavis magnificens*, a Miocene bird belonging to the extinct clade Teratornithidae. The remains of this animal, although not particularly complete, are spectacular in documenting the enormous size of this bird. With an estimated wingspan of 6–8 m, it is without doubt the largest known bird (fig. 5.24). Its large size led many researchers to wonder how it flew and, indeed, whether it was capable of taking flight at all. To analyze this, Vizcaíno and Fariña (1999, 2000) used previously made mass estimates, including allometric regression equations from measurements available from the specimen. Published values of its body mass ranged from 60 to 160 kg, much larger than the largest current flying birds (the Andean condor *Vultur gryphus*, the wandering albatross *Diomedea exulans*, and the African bustard *Ardeotis kori* all weigh between 10 and 16 kg). Using previous estimates of WL and AR based on geometric similitude with the condor, Vizcaíno and Fariña (1999, 2000) calculated that *Argentavis* would have had to travel at a minimum relative air speed of 40 km/h to generate sufficient lift to get its 80 kg (averaged value of

5.24, Comparison of the reconstructed silhouette of *Argentavis magnificens* with the silhouette of an Andean condor and a human figure. Modified from Vizcaíno and Fariña (1999).

weight estimates) body airborne. On land and in calm air conditions, *Argentavis* would have had to reach this speed by running at a speed achieved only by specialized cursorial birds such as rheas or by executing a full wing stroke of more than 3 m in height.

In light of the extreme improbability that *Argentavis* would have been capable of achieving such behaviors, these authors suggested that to generate lift, this extinct bird would have had to jump from a height in mountainous areas and/or taken advantage of strong winds in its wide range of distribution. Once in the air, it would have engaged mainly in gliding flight, taking advantage of updrafts in the manner of many large birds of prey. Previous studies also suggested that a bird of this size would have been unable to develop sufficient muscular force to flap its wings (Tucker 1977; Pennycuick 1992), but Vizcaíno and Fariña (1999, 2000) calculated a resistance index for the humerus, the value of which, when compared to those of modern birds (table 5.1), was sufficiently high to support the hypothesis that it was capable of withstanding the forces to which it would be subjected in an intermittent flapping flight.

Chatterjee et al. (2007) used a computer simulation model to analyze the flight capabilities of this bird.

Table 5.1. Index of resistance (Z/amgx, in Gpa-1) comparisons

Species	Z/amgx
Argentavis magnificens	36
Teratornis merriami	53
Gymnogyps californianus	54
Diomedea exulans	10

Note: Index calculated for the humerus of *Argentavis magnificens* compared with those of the teratornithid *Teratornis merriami* from the Pleistocene asphalt deposits of La Brea Ranch (California), the condor from California *Gymnogyps californianus*, and the wandering albatross *Diomedea exulans*. Modified from Vizcaíno and Fariña (1999).

Estimating the relevant parameters (wingspan, chord, AR, WL), they compared a series of calculated values (turning radius, maximum speeds, and minimum glide angles; fig. 5.25) with those of several current gliding birds and reached conclusions very similar to those of Vizcaíno and Fariña (1999, 2000).

Recently, new techniques have revealed preserved remains that were undetectable to study by previous analytical tools. For instance, use of laser fluorescence allowed identification of exceptionally preserved soft tissues forming the wing in addition to preserved feathers, including fleshy phalangeal pads as well as pedal scales and claws, in the basal bird *Confuciusornis* from the Early Cretaceous of China (Falk et al. 2016). This information enabled the authors to infer powered flight and arboreal habits for this bird.

In the case of pterosaurs, the wing parameters of which can be estimated from the proportions of their forelimbs, body mass has traditionally been estimated by modeling body volume, using a density similar to that of the average of modern birds. This method yields comparatively lower body masses with respect to flying birds of similar wingspan and mean chord, suggesting that pterosaurs were exceptionally light flying forms, a hypothesis also supported by the pneumaticity of their bones. In contrast to traditional methods for modeling body mass, Witton (2008) proposed using the ratio of dry bone mass to total mass, which is identical in birds and flying mammals, regardless of size, morphotype, and clade. Based on this method, Witton (2008) derived mass estimates that are, in some cases, as much as three times higher than previously published estimates (e.g., for the pterosaur *Quetzalcoatlus*, this author calculated a body mass of up to 250 kg, whereas most previous studies estimated its mass as 70–85 kg). Clearly, these estimates have an enormous impact on inferences on wing loading and, therefore, flight capabilities and types. For example, the WL of *Pteranodon* calculated by Alexander (1985), based on a mass of 15 kg, was

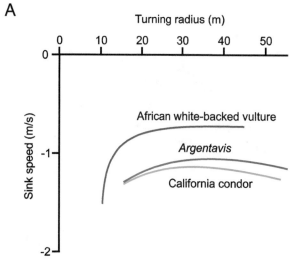

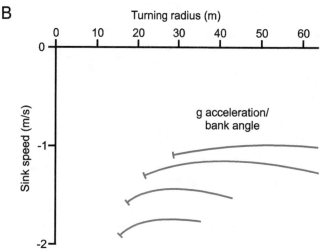

5.25. Turning radius versus sink speed in thermal currents. *A*, variation of speed estimated with respect to the turning radius in *Argentavis* compared with values for two extant vultures. *B*, different speed/radius curves for *Argentavis*, indicating how turning radius could be reduced by increasing wing inclination (that is, banking). According to calculations by Chatterjee et al. (2007), *Argentavis* could fly with a turn radius close to 30 m at a sink speed of 1 m/s. Modified from Chatterjee et al. (2007).

between 33 and 60 N/m², whereas Witton (2008), based on a mass of approximately 36 kg, obtained a WL of 135 N/m². As it is necessary to estimate wing area and wingspan to calculate the AR and, together with the estimated body mass, WL, the shape and size of the lift surfaces, formed by membranes in the case of pterosaurs, have to be reconstructed. However, different hypotheses on the posterior extent of the membranes—for example, to the ribs, hip, thigh, ankle, or foot—have been proposed (fig. 5.26) and result in entirely dissimilar estimates of flight capabilities. Recently, exceptionally preserved specimens from various groups of pterosaurs demonstrate that the membrane extended to the calf or ankle (Witton 2008; Elgin et al.

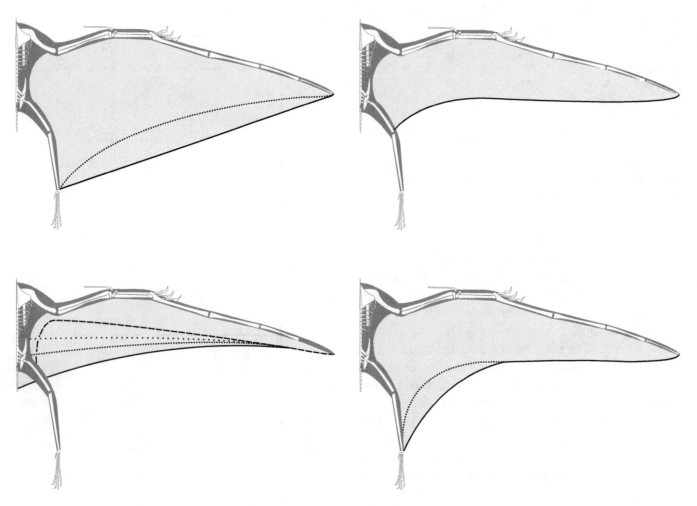

5.26. Different reconstructions of wing membranes (patagia) in pterosaurs. Modified from Elgin et al. (2011).

2011). This, however, does not preclude the possibility of high-AR wing configurations among pterosaurs, as the trailing edge of the membranes did not necessarily have to be straight (fig. 5.26).

Based on body mass, wing area, and AR and WL values, Witton (2008) analyzed the diversity of flight types in pterosaurs by constructing a morphospace based on WL versus AR, including the values of a variety of birds and bats. The range for AR among the pterosaurs analyzed by Witton (2008) is 15 (between 6.60 and nearly 23), similar to the range observed in current birds (4.5 to 18). The morphospace occupied by pterosaurs is shared with that of birds and bats, which led this author to propose a much greater diversity of flight types than traditionally interpreted based on lower estimates of masses, although no pterosaurs were found to have flight characteristics comparable with diving birds or water birds (fig. 5.27).

With regard to takeoff, it was traditionally proposed that pterosaurs, especially the larger ones, would launch themselves from a height. This inference was based in part on the hypothesis that they would have had a very poor capacity for locomotion on the ground. However, Padian and Rayner (1993) suggested that pterosaurs would be able to run or jump on their hind limbs to facilitate takeoff.

Other fossil vertebrates for which paleobiological studies have been performed include bats and gliding mammals. For instance, Amador et al. (2019) analyzed the exceptionally preserved early Eocene bat *Onychonycteris* and reconstructed its wing parameters (WL and AR, along with other quantities such as body mass) and inferred its flight performance. Amador and collaborators found that *Onychonycteris* exhibited a unique combination of a very high WL and low AR, which is not observed in any modern analog among chiropterans, making this animal an obligate fast flier with low maneuverability, probably alternating between flapping and gliding stages. The hypothetical biological role for flight in this animal was one of traveling

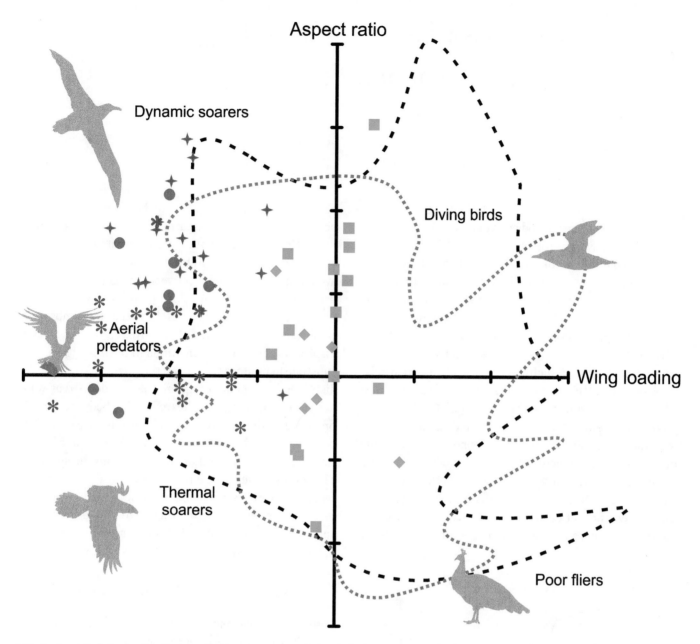

5.27. Scatterplot of wing loading (*horizontal axis*) versus aspect ratio (*vertical axis*) in pterosaurs, birds, and bats. The *dashed black line* corresponds to the morphospace of bats, and the *dotted gray line* corresponds to the morphospace of birds. The *symbols* correspond to different pterosaur clades. Modified from Witton (2008).

between roosting places or foraging sites. Grossnickle et al. (2020) analyzed exceptionally preserved skeletons of several extinct mammals inferred to be gliders and compared them with extant gliding mammals such as the marsupial feathertail gliders (Acrobatidae), the colugos (Dermoptera), and the flying squirrels (Anomaluridae), and also with other extant nongliding arboreal mammals.

Using both cranial and appendicular linear measurements to perform several phylogenetic-oriented analyses, these authors found that some of the fossil taxa exhibited unique gliding morphologies not present in modern analogs, whereas others showed mixed traits between arboreal nongliding and gliding mammals, and others still more closely resembled extant nongliding arboreal mammals.

6

Substrate Preference and Use: Terrestrial Locomotion

As described in chapter 5, the evolutionary transition to land involved changes in body movement and support that were achieved by means of important modifications of the vertebral column, girdles, and limbs. Even allowing for a body sufficiently small and light that it would not have caused the lungs to collapse and prevent ventilation, air is not sufficiently dense for a caudal fin to produce thrust and provide propulsion. The solution to the issues of support and locomotion was foreshadowed in the pectoral and pelvic appendages of such basal Paleozoic tetrapod relatives as *Tiktaalik* (see Shubin et al. 2014), *Elpistostege* (see Cloutier et al. 2020), and *Acanthostega* (see Clack 2012).

In fish, the fins are generally moved in coordination with the trunk, so that diagonally opposite pairs of fins move more or less synchronously. This movement is used by several groups that walk on the sea bottom (e.g., epaulette shark, the teleostean frogfish; coelacanths share this movement, but do not use it for walking; see Clack 2012). As in these animals, the chiridium of Paleozoic

basal tetrapod relatives was capable of acting as a support, aiding propulsion and maneuvering in a complex, three-dimensional aquatic environment with abundant submerged vegetation. But out of the water, the effort of overcoming the friction between the substrate and the trunk and limbs dragging on the ground would be extremely energetically expensive; instead, the limbs must act to raise the trunk and themselves for the body to be displaced. The action of the limbs in pushing against the substrate to produce thrust and the lifting of the limbs for displacement without dragging them is defined as walking in steps. In tetrapods, the undulating body movements produced by the metameric axial musculature (as in fish) move the transversely oriented limbs forward in a succession that allows the trunk to be lifted from the substrate. The body's center of gravity remains within a triangular region, the vertices of which are defined by the three limbs resting on the substrate, while a fourth limb is raised. The appendicular musculature helps in forward displacement of the limbs and then in pulling the body forward (fig. 6.1).

Posture and Energetic Cost of Locomotion

Posture is defined as the disposition of an animal's body when it is motionless or at rest, particularly with respect to how the body is supported by the limbs. As we have already mentioned, some configurations are more efficient for support and locomotion than others. In the sprawling postures of urodeles and saurians, and to a lesser extent crocodiles, the limbs are oriented transversely to the anteroposterior (or longitudinal) axis with the stylopodium (humerus or femur) held in a nearly horizontal attitude. The torque produced by the moment arm between the pivot (the position where the limb rests on the substrate) and the center of mass, which for simplicity we will take as the length of the stylopodium (fig. 6.1), must be continuously counteracted if the body is to be held off the ground; the animal would otherwise drag its belly on the substrate. Clearly, the latter option would be very energetically expensive due to the friction produced with the substrate. Lifting the body to prevent this involves a great

6.1. Body support and locomotion in tetrapods with a sprawling posture. *A*, maintaining the body raised requires that body weight, acting through the center of gravity (*red arrow*), must be counteracted by the contraction of the limb adductor musculature (*blue arrow*). *B*, muscle groups in *red* are those that pull limbs backward, propelling the body forward, whereas the muscle groups in *blue* are those that move limbs forward to start a new step.

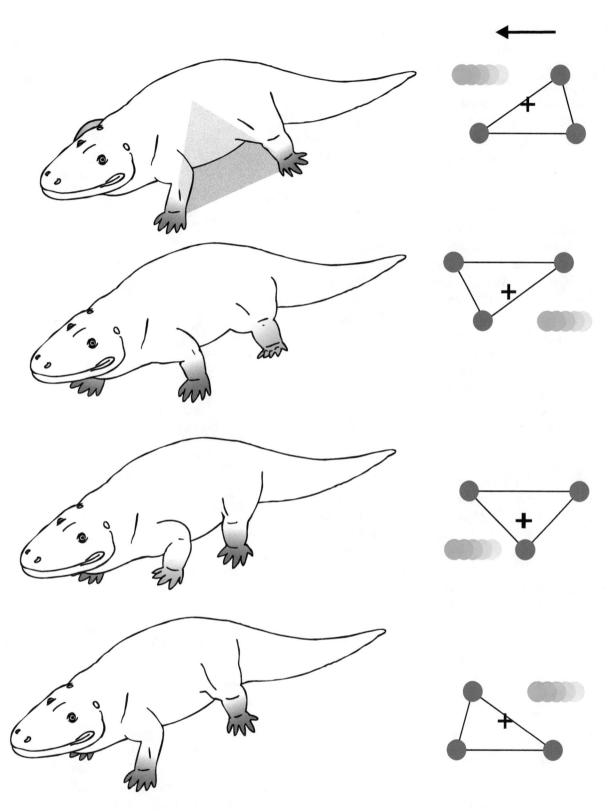

6.2. Walking in tetrapods. The *left-side* diagrams illustrate the sequence, from *top* to *bottom*, of limb movement as the animal moves forward. *Red shading* indicates a foot that is in contact with the ground, whereas *blue* indicates a foot that is raised and being moved forward. The *right-side* diagrams indicate the triangle of support formed by the limbs that are in contact with the ground (*red circles*) and the limb that is raised (*blue circles*) as the animal proceeds forward (*arrow at the top* of the sequence of figures). The animal's center of mass (*cross*) is always contained within the triangle of support. Modified from Liem et al. (2001).

development of the humeral and femoral adductor musculature and of the skeletal elements (e.g., coracoid, pubis) that serve to anchor this musculature. Its contraction results in the pulling of the stylopodium ventrally toward the sagittal plane, consequently raising the trunk (such as occurs in executing "push-ups"). In this posture, one limb at a time is generally moved during locomotion, resulting in great stability, as the center of mass is always located at a low height within a wide triangle formed by the support of the other three limbs (figs. 6.1, 6.2).

Posture is also related to ventilation and metabolism. In tetrapods with transversely oriented limbs, as noted by Drew (2017) and Pough and Janis (2019), inhalation and exhalation are produced as a consequence of trunk musculature contraction. When the animal walks, the body's alternating lateral flexion, resulting from the undulating movements, compresses first one side of the body (and its contained lung) while expanding the other side and then vice versa, forcing air to pass from the compressed lung to the expanded lung, and, as a result, severely restricts the exchange of fresh air (fig. 6.3). During such undulatory movements, then, ventilation is effectively curtailed and gas exchange severely reduced. This results in insufficient oxygen availability in the blood for glucose catabolism, meaning that the animal cannot produce sufficient energy, in the form of ATP (adenosine triphosphate, the basic biomolecular energy source for cellular metabolism), and is compelled to stop to renew its oxygen and thereby return to aerobic metabolism (Pough and Janis 2019). Although lizards with short bodies (such as many agamid and corytophanid lizards; see Xu et al. 2013a, 2013b; Clemente and Wu 2018) can rise up on their hind limbs during running, this is not possible for those with long bodies. Crocodiles hunt by stalking, with sudden attacks at fast speed that they cannot maintain should their initial attempts be unsuccessful. These outbursts of rapid activity produce and hydrolyze great quantities of ATP, wear the tissue's buffers systems, and produce acidosis by lactic-acid accumulation, so that the animal has to rest while blood is oxygenated and the lactate transformed to glucose.

In the history of tetrapods, a parasagittal posture was adopted within at least two distantly related amniote lineages, archosaurians (dinosaurians and crocodylians) and therapsids, and resulted in a great diversity of morphological specializations. Today, however, parasagittal tetrapods are represented only by some members of these two lineages, the birds and mammals, respectively (see later for modern crocodiles). In a parasagittal posture, the vertical component of weight is transmitted directly through the diaphysis of the long bones of the limbs, which nearly eliminates a transverse moment. The support triangles

6.3. Ventilation during locomotion of a tetrapod with a sprawling posture. Flexion of one side of the body increases the pressure on the lung of that side, while the extension of the other side of the body decreases pressure on the other lung, forcing air to pass from the former to the latter lung.

are much narrower with respect to the height of the center of mass (fig. 6.4) than in animals with a sprawling posture. However, their dynamic stability allows them to rise and descend at each step without dragging the body on the substrate, which is energetically less expensive. As well, they can reduce their duty factor (see below) and have fewer limbs resting on the substrate at the same time, which allows diversity in gait type and increased maneuverability. In the girdles, the glenoid fossa and the acetabulum are displaced ventrally, accompanying the change in the posture of limbs. Particularly in the pectoral girdle of mammals, the position of limbs directly beneath the scapula

6.4. Parasagittal posture in amniotes. Note, in comparison with figures 6.1 and 6.2, the markedly narrower support triangle (*shaded in gray, left-side* image) and the shorter lever between the shoulder joint and the line of action of the animal's body weight (*red arrow, right-side* image) as a result of the positional change of the limbs toward the midline, and thereby requiring relatively decreased contraction of the limb adductor musculature (*blue arrow*) to produce a parasagittal posture.

displaces the forces acting on the limbs over the scapular midline. Consequently, the scapula's roles in locomotion and weight bearing are increased, whereas the medial elements (clavicle, interclavicle, coracoids, and procoracoids) of the pectoral girdle are reduced. The humerus does not rotate around its longitudinal axis during its retraction in the propulsive phase. During this phase, the scapula rotates backward, so that the glenoid cavity moves posteriorly, increasing the stride length. In the pelvic girdle, the origin sites of the adductor musculature are also reduced, and muscle orientation changes to favor forward thrust. The elbow is directed backward and the knee forward, while the manus and pes point forward.

A parasagittal posture in mammals is linked to the evolution of a more efficient respiratory system (Pough and Janis 2019). As noted, tetrapods with a sprawled posture experience reduced ventilation during locomotion effected by lateral trunk undulation. Mammals, however, bend their vertebral column in the vertical (sagittal), rather than horizontal (frontal), plane, which improves ventilation during locomotion. With each step, the viscera move forward (pushing on the diaphragm, see below) by inertia,

increasing the pressure in the thoracic cavity, which forces air out of the lungs. When the animal's trunk is extended during its stride, the viscera move backward, increasing the volume of the thoracic cavity, thereby resulting in decreased pressure in the cavity and thus on the lungs, so that air consequently and largely passively enters the lungs. Movements of the viscera are modulated also by the muscular diaphragm, which normally functions in ventilation and is essential for the high metabolic rates that mammals develop during their activity. The diaphragm is the muscular partition between the thoracic and abdominopelvic cavities. When relaxed, the diaphragm is anteriorly convex, resembling a sideways dome. In inspiration (or inhalation), the diaphragm contracts and flattens, thus pulling the floor of the thoracic cavity backward. This expands the volume of the cavity, thus reducing the pressure within the cavity and on the lungs, and air enters the lungs passively. In exhalation (or expiration), the diaphragm relaxes and returns to its dome-like shape, so that the thoracic cavity decreases, thereby increasing pressure within the cavity and on the lungs, so that air is forced out (fig. 6.5). As a consequence of these high rates of ventilation

6.5. Ventilation during running of a tetrapod with a parasagittal posture. When the vertebral column is vertically flexed (*top* image), pressure increases on both lungs, forcing exhalation. When the trunk extends (*bottom* image), pressure on the lungs decreases, promoting passive inhalation.

during activity, increased desiccation of the lung mucosa, resulting in enormous water loss, would occur. Mammals have solved this problem by developing laminar bony expansions (turbinates) in the nasal vestibule covered with richly vascularized epithelium, which heats and saturates the air that enters the lungs with water vapor, facilitating exchange. The epithelium of the turbinates also reclaims this water as air is exhaled (Drew 2017).

Limb and Trunk Specializations in Tetrapods with Parasagittally Oriented Limbs

Throughout their evolution, mammals have developed a diversity of designs that have allowed them to adapt to the most varied environments and ways of life and to assume the widest range of sizes within vertebrates (from minute shrews to the great whales). Some of these specializations have also been recognized, to nearly the same degree, in archosaurs with parasagittal limbs; among them, nonavian dinosaurs exhibit remarkable postural and locomotory diversity, from quadrupedal giants with columnar limbs to small, agile bipedal forms (Carrano 1999). Hence, understanding these designs is relevant for their paleobiological interpretation, especially in nonavian dinosaurs (Carrano 1999). This diversity is briefly presented here, but several more specific aspects will be dealt with in the following sections.

Historically, the knowledge of the appendicular skeleton of earlier mammalian relatives that lived during the

Mesozoic (i.e., stem mammals) was rather scarce. These stem mammals were inferred as relatively small, with the limbs commonly adopting a flexed stance. The manus and pes were plantigrade—that is, with their underside resting almost completely on the substrate. Such a conformation allows achievement of high accelerations that are, however, of short duration due to high energetic costs. In the past few decades, however, several important findings have expanded the knowledge of these Mesozoic mammalian relatives (see Luo 2007). The traditional view that they were homogeneously small, terrestrial, and ecological generalists has been revised, with several lineages showing burrowing (e.g., multituberculates and the theriimorph *Fruitafossor*), swimming (e.g., the docodonts *Castorocauda* and *Haldanodon*), climbing (multituberculates), and even gliding specializations (e.g., the triconodont *Volaticotherium*; see also Chen and Wilson 2015 for an ecomorphological update on these animals).

Among mammals, in groups that specialize in running, the two distal limb segments (zeugopodium and autopodium) lengthen compared to the proximal segment (stylopodium). Also, the joints between limb segments allow rotation only in an anterior-posterior direction (i.e., in a nearly parasagittal plane) and further, the disposition of the manus and pes changes until they contact the substrate only with their digits or even just with the distal end, capped in hooves. The most extreme extant example is that of the horse and its closest relatives, in which the

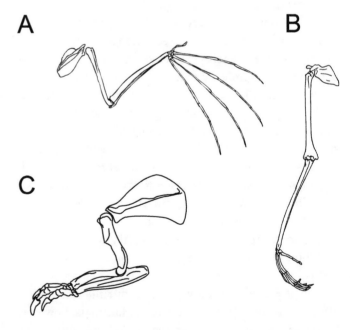

6.6. Specializations of the forelimb of mammals. *A*, bat wing (Chiroptera, flier). *B*, prehensile limb of a gibbon (Primates, climber). *C*, digging limb of an armadillo (Xenarthra, digger).

lateral digits are strongly reduced, effectively disappearing, and all the weight rests on the hoof of the distal phalanx of the third, or middle, digit.

In elephants, the largest extant quadrupeds, the proximal segment lengthens more than the others, and the limbs are arranged so as to be nearly vertical. In this arrangement, all forces are aligned in a bony column that supports weight virtually without muscular effort.

Kangaroos (Macropodidae) and some rodents (e.g., jerboas) are specialized to use their hind limbs in jumping. In these mammals, the distal segments are very elongated, which increases takeoff speed and jump amplitude. In landing, the pedal ligaments store energy that is released to aid in extending the pes in the subsequent takeoff, therefore making it a rather energetically economical mode of locomotion.

In many other terrestrial mammals, the more extreme specializations occur in the forelimbs. They can be long, ending with a hook-shaped manus, as in gibbons (hylobatid primates), for balancing and maneuvering from branch to branch; robust for the insertion of a powerful musculature and equipped with strong claws for digging, as in armadillos (dasypodid xenarthrans); or slender and having long digits that support a membrane that permits powered flight, as in bats (chiropterans; fig. 6.6).

Locomotion in Tetrapods

The Step Cycle

Tetrapods move by a cyclic succession of steps (Liem et al. 2001). The cycle begins when the foot is set on the substrate and, making use of friction, then accelerates the body and moves it forward (propulsive phase). Subsequently, the foot is raised, moved forward, and readied for placing it down again (swing phase). At least one limb must be raised from the substrate and remain so for a momentary period until it is placed down again at its subsequent point of support (fig. 6.7). During this phase, stability is maintained by means of continuous postural changes that involve a considerable part of the body musculature, limbs, girdles, and vertebral column.

A stride is the successive step cycles of all the limbs. In a human (or other bipeds), a stride consists of two step cycles (one of the right leg, followed by one of the left leg). The length of a step is the distance traveled during

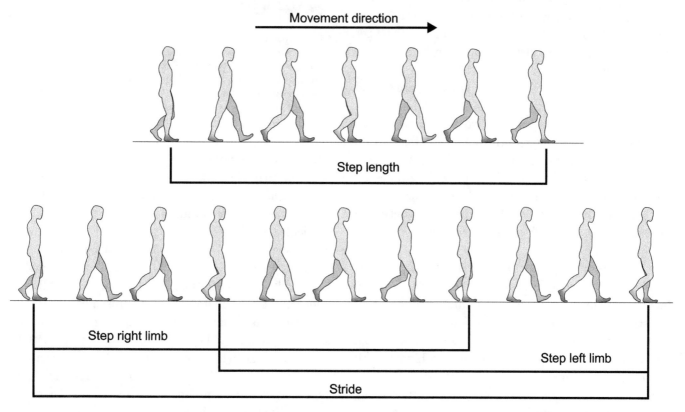

6.7. Step cycle. *Above*, step cycle in a biped: the step includes a propulsive phase (*red*), with the limb of a side resting on the ground, and a swing phase (*blue*), in which the same limb rises for moving to the new place of support. *Below*, stride: this consists of the whole step cycle of all limbs (in this case, only two limbs).

the propulsive phase of one leg, and the length of a stride is the total displacement after the step cycle of both legs (two step cycles). But in quadrupeds, the stride consists of four step cycles (one per limb). Stride length is measured from the position of one foot on the substrate until its next footfall (fig. 6.7). During the stride, there is a compromise between producing a sufficient thrust for motion and maintaining body balance. Anyone who has ridden a bicycle would appreciate, at least intuitively, this compromise between velocity and stability.

Displacement Velocity

There are two ways of augmenting velocity: increasing the stride length (by taking longer steps) and increasing the stride frequency (by taking more steps per time unit). The former is less energetically expensive. A bicycle provides an analogy. In a bicycle with gears, changing to a gear with a larger chain ring is equivalent to increasing stride length (each turn of the pedal carries the bicycle a longer distance), but in a bicycle without gears the only option for increasing velocity is to pedal more quickly—that is, increase stride frequency, which requires more energetic expenditure. Lengthening stride is, therefore, the most usual way to increase velocity. Evolutionarily, the form-function complex for achieving increased stride length is an increase in the length of the limbs; animals with longer limbs can cover a longer distance with each step, and thus their velocity is faster.

From the perspective of body design, there are several strategies to increase stride length and/or frequency. One alternative is to increase the effective limb length, which may occur to varying degrees (fig. 6.8). First, the posture of the autopodium can be modified. The most generalized tetrapods are *plantigrade*: they place their digits on the substrate, as well as the metapodials and, usually, the basipodials. The support surface (the sole) is relatively extensive, maximizing friction and increasing grip and stability; because of such characteristics, many climbing tetrapods are plantigrade. *Digitigrade* animals set only their digits on the substrate. It is a relatively easy way to increase the effective length of the limb and therefore of the stride, but clearly at the expense of grip and stability. Some animals are capable of alternating between foot postures depending on whether they require higher speed or more grip (e.g., many felids that pursue as digitigrades adopt plantigrady when taking down prey). When walking, humans are plantigrade, but they run at high velocity on the tips of their toes. It is likely that digitigrady appeared evolutionarily as an initial response to increasing stride length in forms that needed to move at higher velocity,

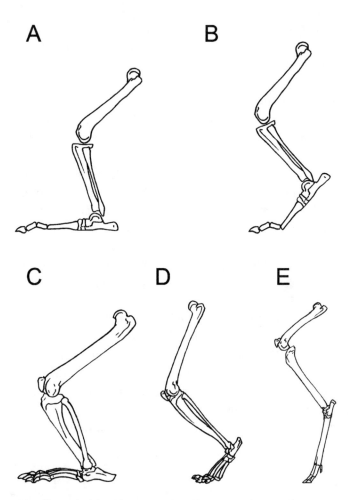

6.8. Effective limb lengthening produced by a postural change from A, plantigrade to B, digitigrade postures in the same limb (schematic); and lengthening due to changes in limb proportions from C, plantigrade (panda bear) to D, digitigrade (puma) and E, unguligrade (South American deer) postures. Images not to the same scale. Modified from Liem et al. (2001).

and an increase in the relative length of metapodials and digits occurred later. When the distal end of the pes applies the output force, its heel serves as the attachment site for the gastrocnemius muscle (i.e., where input force is applied), and its ankle acts as the pivot, so that it functions as a first-class lever (see chap. 2): increase in the relative length of the metapodial and acropodial segments implies (especially in mammals) an increase in velocity at the expense of force. Some forms, taking this specialization to its extreme of placing only the ungual phalanx on the substrate, are referred to as *unguligrades* (e.g., ungulates).

This change in the foot posture is followed by a lengthening of the distal limb elements (zeugopodium and, especially, autopodium; fig. 6.8). This strategy favors designs with elongated limbs and is generally accompanied by a digitigrade posture. Many muscles have their insertions closer to articulations (joints), meaning that a given amount of muscular shortening will produce a greater

6.9. Reduction of limb inertia achieved by the bending of limb segments during the flight or recovery phases.

the stylopodium, rather than closer to the elements they move—and force is transmitted distally by means of long tendons. This arrangement reduces the mass of the distal limb segments and thus decreases inertia, so that movement is effected more easily and efficiently, with less expenditure of energy (figs. 6.8, 6.9). Long tendons also contribute energy efficiency by storing elastic energy. Distal limb lengthening and mass-reduction strategies are often accompanied by reduction in the number of digits, with the more medial and lateral ones being strongly reduced or lost.

Another way of increasing the stride length is to increase the distance that limbs travel while they are not in contact with the substrate (fig. 6.10). When walking, lizards and crocodiles bend their trunk in the horizontal plane (frontal), increasing the distance traveled by the body during the swing phase of each limb. In analogous manner, the strategy of using the trunk to increase distance is observed among many mammals, with one of the most impressive examples being that of the cheetah (*Acynonyx jubatus*), which executes long jumps forward through alternating extreme flexion and extension of the vertebral column in the dorsoventral plane.

From an evolutionary perspective, increasing stride frequency is not as common as increasing stride length. A short limb facilitates oscillation because it decreases limb

distal limb displacement, thus producing longer strides. This also affects the velocity of the limb, according to $v_o/v_i = l_o/l_i$, which, when solved for v_o, is $v_o = v_i l_o/l_i$. Output velocity (v_o) is increased by shortening the distance between the muscle insertion and the joint; this shortens the input moment arm l_i in relation with the output moment arm l_o and thus optimizes velocity at the expense of force, as noted in chapter 2. Thus, the relationship l_o/l_i is often greater in animals that are specialized for running than in more generalized animals. However, limbs maintain a compromise between velocity and force. Tetrapods generally have some muscles with high moment arms (e.g., the semimembranosus and semitendinosus in the hind limb) that act synergistically with others that have rather short moment arms (e.g., iliopsoas and gluteal muscles).

Lengthening limb elements, particularly of the zeugopodium, to increase stride length is linked with a strategy to increase stride frequency: reducing limb inertia by decreasing the mass at the distal end of the limb. In forms adapted for sustained and rapid displacement, the main musculature mass that produces movements of the more distal limb elements is located close to the body—in

6.10. Increase in the total length traveled by limbs due to trunk undulation. *Above*, lateral undulation in a lizard. *Below*, dorsoventral undulation in a mammal. The *red arrows* represent a simplified trajectory of each limb, and the *blue arrows* represent the additional length added by trunk undulation. Based on Hildebrand and Goslow (2001).

inertia, which promotes increase in stride frequency, but it shortens stride length and, therefore, compromises velocity. However, during the swing phase, many long-limbed animals flex their limbs, thus shortening the limbs' overall length and enhancing their forward movement (because reducing the radius of swing increases the angular velocity of a limb), and then extend them as far as possible just before they touch the substrate. In general, tetrapods that are adapted for speed simultaneously employ several strategies for increasing velocity, either in design (such as digitigrade posture, zeugopodium and stylopodium lengthening, proximal concentration of musculature, and reduction in the number of digits; fig. 6.8) or behavior (such as limb flexion in the swing phase; fig. 6.9).

Gait

Gait is defined as the pattern of foot contact with the substrate during a stride. The type of gait that is used in locomotion depends on an animal's size, posture and limb length, velocity, and substrate characteristics. Thus, it is common that most animals modify the type of gait depending on the context.

Limb movement during gait can be expressed in quantitative terms by means of two descriptors. One is the duty factor, the proportion of time during a stride when a foot is on the substrate in relation to the duration of the swing phase. As speed of the animal increases, there is a tendency to decrease the contact time that a foot has with the substrate; this means that the duty factor diminishes as the duration of the swing phase increases. The other quantity is the relative phase, the relative time (expressed as proportion or percentage of the total time of the stride) that each limb is in contact with the substrate with respect to the other limbs. If both limbs (fore- and hind limbs of the same side, for example, or both forelimbs) perform a step cycle in unison, they are completely in phase, but if they have their respective step cycles decoupled, they are out of phase, by a proportion that is determined by the time during the step in which it makes contact with the substrate.

Gaits are classified as symmetrical or asymmetrical. In symmetrical gaits, the limbs of each pair (e.g., anterior and posterior, or both anterior) move with a phase of 0.50; that is, their step cycles are spaced apart by one-half of the stride interval or, in other words, they are set on the

TYPES OF GAITS

Symmetrical Gaits

Diagonal gait: considered a basal gait, it is characterized by the simultaneous support of diagonally opposed feet, as occurs in urodeles. The center of mass is balanced on a line between the supporting points.

Lateral sequence: quite stable; most of the time the body's center of gravity is projected within a triangle with vertices at the three limbs resting on the substrate. Many reptiles move with the belly resting on the substrate, moving a leg at a time (belly crawl). Crocodiles use this gait when in water; during short bursts of acceleration to move rapidly, they transition to a belly run through exaggerated sinusoidal movements of their trunk. However, their usual gait is with the belly raised from the substrate, the hind limbs practically beneath the body and the pes rotated forward; small species can gallop (see below). Turtles execute a very stable gait, with three or four feet on the substrate for most of the stride. Many mammals walk in lateral sequence, and at a higher speed they increase the stride length and/ or frequency until a trot (a lateral gait in which there is a flight stage) develops. At very high speeds, they transition to fast gaits (either symmetrical, such as jump, leap, and

bound, or asymmetrical, such as half-bound and gallop) that are defined below.

Pace: the animal oscillates the leading and trailing limb of the same side more or less in unison (in lateral, rather than diagonal, phase), preventing limbs of the same side from interfering with each other. It is characteristic of animals with long limbs in relation to the trunk, particularly when they move at low speed (e.g., giraffes, camels, and the maned wolf).

Jump, leap, and bound are fast gaits used by animals that move when dodging obstacles or during display (e.g., antelopes). The four limbs touch the substrate at the same time and are raised from the substrate during most of each cycle. They provide great stability and produce abrupt deceleration with each landing.

Asymmetrical Gaits

Half-bound: the hind limbs make contact more or less simultaneously and the forelimbs do as well, but with a different leading-trailing pattern.

Gallop: each of the forelimbs and hind limbs have different leading-trailing patterns. At lower speeds, the gait is referred to as a canter.

6.11. Duty factor and relative phase in tetrapod locomotion (domestic dog). On the *left*, symmetrical gait at low velocity: limbs rest at regular intervals, and the relative proportion of limbs resting on the substrate (*red*) and in flight phase (*blue*) is similar. On the *right*, asymmetrical gait at high speed: limbs do not rest at regular intervals (pairs of limbs move at different phases), and the proportion in which they rest on the substrate is less (less total duty factor).

substrate at regular time intervals. In asymmetrical gaits, pairs of limbs have different phases. This does not necessarily imply that limbs within a pair are in phase; that is, the limbs are not set on the substrate at regular intervals of time during a stride. For example, when a pair of limbs approaches the substrate, the leading limb can make contact later and in front of the trailing limb. Asymmetrical gaits can be less stable than symmetrical gaits, and always include a flight stage; that is, an interval during which all four limbs are raised above the substrate and the stride

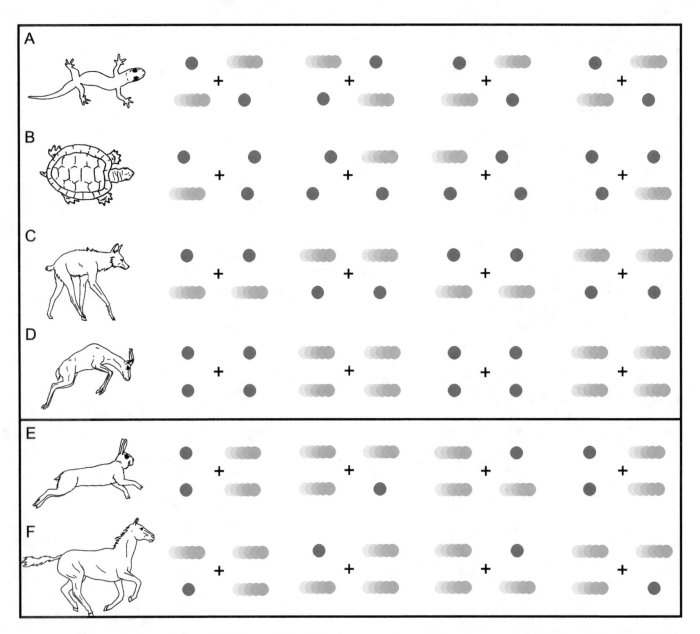

6.12. Types of gaits in vertebrates. Symmetrical: *A*, diagonal gait; *B*, lateral sequence; *C*, pace; *D*, jump. Asymmetrical: *E*, half-bound; *F*, gallop. *Red circles* indicate limbs resting on the ground; *blue circles* indicate limbs raised from the ground. The movement direction is from left to right. The center of mass is represented by a *black cross* (+). Modified from Liem et al. (2001).

length is significantly increased (fig. 6.11). Obviously, during the flight stage the duty factor decreases to 0. Several symmetrical and asymmetrical gaits, summarized in the text box "Types of Gaits" and figure 6.12, are recognized.

Locomotory Modes

In biology, animals have often been classified according to categories based on locomotory mode. However, as Oxnard (1984) pointed out, most animals are capable of performing several locomotory modes depending on circumstances; therefore, such categories are not exclusive, which

diminishes their usefulness in classifying animals (see chap. 5). Considering the definitions provided in chapter 1, it is simpler and less arbitrary to view these locomotory modes as faculties, rather than as fixed categories to classify animals. Thus, animals can exhibit a spectrum of faculties, which in turn can be aligned with a particular spectrum of biological roles. There are animals that possess the faculties of jumping and digging (e.g., toads, rabbits), for which the biological roles are, respectively, locomotion and to build shelters or search for food; other animals are capable of jumping and flying or of digging and climbing, among many other possibilities.

The term *cursorial* is an excellent example of a frequently used word that is not unambiguously defined. In general, it is said that animals that cover great distances and/or achieve high velocities on land are cursorial animals. Cursorial is derived from the Latin word *cursor*, for runner. In an attempt to clarify this term, Stein and Casinos (1997) defined cursorial mammals based on two criteria, as terrestrial quadrupeds with (1) limbs oriented vertically (posture) and (2) moving in a parasagittal plane (motion). We could extend this definition to other groups with parasagittal limbs, like many Triassic archosaurs (e.g., rauisuchids and basal ornithodirans) and many dinosaurs. However, as the definition applies only to quadrupeds, birds and some bipedal mammalian groups would not be recognized as cursorial (e.g., kangaroos and other bipedal jumpers; humans). Further, in some mammalian groups such as bears, primates, and kangaroos, the animal can alternate between bipedal or quadrupedal postures, depending on the circumstance (fig. 6.13).

Carrano (1999) used a different definition of cursoriality, with the aim of avoiding problems that could arise from assigning a particular animal to a single category of posture and locomotion. According to this author, cursorial animals are those that maximize locomotion velocity using a design that minimizes energetic consumption (by using strategies explained above, such as elongation of limb distal segments and reduction of their inertia). Such animals would constitute the end of a functional continuum, the other end of which would comprise graviportal animals, defined by a design that affords slower locomotion velocity but a greater force applied per stride (fig. 6.14). Although this definition of cursoriality may not allow, due to its comparative and relative nature, classifying animals quickly and easily, it is useful because it is based on clear physical concepts related directly to movement; it is adopted here for this reason.

The horse and the cheetah are examples of cursorial animals that take advantage of their velocity in different ways for different faculties: endurance versus explosive acceleration (fig. 6.15). Their anatomical designs clearly

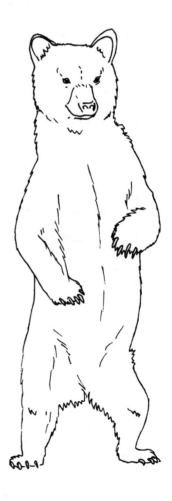

6.13. Mammals capable of bipedal locomotion: kangaroo, brown bear, and lemur.

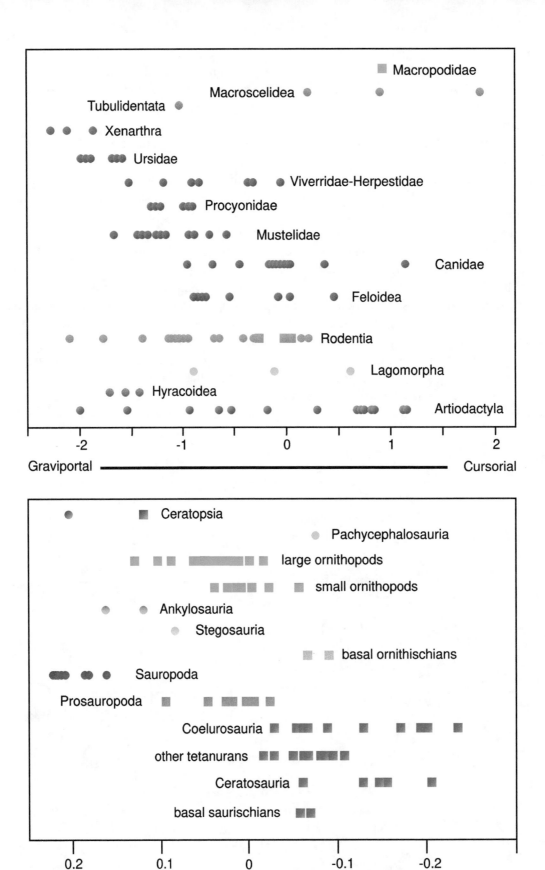

6.14. Gradient of cursoriality according to Carrano (1999). Relative proportions of the limb segments in mammals (*above*) and dinosaurs (*below*). In each box, the relative length of the limb distal segments (especially the autopodium) increases to the *right*. That is, toward the *left* there are morphotypes with comparatively long stylopodia and short zeugopodia and autopodia, able to apply greater force in each step though at a lower velocity (graviportality) than the morphotypes that are on the *right*, with short stylopodia and elongated zeugopodia and autopodia (cursoriality). Modified from Carrano (1999).

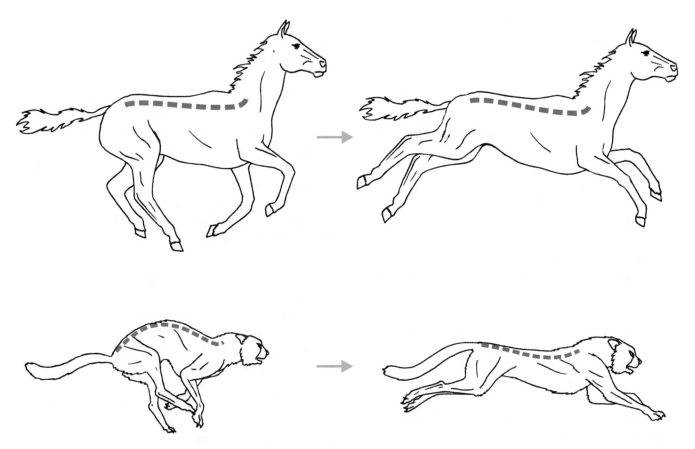

6.15. Cursorial mammals specialized for cruising velocity (horse, *upper* images) and for explosive acceleration (cheetah, *lower* images). Note the difference in the degree in undulation of the vertebral column. Modified from Liem et al. (2001).

reflect the influence of body size. The horse is a cruising runner, able to maintain a velocity of 30 km/h over 30 km. The vertebral column is quite rigid, and the displacement of its body from the vertical plane is negligible. It requires relatively little energy for lifting its mass, and the body is propelled almost exclusively along the direction of movement. The cheetah, on the other hand, is a sprinter able to run short distances, but for some stretches can reach approximately 90 km/h. This animal increases the stride length during the flight stage thanks to the great dorsoventral flexibility of its column, but this requires considerably more displacement of that body mass both vertically and in the direction of movement. Thus, maximum body size is strongly restricted; if the animal were larger, it would have to spend more energy raising its body mass during each stride.

JUMPERS

Many tetrapods are jumpers, among them anurans, some lizards, birds, and many mammals. And most cursorial animals are at least somewhat proficient at jumping. As jumping produces an abrupt acceleration from rest, it is useful both for escaping and for capture. Jumpers are often bipedal or can adopt a bipedal stance, and they can move easily on land or in arboreal environments. Animals that use their hind limbs in unison for successive jumps are called ricochetal (e.g., kangaroo rats, jerboas); the tail plays an important role in balance. McGowan and Collins (2018) hypothesized that among mammals, the evolution of bipedal jumping was linked to predation avoidance.

This mode of locomotion imposes several design restrictions that led many forms to develop convergently similar morphofunctional features for marked acceleration from a resting position. Such animals can also modify both their velocity and direction of movement more quickly than animals that do not jump. With each landing, the ligaments of the feet store elastic energy, which is released when extending the feet in the subsequent takeoff, making it an energetically economical mode of locomotion.

The height (h) that an animal can jump is determined by its vertical velocity and the angle (θ) of takeoff:

$$h = \mathbf{v}^2 \sin^2\theta / 2\mathbf{g}.$$

Note that for maximizing height, the animal must jump at an angle of 90° (perpendicular to the substrate).

The range (distance R, analogous to the stride length) that an animal jumps depends on velocity $\mathbf{v}$ and takeoff angle (θ):

$$R = (\mathbf{v}^2 \sin\theta)/\mathbf{g}.$$

In theory, for the same takeoff velocity, the maximum range is reached when θ is 45°, the same angle as in a parabolic trajectory.

These equations are better known as "ballistic formulae" (Emerson 1985). The necessary acceleration for reaching the takeoff velocity is $\mathbf{v}^2/2s$, where s is the distance through which force is applied. In turn, this is the difference in the functional length of the hind limb between its positions of initial flexion and full extension (fig. 6.16). Therefore, jumpers benefit from having long hind limbs.

Many anurans are excellent jumpers. At rest, they lie with their hind limbs folded and pelvis depressed. For jumping, they propel forward by stretching their hind limbs and straightening their pelvis to its maximum extension by contracting the powerful musculature that is inserted between the pelvis and urostyle (fig. 6.16).

Among mammals, jumping abilities evolved independently in several lineages. Kangaroos are considered jumpers par excellence, although they can also adopt an entirely different gait (fig. 6.17). At low speeds, they walk with a kind of shuffling gait that is apparently unique among tetrapods, using all four legs and tail and barely lifting their feet. They first support their weight by placing their forelimbs and tail on the substrate and move their hind limbs forward; then, they move the anterior part of the trunk forward while keeping their hind limbs set. At higher velocities, of course, they jump on their hind limbs. Kangaroos, as noted above, are known more formally as macropodids (derived from the Greek for "long foot") due to their long and narrow hind feet, which have a particular arrangement of digits. Generally, digit I is missing, II and III are fused, IV is quite large, and V is moderately developed. The unusual development of the hind limbs is optimal for traveling long distances at relatively high velocities at low energetic cost because of the great amount of elastic energy stored by their tendons; indeed, most of the energy required for each jump is supplied by the springing action of the tendons, rather than by muscular effort. The elongated feet provide an extensive lever, and the key factor is the ability to extend the hind-limb joints before takeoff (fig. 6.17).

In addition, as in running, there is a close link between jumping and ventilation. When the feet leave the substrate, air is expelled from the lungs. While moving forward for landing, the lungs fill again, providing the oxygen necessary for producing energy. Thus, successive

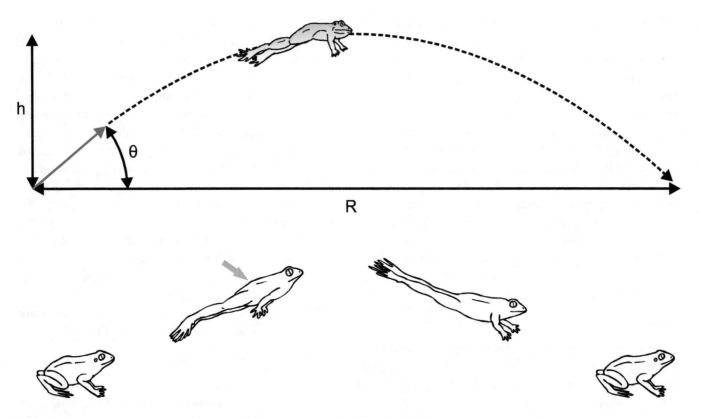

6.16. Jump in a frog. *Above*, trajectory for a maximum range, R, which is a function of the takeoff angle, θ; the *red arrow* indicates the output velocity. *Below*, jump sequence; note the maximum extension of hind limbs and dorsum (*blue arrow*). Modified from Liem et al. (2001).

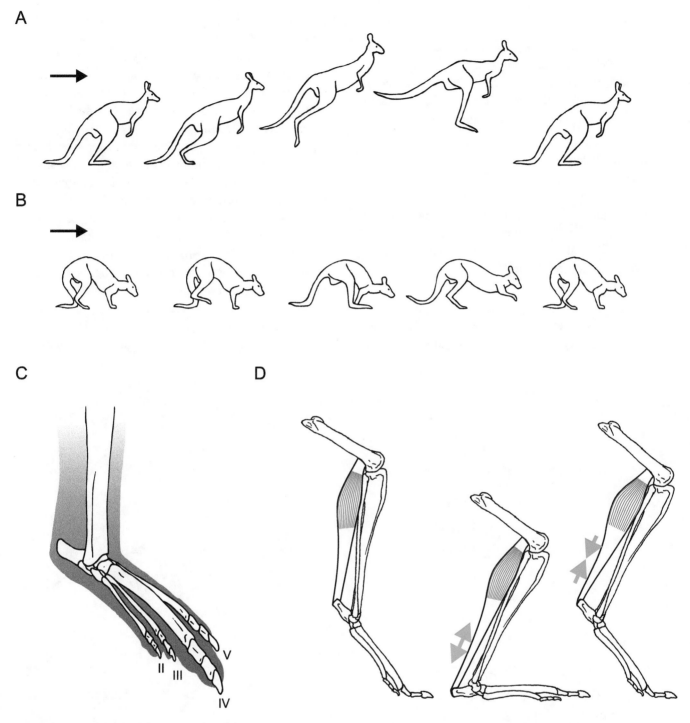

6.17. Locomotion in kangaroos. *A*, jump. *B*, shuffling gait. *C*, kangaroo foot (note the reduction of digits II and III and the hypertrophy of IV). *D*, extension of the tendon of the calf musculature (*blue arrows* in *middle* image) stores elastic energy that, when released, contributes to the force for extension of the foot (*blue arrows* in *right* image). Modified from Hildebrand and Goslow (2001).

increments in velocity require much less effort than the equivalent activity in horses, dogs, and human beings.

Cursoriality and jumping offer advantages for:

· Covering large areas in search of food or water sources.
· Migrating seasonally.

· Pursuing prey or escaping predation.
· Dodging obstacles when moving.

Cursorial animals and jumpers have different abilities with regard to speed, resistance, acceleration, and maneuverability. Within each of these locomotory modes, there are different designs that favor some of these over the others.

CLIMBERS

Tetrapods adapted to climbing are called climbers or scansorial (fig. 6.18). Some authors, such as Hunt et al. (1996) and Gebo (1996), defined climbing as the act of moving over a substrate with a slope equal to or greater than 45°. Climbing faculties are advantageous for:

· Obtaining food.
· Escaping predation.
· Finding shelter for resting, nesting, or breeding.
· Displacement in areas of dense vegetation or where the land is rugged or flooded.
· Accessing higher elevations from which gliding or flying forms can take off.

In turn, climbing, particularly as it applies to arboreal substrates, has its own difficulties, related to two important mechanical issues: traveling over a discontinuous tridimensional substrate and avoiding falls. Body mass imposes limits to locomotion on an arboreal substrate, because an animal can progress only on branches that can sustain its weight; therefore, heavier climbing animals (e.g., orangutans, gorillas, and bears) have restricted access to thinner branches. One strategy for such animals to gain access to thinner branches is to distribute body weight among several supports, which requires longer limbs, as is the case in many primates and extant sloths.

There are diverse modes of locomotion in an arboreal substrate. Animals that walk or run on more or less horizontal branches have the same types of propulsion as nonarboreal forms, but their manus and pes, and sometimes tail, may be modified to enhance support.

Some jump from a support to one that is at a lower or nearly the same level. In these, the trunk of the body is often relatively long, strong, and flexible (e.g., the lizards *Anolis*, tree kangaroos *Dendrolagus*, squirrels *Sciurus*, capuchin monkeys *Cebus*, and langurs *Semnopithecus*),

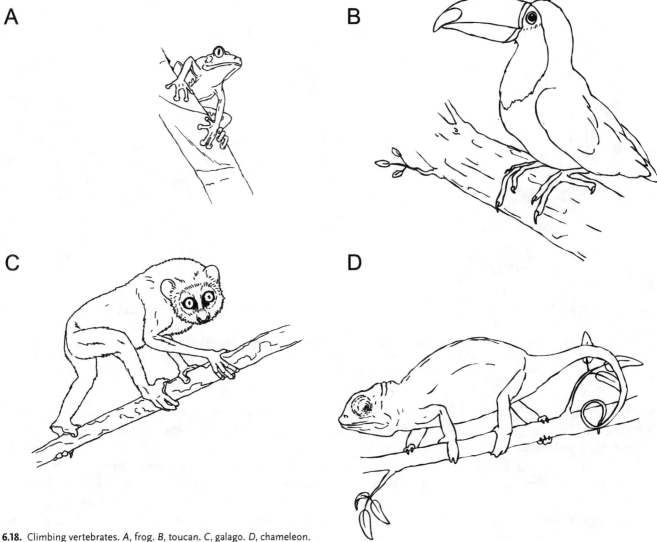

6.18. Climbing vertebrates. *A*, frog. *B*, toucan. *C*, galago. *D*, chameleon.

tending toward long limbs, slender bones, and muscular mechanisms similar to those of cursorial animals, although the proximal limb segment, unlike in running animals, also tends to be relatively long. The body is often approximately horizontal when taking off.

Other primates jump more or less vertically from a stationary takeoff position. They have elongated hind limbs, kept bent at rest, and hind-limb extension mechanics similar to those of jumpers, although the femur is not as shortened (e.g., the tarsiid *Tarsius* and galagid *Galago*; fig. 6.18) and the elongation of the pes is concentrated on the calcaneum and navicular instead of digits (Schmidt 2010).

Other mammals are propelled by reaching for supports and pulling from them (orangutan *Pongo pygmaeus* and sloths *Bradypus* and *Choloepus*). They are agile and flexible animals, with long limbs and thorax. Hands and feet are also elongated, possibly for improving grip rather than for propulsion; girdles enable freedom of movement; and ulna and radius are free and of the same length, ensuring maximum pronation and supination. This type of locomotion does not require emphasizing force, which is why muscles and their bony insertions are not prominent and bones are thin and light. Some monkeys move in this mode but preferentially use their forelimbs, and they are thus often referred to as brachiators (e.g., gibbon *Hylobates*).

Avoiding falls requires overcoming two issues. One is weight transference from one support to another (fig. 6.19). Some animals jump, soar, or fly between supports

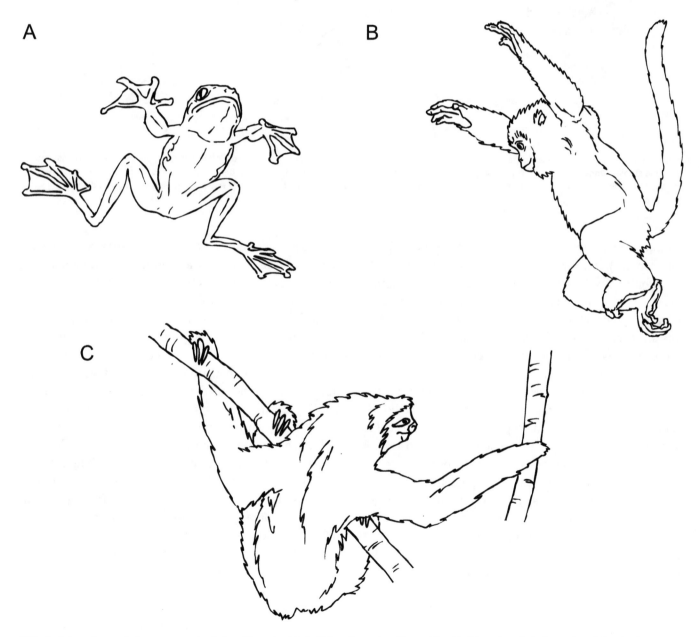

A

B

C

6.19. Body transference between supports. *A*, frog (Racophoridae) gliding. *B*, monkey (Cercopithecidae) jumping. *C*, sloth (Bradypodidae) bridging.

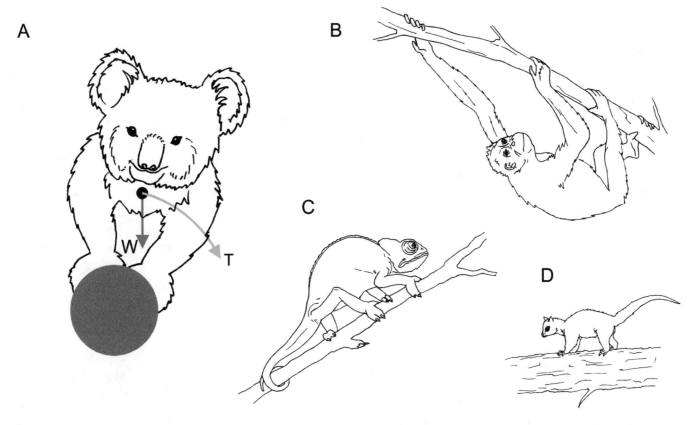

6.20. The problem with torque around a support. *A*, in *red*, weight (**W**); in *blue*, the torque (**T**) produced. Solutions: *B*, adopting a position of maximum dynamic stability under the support in an orangutan; *C*, application of muscular forces produced by prehensile limbs opposed to torque in a chameleon, in conjunction with bent postures that decrease the torque radius; *D*, dynamic stability favored by a body size smaller than the support diameter in a mouse opossum.

(arboreal snakes and lizards, several primates, marsupials, and rodents), whereas others grasp the new support to transfer body weight, a maneuver referred to as bridging; examples of vertebrates that bridge include chameleons (Chamaeleonidae), numerous snakes and lizards, sloths, anteaters, pangolins, primates, and marsupials. Most of these animals (except for snakes) have long limbs and/or a prehensile tail for increasing reach.

The second problem is the difficulty of keeping the body on the substrate (fig. 6.20). Provided that the body's center of mass is above the support where the animal is located, a torque or rotational moment is produced around the support (which acts as a pivot point around which the animal will tend to tilt). Rotational moment depends on two factors: body weight and distance to the support (or turning radius, analogous to a lever arm), the latter depending, in turn, on limb length and body posture. Four main strategies are recognized for overcoming these difficulties (Cartmill 1985). First, chameleons, some primates, and sloths can adopt a suspensory posture below the support. The rotational moment is eliminated if the animal adopts a suspended or hanging position, with its center of

mass below the support, where it achieves a position of dynamic balance. Second, other animals produce a rotational moment that counteracts that of body weight; this opposing torque is produced by the active application of muscular force by the limbs, more specifically by prehensile hands and feet, and in some cases also by a prehensile tail. The third and fourth strategies consist of modifying one or both variables that influence the rotational moment: decreasing distance to the support (by flexing the limbs and/or reducing their length) and/or decreasing body size. Some small animals can move on a branch as though walking or running on land; in this context, their locomotory habit is similar to that of a ground-dwelling form.

Frictional forces play a very important role in climbing faculties (fig. 6.21). Given that frictional force depends on the coefficient of static friction of the contact surface and on the component of the force of weight orthogonal to the contact surface (Cartmill 1985; Hildebrand and Goslow 2001), an increase in the slope of the support surface results in a decrease of the component of the force of weight orthogonal to the contact surface. This means that, whether during locomotion or at rest, an animal must, in

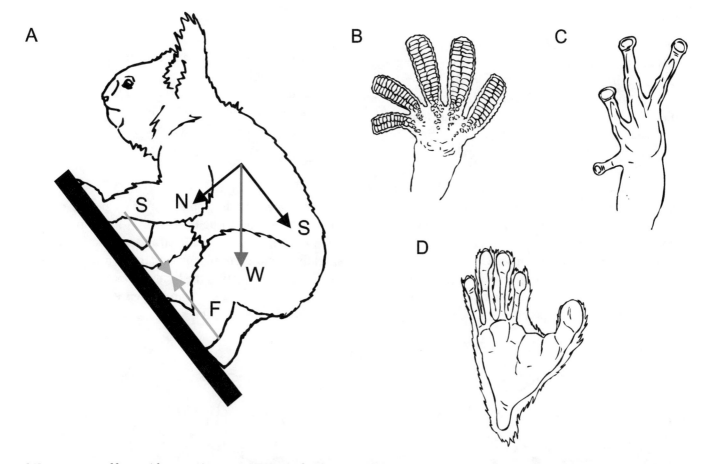

6.21. Importance of frictional forces. *A*, diagram indicating the forces involved in locomotion on an inclined substrate; weight (**W**) has a component normal (**N**) to the surface and a component parallel (**S**) to it; the latter must be counteracted by a frictional force (**F**) of opposite direction and magnitude to prevent the animal from sliding downward. *B*, digital pads of a gecko. *C*, adhesive pads of a frog. *D*, prehensile foot of a lemur. Modified from Hildebrand and Goslow (2001).

addition to applying forces intended to keep the body lifted from the substrate, also maximize the component orthogonal to the surface. A possible strategy to achieve this is to increase the friction coefficient by means of adhesive or rugged surfaces, as is the case in many amphibians and lizards. Another strategy is to apply additional muscular force perpendicular to the support. This can be achieved with prehensile limbs by opposing the contralateral autopodia on either side of the support and/or by using claws to produce new contact surfaces perpendicular to the surface of the support and, thus, new forces with orthogonal components. Many climbing animals have strong claws and prehensile or, at least, opposable autopodia. These strategies are not exclusive; different climbing animals exhibit several combinations of these functional solutions to the problem of stability and movement.

Many climbing tetrapods are not markedly different from other terrestrial tetrapods and are able to move on other substrates using different types of gaits and perform other activities (e.g., running, jumping, digging,

swimming). Indeed, many animals with generalized anatomical designs are able to perform multiple types of locomotion in or on a variety of substrates, without exhibiting evident specializations for any of them (and are, therefore, difficult to categorize; see Miljutin 2009). Polly (2007) used the term *ambulatory* for these animals, particularly mammals, an example of which is the opossum *Didelphis*.

All groups of terrestrial vertebrates have climbing representatives (fig. 6.18). Among amphibians, many frogs and salamanders are outstanding climbers. Also, many lizards and snakes are excellent climbers, whether on plants or rocks, and some are completely arboreal, such as chameleons and geckos (Gekkonidae). Many birds move on branches or rocks using their legs, tail, beak, and even wings. Most mammals are able to climb to some degree, and there are excellent climbers even in mainly cursorial lineages (e.g., Carnivora). Some clades include forms especially adapted to locomotion on arboreal substrates (primates, marsupials, sloths and anteaters, pangolins, rodents, dermopterans, and bats).

DIGGERS

Most terrestrial tetrapods include digging—that is, actively and purposefully removing substrate—among their activities. This intentional alteration of the substrate has varied purposes, such as moving from one point to another, building shelters for resting, obtaining food, avoiding predation, and protecting offspring. Some dig their own shelters, whereas others use natural cavities or shelters built by other animals and would thus not be "diggers." Tetrapods that spend all or nearly all of their life beneath the surface are termed *subterranean* (e.g., moles, fairy armadillos, mole rats, prairie dogs, tuco-tucos). Those that perform activities on the substrate, but which are well adapted for digging in search of food and/or shelter, are known as *fossorial* (e.g., many toads, some turtles, the platypus, tenrecs, wombats, the aardvark, rabbits, marmots, badgers, dogs, bears, anteaters).

Digging habits offer advantages for:

· Producing microhabitats for resting, aestivating, or hibernating.
· Obtaining food (insects and their larvae, earthworms, roots and tubers).
· Accumulating and hiding food.
· Escaping from predation.
· Protecting offspring.

Vertebrates can develop several strategies for digging (fig. 6.22), which can be combined in the same animal. Some caecilians (Gymnophiona) and some tetrapods with short limbs move through ground that is generally soft or sandy but firm by using the head as a tool for digging. In relation to their subterranean existence, caecilians possess remarkably compact, wedge-shaped, well-ossified,

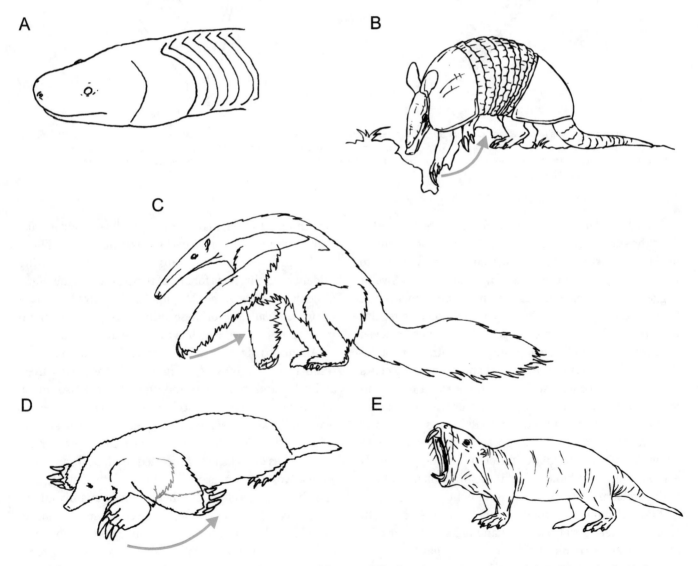

6.22. Digging vertebrates. *A*, caecilians use their solid head. *B*, armadillos scratch-dig the substrate. *C*, anteaters sink their claws and pull with the forelimb (hook-and-pull digging). *D*, moles rotate the entire forelimb (humeral-rotation digging). *E*, mole rats use their incisors (chisel-tooth digging).

and solidly roofed skulls, an architecture that has been considered as particularly well suited to the functional demands of burrowing (Duellman and Trueb 1986; Wake 1993; Kleinteich et al. 2012). The bones exhibit tight sutures, and many of the bones are fused into larger compound elements (Wake 1993). There are two skull types: the stegokrotaphic condition, in which the temporal region and jaw-closing muscles are covered by the squamosal, and the zygokrotaphic condition, with incomplete coverage of the temporal region by the squamosal (Kleinteich et al. 2012).

Most tetrapods loosen and rip apart the substrate by scratching at it with their claws by means of alternating flexion and extension of their limbs (scratch-digging). In many of them, a powerful forearm extension is crucial, which is why the forearm extensor musculature is well developed. The humeral-rotation digging executed by moles is different. These mammals cannot pronate and supinate their forelimbs, and the force, generated by powerful humeral retractor musculature, is exerted by rotation along the axis of the robust humerus; the forelimb only positions the manus. Anteaters, with their powerfully developed flexor and supinator musculature, are able to sink their strong claws into termite mounds and pull with the elbow bent (hook-and-pull digging). Many rodents use their incisors for digging (chisel-tooth digging; e.g., tuco-tucos *Ctenomys* and mole rats *Heterocephalus*).

Tetrapods that dig on a solid substrate must be able to exert considerable force on it, and their musculoskeletal design is such that it emphasizes output force. One way of enhancing this force, perhaps most obviously, is to increase the amount of musculature, but shortening the output arm (i.e., the distal segments of limbs) or lengthening the input arm (by inserting the musculature farther from the articulation) produce the same effect. These last two arrangements, however, decrease the velocity of the system, which is why tetrapods that are well adapted for digging are not good runners. For the same reason, on the other hand, good diggers are often excellent climbers as well.

LOCOMOTION WITHOUT APPENDAGES

Many amphibians (caecilians) and squamates (amphisbaenians and snakes) have secondarily lost their limbs during the course of their evolution. Four main modes of traveling without appendages are recognized: lateral undulation, sidewinding, concertina, and rectilinear (fig. 6.23; however, several distinct variations within these categories have been recognized by Jayne 2020, who listed, for example, five distinct modes of lateral undulation). The most common is lateral undulation, in which a lateral wave is

propagated from the front to the back of the body. In the posterior part of each curvature of the wave, the body has a sliding contact with the surroundings and presses against the substrate (generally against pebbles or other more or less firmly set objects) obliquely to the direction of progress, producing a propulsive vector component in the same direction as the anteroposterior axis of the body (Summers and O'Reilly 1997). As the animal moves forward, each point along the body passes through the path followed by the head.

Snakes that travel on loose or slippery substrates do so with a sidewinding locomotion. This kind of locomotion differs from lateral undulation in that the animal does not progress by a sliding contact with its surroundings. Instead, it raises parts of its body, while other parts maintain static contact with the substrate, providing considerable grip. Those parts that are not in contact are raised and are thrust forward in stepwise fashion. When these are brought back into contact with the substrate, the head extends and the cycle restarts. All points along the snake's body follow different though nearly parallel paths, and more posterior parts of the body are consistently in front of more anterior parts within a region of static contact (Jayne 2020). The resulting general movement is diagonal with respect to the snake's anterior-posterior axis and leaves characteristically elongated, discontinuous, and parallel tracks that are aligned obliquely with the snake's movement (Jayne 1988).

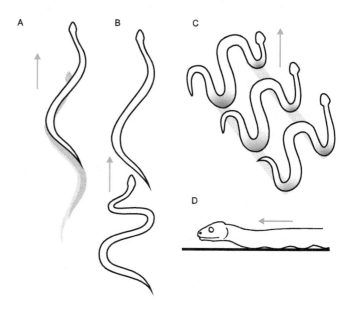

6.23. Locomotion in snakes. *A*, lateral undulation. *B*, concertina. *C*, sidewinding (in *red*, parts of the trunk resting on the substrate; in *gray*, characteristic tracks). *D*, rectilinear; note the production of ventral waves. *Blue arrows* indicate forward movement. Modified from Hildebrand and Goslow (2001).

In concertina locomotion, the body is folded laterally like an accordion. With static contact being maintained by the posterior part of the body, the anterior part is extended forward, and then the body is folded again. Snakes employ this type of locomotion in several circumstances, such as for moving through tunnels or climbing. Jayne (2020) recognized four modes: (1) flat-surface concertina, (2) tunnel concertina, (3) arboreal concertina with alternate bends, and (4) arboreal concertina with helical wrapping.

Larger snakes such as boas and pythons can perform rectilinear locomotion. In contrast to lateral undulation, sidewinding, and concertina modes, which use lateral bending of the vertebral column to generate propulsive movement, rectilinear locomotion does not involve such movement (although the body does not have to remain perfectly straight). Instead, rectilinear locomotion results from movement of the skin relative to the underlying skeleton to propel the snake forward through the action of specialized costocutaneous (extending between the ribs and skin) and interscutalis (extending between adjacent ventral scales) musculature. This movement is accomplished by periodic shortening and lengthening of the ventral and ventrolateral parts of the skin, to which are attached the ventral scales, via the interscutalis muscle. This muscle, however, is not responsible for propulsion; rather, it keeps the ventral scales shortened while they maintain static contact with the substrate, and the costocutaneous inferior muscle pulls the skeleton (the ribs remaining immobilized) forward relative to the skin. The skin musculature thus modulates the stiffness of the skin, allowing it to assume an important role in the transmission of forces (Jayne 2020).

Posture, Preference, and Use of Terrestrial Substrate in Paleobiology

As noted above, the morphology of extinct forms is often so different from that of their extant relatives that inferring paleobiological features by means of homology and parsimony often leads to incorrect reconstructions (Vizcaíno 2014; Vizcaíno et al. 2018). An illustrative example is the case of nonavian dinosaurs. Their most closely related extant forms (birds and crocodiles) differ so markedly in morphology, function, and ecology that, to a great extent, they are not useful as functional models, and it is necessary to turn to forms that, though distantly related, are more similar in size and morphology. Thus, Bakker (1971) and Alexander (1989), among other authors, used elephants, rhinoceroses, and hippopotamuses as functional models of substrate preference and use for inferring such aspects of sauropod and ceratopsian paleobiology.

Moreover, Carrano (1999) proposed that mammals and dinosaurs share similar variation patterns along the cursorial-graviportal functional continuum.

Body size imposes certain restrictions on the types of locomotion in different substrate uses. For instance, in inferring substrate preference and locomotory habit of sloths of the Santa Cruz Formation (Santacrucian age), Toledo et al. (2012, 2013, 2015) and Toledo (2016) considered inferred functional properties with respect to body-mass estimates, taking into account the limitations that size imposes on locomotion on arboreal substrates. The hypotheses of climbing abilities were considered based on the knowledge of the largest extant climbing mammals. Given that, independently of the functional properties of limbs, body mass restricts locomotion to trunks and branches that can support the animal's weight, these authors concluded that the larger climbing Santacrucian sloths (e.g., *Nematherium, Analcimorphus*) were limited to moving on trunks or the thicker branches at the base of a tree's crown.

Body size also imposes restrictions on cursorial animals; as we have seen, the design solutions for velocity differ depending on an animal's mass. For instance, Cassini et al. (2012a), using indices for inferring functional properties, found that typotheres (Notoungulata) from the Santa Cruz Formation were faster and more agile than proterotheres (Litopterna). However, in their conclusions, the authors pointed out that these results may be due to the effects of size (see below): typotheres were smaller (from 1 to 10 kg) than proterotheres (from 20 to 100 kg).

Functional morphology, as already noted, can provide considerable information on the functional properties of the locomotory system, especially with regard to mobility and stability of articulations and musculature configuration. Toledo et al. (2013, 2015) made use of this methodology in analyzing the appendicular skeleton of Santacrucian sloths to infer their substrate preference (terrestrial or arboreal) and locomotory strategies. These authors' methodology is an example of a qualitative functional morphology approach that also (in contrast to most such efforts) differentiates between function and biological role; it may also be considered an example of Currie's (2015) investigative scaffolding approach. Like Candela and Picasso (2008) in a work on porcupines (Erethizontidae) from the Santa Cruz Formation, Toledo and collaborators first analyzed and thoroughly described the limb and girdle skeleton, emphasizing features of functional relevance as recognized in extant taxa, and then reconstructed the musculature based on the better-preserved fossil remains. For this latter component, they drew on the knowledge of the muscular anatomy of closely (sloths and anteaters) and more distantly (marsupials, rodents, primates, and carnivorans) related forms. The following step was a mixed approach of inference by comparison and qualitative application of mechanical concepts. Thus, they developed a mechanical profile of the sloth limbs, listing the functions for which they were better suited (table 6.1). These functions were evaluated through comparison with the biology of extant forms. The functional profile of sloths from the Santa Cruz Formation, deduced in this way, is more similar to that of tamanduas and other slow climbers such as koalas than that of extant sloths and arboreal primates. As just noted above, the authors concluded that although many limb capacities enabled these sloths to climb, this ability was likely limited, given their medium to large size (from 40 to 120 kg).

As explained in chapter 2, the tools of Newtonian mechanics can be used to consider parts of the organisms as lever and pulley systems moved by muscles for evaluating whether a particular arrangement of these elements favors force or velocity. Such biomechanical analysis requires the identification of pivots (represented by the joints) and the attachment sites of muscles for evaluating lever arms at the points of force application. An example of the use of such analysis is that by Fariña (1995), who analyzed the posture of a Pleistocene glyptodont (*Glyptodon*). This author modeled the animal in static and stable bipedal postures, using the estimated center of mass for calculating the moment (torque) exerted by body weight with the

Table 6.1. Mechanical-functional profile of the limbs of Santacrucian sloths

Forelimb	Extensive glenohumeral mobility
	Powerful humeral retraction and adduction
	Powerful antebrachial flexion and extension
	Limited pronation and supination
	Mechanical loads transmitted mainly through the humeroradial articulation
	Powerful carpal and digital flexion
Hind limb	Extensive acetabulofemoral mobility, especially in abduction
	Powerful crural retraction and adduction
	Posture with knee usually flexed
	High mechanical diaphyseal loads
	Powerful foot extension and flexion of toes
	Cruroastragalar mobility restricted to plantar-flexion
	Slight capacity for inversion of medial toes

Note: Profile based on analyses of qualitative functional morphology, muscular reconstruction, and biomechanical indices by Toledo (2016) and Toledo et al. (2013, 2015).

hip joint as the pivot. Then, Fariña (1995) estimated the necessary force of the muscles responsible for lifting the vertebral column to counteract this moment and stabilize the animal (it was also necessary to have the body-mass estimation and at least one superficial reconstruction of musculature). For estimating this force, the author assumed that the muscles had parallel fibers and were approximately of the same diameter, and that their force corresponded to the isometric tension of extant mammal

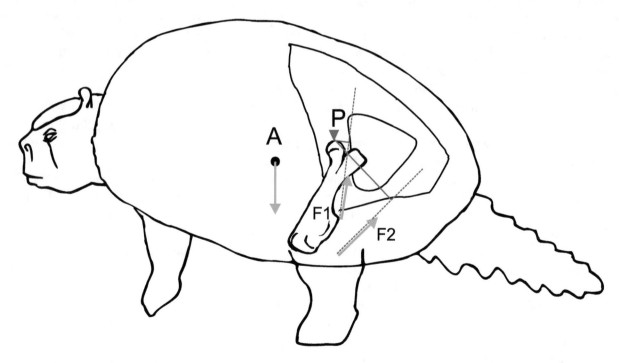

6.24. Biomechanical analysis of posture in glyptodonts by Fariña (1995). *A*, center of gravity; *P*, pivot; **F1**, force produced by the gluteus maximus; **F2**, force produced by the femoral biceps and semitendinosus muscles; *blue arrows*, direction of forces; *dashed red lines*, line of action of the muscles; *solid red lines*, mechanical advantage of the muscles. Based on Fariña (1995).

muscles. Fariña proposed that glyptodonts could adopt a bipedal posture, given that the system design formed by their hind limbs, pelvis, and analyzed muscles was able to stabilize the moment produced by the mass of the anterior part of the body (fig. 6.24). As pointed out in chapter 2, Fariña's analysis could produce hypotheses on bipedalism as a potential faculty; however, making inferences on the biological role of bipedalism (e.g., self-defense, copulation, locomotion) requires an additional analytical step.

A similar approach was used by Hutchinson (2004) for estimating forces implied in the bipedal locomotion of *Tyrannosaurus* and other theropods by calculating the moments around each pivot point (pelvic girdle, knee, ankle, and metapodium-digit articulations). Another less sophisticated approach involves modeling individual bones or their parts as rigid solids subjected to forces. Fariña (1995) and Vizcaíno et al. (2011a) analyzed limb posture and function in glyptodonts using the cross section of the humerus and femur and the proportion of body mass supported by each limb. These authors used a strength indicator (SI) formulated by Alexander (1983, 1985) that provides a measure of the strength of a bone in relation to its expected loading. The SI is based on a parameter used in engineering, the second moment of area of the cross section (also called the section module), and quantifies mechanical strength in flexion that the bone presumably should bear (again, an estimation of mass is required). Fariña (1995) and Vizcaíno et al. (2011a) found that the hind limbs of *Glyptodon* had a mechanical strength capable of supporting most and even all of this animal's body mass, leading them to conclude that *Glyptodon* had the capacity of adopting a bipedal posture. Analyses similar to these have been carried out in diverse groups of dinosaurs to infer not only posture and type of locomotion, but also maximum velocity (e.g., Alexander 1989; Hutchinson and García 2002).

Several methodological approaches for inferring substrate use are based on indices (see the text box "Functional Indices" for several examples), calculated as the ratio between two measurements of the postcranial elements. Generally, they are not merely ratios or proportional relationships between measurements, but (as we will see further below) they are intended to reflect some biomechanical aspect (e.g., in-lever/out-lever ratios). Their application has been criticized based on arguments of a statistical nature (e.g., lack of independence of measurements, issues related to nonnormality and heteroscedasticity, and difficulty in differentiating the effects of the variables involved, among others). However, several articles have demonstrated that indices are indeed useful (e.g., Samuels

and van Valkenburg 2008; Chen and Wilson 2015) and have many advantages:

- They are dimensionless quantities.
- They are easily calculated.
- As many indices as necessary or as many as the available information allows can be generated.
- In general, they involve traditional linear measurements that can be easily taken with calipers or other methods.
- They can be applied to incomplete or fragmentary specimens.
- They can be used in different ways, from simple comparative tables and boxplots to variables for multivariate analyses such as PCA and clustering analysis, among others. Some authors, for example Carrano (2006), have even used indices as continuous characters for analyzing phylogenetic patterns.

The literature provides many examples of the use of indices and their biomechanical interpretation (fig. 6.25). For instance, the index of fossorial ability (IFA), also described in previous chapters, is useful not only for comparisons between animals that dig with forearm extension and those that do not dig, but also for every functional context where forearm extension is relevant (e.g., Vizcaíno et al. 2006b), because the index describes a function (forearm extension) rather than a faculty (digging; see fig. 1.7, chap. 1). In their work on glyptodont posture and locomotion, Vizcaíno et al. (2011a) used functional indices in two ways: the indices were derived as the quotient of two linear variables that represent lever arms and then analyzed comparatively based on knowledge of closely related extant forms (armadillos) to test their correlation with functional features and faculty. Cassini et al. (2012a) carried out a paleobiological study on native ungulates of the Miocene of Patagonia, using indices to infer the functional capacities of the limbs of the groups under study. In addition, they discussed these functional inferences through comparison of the value of indices with those of a sample of extant mammals such as rodents, carnivorans, elephants, and modern ungulates. As mentioned above, Chen and Wilson (2015) used a number of indices to assess substrate preference, uses, and locomotory strategies in Mesozoic stem mammals compared with a wide sample of extant mammals encompassing several orders.

Ecomorphological studies, as noted in chapter 2, use statistical analyses to assess the correlation between morphometric variables and the ecological traits (e.g., posture,

FUNCTIONAL INDICES USED FOR CHARACTERIZING SUBSTRATE USE

Intermembral index (IMI): sum of the lengths of the humerus and radius (or ulnar functional length) divided by the sum of the lengths of the femur and tibia; (HL+RL)/(FL+TL) × 100. This index reflects the relative proportions of the fore- and hind limbs (Vizcaíno and Milne 2002; Samuels and van Valkenburgh 2008).

Shoulder moment index (SMI): distance between the proximal end of the humerus and the deltoid tuberosity divided by the total length of the humerus; DLH/HL × 100. SMI indicates the mechanical advantage of the deltoid and pectoralis major muscles acting at the shoulder joint (Vizcaíno and Milne 2002).

Humerus robustness index (HRI): minimum transverse diameter of the humeral diaphysis divided by the functional length of the humerus; TDH/HL × 100. It indicates the diaphyseal robustness of the humerus (Elissamburu and Vizcaíno 2004).

Olecranon index (OI): length of the olecranon divided by the functional length of the ulna; OL/(UL − OL) × 100. Functional length is the distance from the elbow joint to the distal end of the ulna, being the difference between ulnar length and olecranal length. It gives a measure of the mechanical advantage of the elbow extensors (e.g., triceps) and is considered a measure of digging ability (Vizcaíno and Milne 2002; Elissamburu and Vizcaíno 2004). OI is also referred to as index of fossorial ability (IFA; see chap. 2).

Epicondylar index (EI): epicondylar width (EW) divided by the functional length of the humerus; EW/HL × 100. It reflects the relative available attachment area for the origin of the flexor, pronator, and supinator muscles of the forearm. It is also considered as an indicator of digging ability (Lessa and Stein 1992; Hildebrand and Goslow 2001).

Brachial index (BI): functional length of the forearm (the difference between ulnar length and olecranal length) divided by the humeral length; (UL − OL)/HL × 100. It indicates the extent to which the forelimb is designed for fast movement (i.e., relative extension of the forearm; Howell 1944; Fleagle 1979; Vizcaíno and Milne 2002).

Ulna robustness index (URI): transverse diameter of the ulna at the middle of its diaphysis divided by the functional length of the ulna; TDU/(UL − OL) × 100. It provides an indication of forearm robustness and the relative available area for attachment of muscles involved in pronation and supination and digital flexion (Elissamburu 2010).

Manus proportion index (MPI): length of the proximal phalanx of the third manual digit divided by the length of the third metacarpal; Mph3p/MC3L × 100. This index describes the relative proportions of the distal and middle segments of the manus (i.e., the fingers and palmar surface; Samuels and van Valkenburgh 2008).

Femur robustness index (FRI): transverse diameter of the femur at the middle of its diaphysis divided by the functional length of the femur; TDF/FL × 100. It reflects the capacity for supporting body mass and resisting vertical forces associated with an increase in velocity (Biewener and Taylor 1986; Demes et al. 1994; Elissamburu 2010).

Tibial spine index (TSI): proximal length of the tibia (length of the cnemial crest) divided by tibial length; PTL/TL × 100. It reflects leg force and the available area for attachment of the gracilis, semitendinosus, and semimembranosus muscles, as well as the pedal flexors, which are important during the first phase of the step. The origin (i.e., the proximal attachment site) is related to increased speed during initial phases of propulsion (Elftman 1929; Elissamburu and Vizcaíno 2004).

Tibia robustness index (TRI): transverse diameter of the tibia at the middle of its diaphysis divided by tibial length; TDT/TL × 100. It gives an indication of the leg strength and the relative available area for the attachment of muscles that act along the ankle (pedal and toe flexors and extensors; Elissamburu and Vizcaíno 2004).

Crural index (CI): length of the tibia divided by the functional length of the femur; TL/FL. Like the BI, it reflects the extent to which the hind-limb proportions are optimized for speed (Howell 1944; Fleagle 1979; Bond et al. 1995; Vizcaíno and Milne 2002).

Pes length index (PLI): length of the third metatarsal divided by the functional length of the femur; MT3L/FL × 100. It describes the relative proportions of proximal and distal elements of the hind limb, and also the relative size of the pes (Samuels and van Valkenburgh 2008). These hind-limb proportions are important for assessing cursoriality (e.g., see Carrano 1999).

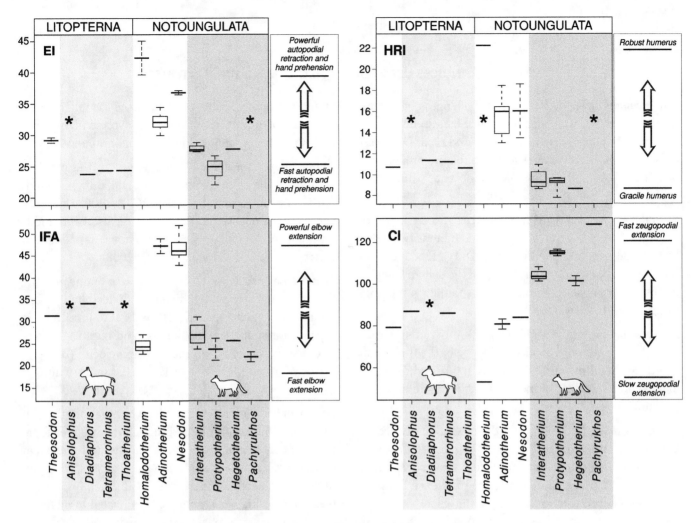

6.25. Use of some biomechanical indices for assessing functional inferences in litopterns and notoungulates of the early Miocene Santa Cruz Formation; *asterisk*, missing data. Modified from Cassini et al. (2012a).

substrate preference) to be inferred. A specific problem is that most of the statistical tests (e.g., ANOVA, discriminant analysis, and canonical analysis) require fixed categories for the comparison sample. This in turn requires classification of sample taxa, an exercise that, as already explained, is often difficult to accomplish unambiguously. One way of dealing with this is to consider substrate preference and analyze whether the extinct taxa are arboreal, live on land, or are aquatic. Another possibility is to classify according to posture as bipedal or quadrupedal, plantigrade or digitigrade, and sprawling or parasagittal, always taking into account that many organisms are able to change posture according to the activity being performed. Classifying by type of gait or locomotory mode is not as straightforward because, as we have already seen, most animals can use different gaits and modes depending on circumstances. The analysis can be repeated by recategorizing taxa. Thus, a test may be carried out that categorizes the lesser anteater *Tamandua* as terrestrial, and another

test may categorize it as arboreal; one test may deem it a quadrupedal walker, another a slow climber or a digger. Difficulties posed by categorization can be curbed by using continuous variables, such as locomotion speed, duty factor, and stride length, among others (although it should be noted that these measurements provide only indirect information about substrate preference and use), or by transforming categorical data into continuous variables (e.g., a matrix to use in partial least squares analysis; see chaps. 7 and 8).

Croft and Anderson (2008) provided an example of using extant mammals to categorize extinct forms. With the aim of inferring locomotory habits of *Protypotherium* (Typotheria), a notoungulate from the Miocene of Patagonia, these authors used estimates of body mass published in the literature and compared them with a sample of extant mammals of similar body size. They employed the morphometric measurements and indices developed by Elissamburu and Vizcaíno (2004) for studying the limb function in caviomorph rodents as a model for typotheres. Both the measurements and indices were applied in principal components analysis (PCA) and discriminant analysis (DA), with the extant taxa categorized as arboreal, jumper, cursorial, and semifossorial, and *Protypotherium* as "unknown." This ecomorphological approach by Croft and Anderson (2008) tested the hypotheses of locomotion previously developed by Elissamburu (2004) through analysis of functional morphology, in which features of the skeletal elements of the limbs (particularly of the joints) were qualitatively analyzed. Note that in the case of PCA, a priori categorization is not required, given that it is an exploratory, variable reduction analytical method (see chap. 2). Samuels and van Valkenburg (2008) also used an approach with indices and DA for finding common morphological patterns (morphotypes) in a sample of extant rodents with similar locomotory habits (e.g., jumpers, cursorial). They then used these results to infer locomotory habits in extinct beavers based on their similarity with semiaquatic and semifossorial morphotypes.

Another example of a protocol using PCA is the work by Toledo et al. (2012). These authors used a series of measurements of the forelimb of a sample of extant mammals of different lineages, body sizes, and locomotory habits for comparison with several sloth genera from the Miocene of Patagonia (fig. 6.26). They found that, within the morphospace defined by forelimb measurements, these extinct sloths had an affinity in shape with the diggers of the sample (either arboreal or terrestrial) and were clearly separated from extant sloths, which exhibited greater morphometric affinity with the primates analyzed. The functional inferences of this similarity in form were weighted using indices such as IFA, among others, that allowed the authors to suggest that Miocene sloths were able to dig.

Alternatively, a modular analysis of the articular surfaces at joints can be carried out by means of geometric morphometric methods. Articular surfaces reflect the relative movements between adjacent bones, and those of limb joints provide information about their movements and may be useful for making inferences about substrate use. Muñoz et al. (2017) followed this approach for interpreting the paleoecology of two small mammals from the Miocene of Patagonia, the typotheres *Hegetotherium mirabile* (Hegetotheriidae) and *Interatherium robustum*

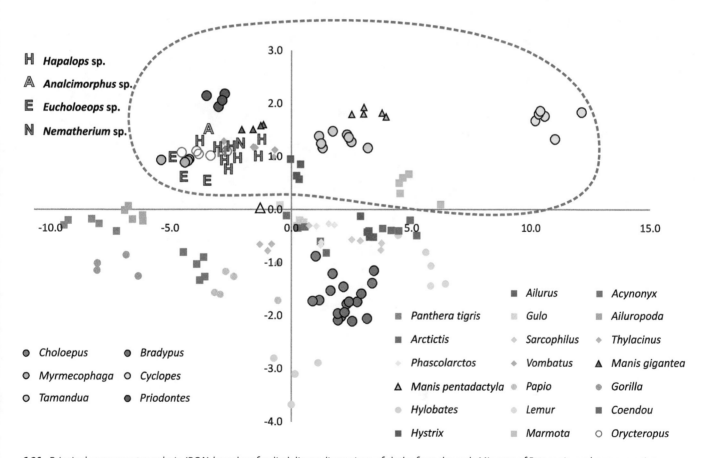

6.26. Principal components analysis (PCA) based on forelimb linear dimensions of sloths from the early Miocene of Patagonia, and a comparative sample of extant mammals. Miocene sloths (*Eucholoeops, Hapalops, Analcimorphus,* and *Nematherium*) share a morphospace with digging forms (*dotted red line*), whereas the extant sloths *Bradypus* and *Choloepus* share a morphospace with primates such as the gibbon, mandrill, and lemur. Modified from Toledo et al. (2012).

(Interatheriidae). The authors compared the proximal articular surface of the ulna of these extinct forms with that of 22 extant small mammals belonging to Rodentia, Carnivora, and Primates using 45 landmarks. While the results of the PCA using the entire surface were inconclusive, the examination of the subunits ("modules") led the authors to recognize morphospaces that allow associating the morphology of the subunits with functional/biomechanics demands. Whereas runners of different clades seem to have different proximal ulnar morphologies, the climbers of the three clades have flattened proximal surfaces that allow the wide range of movements at the elbow joint required for grasping new supports and facilitating directional changes during climbing. In most forelimb diggers (all belonging to Rodentia, but of different clades), the mediolateral convexity of the proximal part of the trochlea would improve the stability and minimize possible dislocations at the elbow joint by enhancing the congruence between the humerus and ulna. In addition, both climbers and diggers have a wide trochlea that increases the contact surface at the elbow joint, which would be effective for resisting the forces generated during these activities. Muñoz et al. (2017) found that none of the typotheres had a specialization for climbing or digging in the features analyzed.

In addition to anatomy, ichnology (the study of organismal tracks and traces) is a source of information on substrate preference and use, including locomotory modes. Among other things, fossil traces can provide valuable information about posture and locomotion of terrestrial animals, such as bipedalism or quadrupedalism, gait, stride length, duty factor, and speed of locomotion. The contribution of ichnology to paleobiology will be described in further detail in chapter 9.

Feeding: Food Diversity and Buccal Apparatuses in Vertebrates

7

We noted in previous chapters that the analysis of substrate preference or use should consider the varied life activities of animals, among them the search for and obtaining of food. By means of these activities, organisms actively extract nutritive substances from the environment, including other organisms. The acquisition by an organism of most of its required biomolecules (e.g., proteins, carbohydrates, lipids) through the ingestion of organic matter produced by other organisms is known as *heterotrophy*; by contrast, organisms that produce organic molecules using inorganic compounds are termed *autotrophs*. The nature of the material in which these biomolecules are packaged define diet or feeding habits of organisms. In this and the following chapter, we will focus on this third essential biological attribute, without losing sight of its close relationship with the other two (size and substrate preference and use).

Vertebrates display quite varied feeding habits, which can be characterized by the size of the item ingested (from microscopic organic particles to organisms that can exceed their own size), by their activity (from sessile or passive organisms such as plants to very active ones such as arthropods and vertebrates), by their nutritive value (from those rich in nutrients such as animal tissues to poor ones such as plant stems), as well as by their availability in the environment or their degree of toxicity, among other features. In fact, any biologist might continue adding to the list depending on the perspectives from which the issue of feeding is approached (e.g., type of food ingested vs. mode of acquisition). For this reason, we devote a significant part of this chapter to the issue of different food classifications.

In vertebrates (and most animals), the part of the digestive system that first comes into contact with food is the mouth, and thus the study of feeding elements (i.e., the mouth and other parts, such as the hyoid apparatus, that contribute to feeding) is highly important for understanding feeding. A good part of this chapter is a summary of the diversity of types and functions of these elements in vertebrates in an evolutionary context, analogous to the treatment in chapter 5 of locomotory types. Given the scope of analysis already covered in this book,

it should come as no surprise that variation in feeding is reflected by multiple designs of feeding systems and that there is both a great diversity of designs within a clade and common solutions in phylogenetically distant clades. The anatomical elements preserved in the fossil record that are useful for inferring the aspects of feeding of extinct organisms are almost exclusively those of skulls and teeth.

The skull and mandible perform a great variety of biological roles, in addition to feeding, that result in different functional requirements that affect their design, such as the relationships with sensory organs (e.g., day or night vision, binocular vision), warming of inhaled air, and defense and/or intraspecific competition (e.g., horns and antlers), among others. These potentially conflicting requirements especially affect the complex formed by the neurocranium and upper jaw bones; the mandible, by contrast, usually exhibits features better reflective of feeding requirements (Vizcaíno and Bargo 1998; Cassini 2013). Teeth, due to their role in obtaining and processing food and resistance to the abrasion and wear caused by their use, allow inferences on feeding, including type of food ingested, masticatory movements, digestive physiology and metabolism, and other specific aspects of habitat use (e.g., digging, grooming).

Classifications of Food and Types of Feeding

In vertebrate paleobiology, the more common initial approaches in morphofunctional and ecomorphological analyses involve the assignment of organisms to one of several possible broad categories or guilds, such as autotroph or heterotroph, primary or secondary feeder, and herbivore, carnivore, or omnivore depending on context, and then inferring the trophic nets of the paleocommunity. This is often followed by consideration of how species in a particular guild partition the resources in that paleocommunity. This requires the use of other dietary categories within each guild, but based on criteria different from those on which the higher-level category or guild is distinguished, such as the mechanical properties of food (soft, hard, turgid, crumbly, fibrous, etc.), the way it is procured (predator, scavenger, etc.), or feeding time (diurnal, nocturnal), among others.

Different disciplines and, indeed, branches within each discipline, including ecology, from which ecomorphological studies are drawn, follow dissimilar criteria. There is, therefore, no single classification for types of food or feeding.

Langer and Chivers (1994), needing to categorize diets as accurately as possible for studies on the mammalian gastrointestinal tract, compiled extensive bibliographic data from different disciplines (ecology, nutrition, functional anatomy, physiology, and evolutionary biology). The main result of this research was a list of terms for classifying food. The authors recognized that they could not compel other researchers to follow their classification, but they expected that it would contribute to the establishment of a widely applicable terminology. As far as we can determine, however, consensus has not been achieved yet. In alignment with the sentiments expressed by Langer and Chivers (1994), the following summary is based on their work (see the text box "Terms Based on Food Types").

A first approach—although one not commonly used in vertebrate paleobiology—involves recognizing terms that can be applied to organisms based initially (or broadly) on the spectrum of food items an organism ingests, and then more narrowly based on the specific food items ingested.

Generalist Feeder—polyphage: refers to animals that consume a wide variety of food; the term *omnivore* is a synonym for polyphage.

Specialist feeder— oligophage: refers to animals that specialize on or make use of a smaller variety of food resources; within this category, we could include *monophage* to refer to extreme cases of organisms that tolerate only one kind of food, such as the koala *Phascolarctos cinereus* and the giant panda *Ailuropoda melanoleuca*, which consume exclusively only eucalyptus leaves and bamboo shoots, respectively.

Regardless of whether an organism is a polyphage or oligophage, the most common procedure in feeding studies is to categorize organisms according to the type of food they consume, in reference to the food's taxonomic affinity or to the portion of the organism (e.g., particular tissue of an animal or portion of a plant) that is consumed.

A common practice in dietary studies in invertebrates is to distinguish between the consumption of nonliving matter (including decaying dead organisms) and living organisms (either vagile or sessile). The consumption of nonliving matter includes feces (coprophage), decaying unstructured organic matter (detritivore), soil with its inorganic component (geophage), bones (osteophage or

ossiphagus), and carrion or dead animal tissues (scavenger). In vertebrates, there are examples of monophages, such as desmodontid bats (*Desmodus, Diaemus, Diphylla*), which are sanguinivores (blood-eaters), but monophagous vertebrates feeding exclusively on nonliving matter are not known, though some of them (birds, mammals) incorporate it to a greater or lesser extent.

When food includes living organisms, they can be of animal or plant origin. For those organisms that feed on other animals sensu lato, Langer and Chivers (1994) found that the most commonly used term is *zoophage* or *faunivore* (as derived from the Greek or Latin root, respectively). However, some published papers apply the term *animalivore* to armadillos (Xenarthra), which prefer consuming animal tissue (animal matter feeder), from insects to the carrion of large mammals (Vizcaíno et al. 2004; Vizcaíno 2009). The reasoning behind this usage is that it differs conceptually from faunivore, as the word *fauna* refers to the whole group of animals of a determined area, and thus faunivory implies feeding on all animals, rather than a particular subset of animals. Beyond this subtle difference, the term *carnivore* is commonly used to refer to a zoophage, though some authors suggest restricting the former term to those that feed on mammals. Zoophage, in turn, can be further subdivided, with the monophage category being the least inclusive; the primary subdivision of zoophages differentiates between those that feed on invertebrates and those that feed on vertebrates.

Microfaunivore: distinctive terms are applied to animals that feed on invertebrates. Those feeding on arthropods may be considered insectivores (insects), myrmecophages (ants or termites), cancrivores or crustacivores (crustaceans), among others; those feeding on molluscs are molluscivores (e.g., snails, clams, mussels) more generally or teuthophages (squids) more specifically; planktonivores are those that feed on zooplankton, which is composed, in turn, largely by the larvae of cephalopods, bivalves, and crustaceans; note that these are not considered molluscivores or crustacivores.

Macrofaunivore: term applied to animals that feed on vertebrates; for example, piscivore (fishes) and ophidiophage (snakes).

Zoophages often feed only on a particular type of animal tissue or component and are classified on this basis; for example, a hematophage feeds on blood, a lepidophage feeds on fish scales, and an oophage feeds on eggs. Those that feed on animals with shells or animals with a resistant exoskeleton are called durophages (thus, a molluscivore can also be a durophage, but not necessarily); in this case, the term is not derived from the hardness of the food, but to the barriers (which are prey specific) that the feeder must overcome to gain access to tissues and organs that are ingested (e.g., muscle, fat, viscera).

Langer and Chivers (1994) used *phytophage* or *florivore* (in relation to flora) for animals that feed on plants in a broad sense. They are plant-matter feeders. Included among phytophages are those that also feed (but not exclusively) on fungi (mycophages), lichens (lichenophages), and moss (bryophytophages), due mainly to a bias in the dietary classification in mammals. Phytophages may feed on one or more parts of the plant: wood (xylophages); bark and branches; roots and organs of underground storage such as bulbs or tubers; diaspores (or disseminula) such as fruits (frugivores); nuts and seeds (granivores); exudates such as sap, gum, and resins (exudativores); blossoms and flowers (florivores, in this more restricted sense, in relation to flower rather than flora—*anthophages* could be used instead to avoid confusion); pollen (pollinivores); or nectar (nectarivores). Langer and Chivers (1994) grouped the remaining phytophages as browsers, including herbivores; that is, those that eat herbaceous plants. This particular case highlights our message that it is important to take note of precision in defining terms; as used more loosely, *herbivore* can have a very different meaning, and unless attention is paid to definitions, confusion is an inevitable result. Browsers are distinguished in two large groups: nongraminoid herb feeders, including leaves, shoots, and stalks; and grazers or graminivores, which consume grass (Gramineae), including leaves, sheaths, and stems. For the latter, Werner (1961) proposed the term *poëphage*, derived from the taxonomic group Poaceae, in turn derived from the Greek πόα for "meadow grass."

In the published literature on phytophagous mammals (i.e., those that feed on plant matter), especially ungulates, the terms *browser* and *grazer* are commonly used but with slightly different meanings from the usage of Langer and Chivers (1994). In this literature, the terms themselves are of ambiguous meaning, as noted in studies of living ungulates (Hofmann and Stewart 1972; Spencer 1995; Fariña et al. 2013), because they have been used to describe both the mode of food acquisition (i.e., degree of selectivity) and the type of food ingested (i.e., botanical connotation).

In an attempt to solve the ambiguity of these terms, Hofmann and Stewart (1972) introduced an alternative classification based on the stomach structure and feeding

TERMS BASED ON FOOD TYPES
(according to Langer and Chivers 1994, table 6.3)

1. Generalist feeder: polyphage; omnivore
2. Specialist feeder: oligophage
   2.1. Animal-matter feeder: zoophage; faunivore
   Can be subdivided into several food types:
   - Invertebrates (microfaunivore)
     Arthropods: ants (myrmecophage),
       insects (insectivore)
     Crustaceans (crustacivore; cancrivore)
     Mollusks (molluscivore; malacophage): squid
       (teutophage), clam, mussel
   - Zooplankton (planktonivore)
   - Vertebrates (macrofaunivore)
     Nonmammal vertebrates: fish (piscivore)
     Mammals (carnivore)
     Blood (sanguinivore; hematophage)
   2.2. Plant-matter feeder: phytophage; florivore
   Can be divided into three types of feeders
   according to the food selection degree:
   - Bulk and roughage feeders
   - Concentrate selectors
   - Intermediate feeders

Plant-matter food in question includes:
   Fungus (mycophage)
   Lichen (lichenophage)
   Moss (bryophytophage)
   Wood (xylophage)
   Bark
   Branch
   Root
   Underground storage organ: bulb, corm, tuber
Diaspores: fruit (frugivore), nuts, cereal grains
(granivore); seed, hull
   Gall
   Exudate: sap, gum, resin (exudativore);
   Blossom/flower
   Pollen
   Nectar (nectarivore)
   Browse (browser)
     Legume
     Nonleguminous angiosperm
     Twig
     Herb (herbivore)
       Forb: shoot; leaf (folivore); leaf stalk or
         petiole; bud
       Grass (grazer, poëphage, graminivore):
         stem; sheath; leaf

habits of east African ruminants. They proposed the terms (1) *bulk and roughage eaters* (grass eaters and, among them, roughage grazers, including fresh-grass grazers and dry-region grazers); (2) *selectors* of juicy, concentrated herbage (tree and shrub foliage eaters, fruits and dicot foliage selectors); and (3) *intermediate feeders* (some preferring grasses and others preferring forbs and shrub and tree foliage). However, according to Langer and Chivers (1994), these three categories do not comprise an unambiguous classification for feeding materials.

The diversity in the usage of terms can be appreciated by considering examples from the literature. In a work on the diet of extinct Old World and North American ruminants, Solounias and Moelleken (1993) indicated that the terms *grazer*, *browser*, and *mixed feeder* should properly be used to express types of vegetation eaten, rather than to distinguish between concentrate selectors and nonconcentrate selectors. In studies on extant and extinct ruminant and nonruminant ungulates, Janis (1988, 1995) used *browser*, *grazer*, and *intermediate feeders*, categories that reflect the percentage of grass eaten. Spencer (1995) suggested using hierarchically arranged categories based on the dietary composition of extant African bovids, ordering them first by dietary composition (grass, dicots, mixed diets preferring grass, and mixed diets preferring dicots) and then by habitat preference (e.g., woodland, desert grassland), followed by feeding height preferences (e.g., tall plants, ground level). Mendoza et al. (2002, 2006) and Mendoza and Palmqvist (2006, 2008) subordinated diets to the type of environment (i.e., open, mixed, and closed) and used a combination of dietary categories that could also include body mass. Cassini (2013), following these authors, maintained the grazer (1), mixed feeder (2), browser (3), frugivore (4), and omnivore (5) categories. All these authors indicated clearly that the categories described are based on the taxonomic composition of the items consumed but differ on the proportion of each type of vegetation consumed.

With respect to the scheme of Langer and Chivers (1994), the above-mentioned categories would be subsumed as oligophages (either phytophages or folivores), with the exception of the omnivores (or polyphages or generalist feeders). Note that terms are not equivalent; for example, browsers in the Langers and Chivers framework include all forms that feed on leaf parts, whereas grazers could be called also pöephagous, and there are no equivalents for intermediate or mixed feeders. As noted before in this book, categorical systems have to be clearly defined; superficial similarities in terms generally lead to confusion in concepts.

Verde Arregoitia and D'Elía (2021) classified the great diversity of feeding preferences of rodents (from flexible generalists to highly specialized forms) into eight types (without considering the framework of Langer and Chivers): carnivore, vermivore, folivore, frugivore, granivore, specialist herbivore, omnivore, and unknown. These authors suggested that mammals select their food according to accessibility, mostly as defined by food mobility; plants and fungi are immobile and animals are generally mobile, implying specialized predatory behaviors. They also considered digestibility of food as a major determinant of mammalian diets. Verde Arregoitia and D'Elía (2021) concluded that feeding categories should be constructed based on food items that can be recognized in natural ecosystems (i.e., different parts of plants, fungi, or animal matter) rather than, for example, isotopic profiles. Similarly, for comparative ecological and evolutionary research, it is important to define dietary categories by composition, rather than by morphological adaptations, so that it is possible to assess how morphology relates to diet without circularity.

In the paleontological literature of South America, the terms *browser* and *grazer* have been largely applied to Cenozoic mammal faunas, but their meaning has hardly ever been clearly defined (Fariña et al. 2013). It is worth pointing out that during the Cenozoic there was an extremely low number of ruminant ungulates on the continent, and the role of herbivores of over 1 tonne (megaherbivores) was mainly occupied by xenarthrans and secondarily by native ungulates (Vizcaíno et al. 2012a). In a study of the masticatory apparatus of large Pleistocene ground sloths to infer their feeding habits, Bargo and Vizcaíno (2008) considered and reviewed the discussion on the ambiguity of the terms *browser* and *grazer*, which incorporate two criteria, one of which reflects the manner of acquiring food and the other the kind of food ingested (see also Fariña et al. 2013). That is, browsing may describe selective feeding of any food type, including feeding on dicot plants; grazing implies the eating of grass, but it may be used to mean the eating of forbs as well. Bargo and Vizcaíno (2008) exemplified this ambiguity by describing the feeding behavior of the Pampas deer *Ozotoceros bezoarticus*. This medium-sized cervid feeds mainly on grass but takes only small nutritive parts of the plants of certain species. That is to say, it would be considered a browser from the perspective of item selection and a grazer based only on the systematic group of the food ingested. Consequently, if the meaning of these terms is not unambiguously defined, particularly when they are applied to an extinct organism, it is not clear whether they refer to the way of acquiring food items by form (size and shape, regardless of the taxonomic nature), the specific selection of taxa independent of form, or some degree of a combination of both.

A possible solution may be to restrict the terms *browser* to those feeding selectively (thus, selective feeders) on shoots and branches of grass and/or bushes, and *grazer* to those that feed less discriminatingly on grasses (Gramineae) in general (thus, bulk feeders). In any event, the fact that both groups are defined on different criteria persists. Further, the correlation between feeding morphology and consumption of members of a particular taxonomic entity (e.g., Gramineae) is indirect. What actually is evaluated is the correlation between morphology and certain physical, mechanical, and ecological characteristics of a taxon, and it can only be confirmed by direct observation in extant forms or supported with chemical information in paleobiology (see the section "Isotopic Paleontology," chap. 9). As discussed in chapters 8 and 9, typotheres, small native ungulates from South America, developed an exaggerated form of the morphology considered typical of extant ungulate grazers. However, recent studies on phytoliths (biomineralized particles of vegetal origin) preserved in the geological deposits from which typotheres are recorded show that this morphology was developed before the widespread establishment of grassland (Gramineae) communities on the continent. Therefore, the morphology of typotheres could reflect a combination of environmental factors and a diet based on plant taxa with certain physical properties analogous to those of Gramineae (Strömberg et al. 2013). Bargo and Vizcaíno (2008) preferred using the correlation between skull morphology and physical features of food that would define the style of food acquisition in proposing a hypothesis on feeding of Pleistocene ground sloths of the Pampean region (see chap. 8).

We should keep in mind that, as noted earlier for substrate use, categories and their definitions represent subdivisions of a continuous spectrum, in this case of proportions or frequencies of items consumed among zoophages and phytophages (sensu Langer and Chivers 1994), or carnivores and herbivores in a more familiar sense. For example, consider the review by van Valkenburgh (2007, 148) on carnivorans, in which the three categories (i.e., hypercarnivorous, mesocarnivorous, and hypocarnivorous) "are not entirely discrete and grade into one another to some extent, but are useful for broad analyses."

In this regard, the analysis by Gagnon and Chew (2000) of herbivores, particularly of African bovids, is especially illustrative and provides a case study to exemplify how a continuous spectrum and statistical analyses can help us

to a better understanding of feeding behavior. Gagnon and Chew (2000) reviewed the categories already available in the literature and decided to depart from the mainstream approach in ecomorphology. Instead of classifying the African bovids in a priori categories, they identified groups a posteriori based on cluster analysis of a data matrix of continuous dietary frequencies. They first performed an extensive survey and compilation of proportions of food types, but also included information on seasonal and geographical variation and, perhaps more interestingly, a measure of data reliability. Regrettably, only 42 of the 78 species analyzed have reliable data (about 54%). One outcome, then, of this study is that long-term ecological studies are still needed, which underscores Radinsky's insistence (see chaps. 1 and 2) that form-function studies require that close attention be paid to behavior in living organisms. Gagnon and Chew then performed a series of cluster analyses and, based on varying and clearly defined proportions of food types, recognized the following categories for African bovids: obligate grazers, variable grazers, browser-grazer intermediates, browsers, generalists, and frugivores. A second outcome of this analysis by Gagnon and Chew is that a priori categories do not always provide appropriate proxies of dietary diversity. These conclusions were reinforced more recently by Pineda-Munoz and Alroy (2014) in a work aimed at creating a global feeding classification scheme for all terrestrial mammalian faunas. These authors summarized the dietary preferences of a broad range of terrestrial mammals feeding on eight resource groups that are readily recognizable in terrestrial ecosystems—seeds, invertebrates, vertebrates, fungi, flowers and gum, roots and tubers, green plants, and fruit— and evaluated them through principal components analysis (PCA) to identify the variables that best summarize the different dietary specializations. Due to the broad aims of this study, herbivore diets were grouped in a single feeding adaptation.

A similar effort was made by Olsen (2015) toward a better understanding of herbivory in anseriforms. The innovation in Olsen's proposal was his treatment of widely varied data (e.g., relative proportions of items ingested, in the gut, or in feces; the observed percentage of time spent by animals foraging on particular items) to create a continuous character representing the degree of herbivory. When possible, Olsen (2015) used as many entries for species as available in the literature and, for each entry, grouped food item consumed into nonoverlapping categories to compile a matrix with the relative importance of each category as a percentage, such that all categories summed to 100 percent. Indeed, this approach led Olsen to incorporate several studies that would be considered as nonreliable by Gagnon and Chew (2000). Later, Olsen (2017) used this continuous data matrix (i.e., continuous dietary data compiled from the literature) to evaluate the association between beak shape (3-D curvature of the upper beak collected from museum specimens) and feeding ecology in Anseriformes by mean of partial least squares analyses. Similarly, Cassini and Toledo (2021) evaluated craniomandibular integration in an ecomorphological context in relation to diet in Neotropical cervids. These authors used a quantitative approach to dietary diversity, instead of using categories that would induce an ecomorphological mismatch between a fine-grained depiction of morphological variation and a crude depiction of ecology.

These several examples suggest that if the use of dietary categories as such can be avoided and attention instead focused on approaches that preserve the quantitative nature of dietary proportions, a better understanding of form-function or, more properly, form–biological role correlation/association can be achieved. Despite the extensive efforts noted in the previous few paragraphs that deal with issues involving feeding classification, a consensus on the most suitable classification broadly applicable in vertebrate paleobiology has not yet been, and may never be, reached. Even so, it probably will not be necessary to reach an agreement, since it will always depend on the type of classification a paleobiologist should apply according to the needs of a particular study. What seems to be indispensable is, once more, clarifying the specific meaning given to the terms in works on paleobiology. Besides, it is advisable not to base the interpretation of diet on a single morphological variable, but rather on assessment of the consistency of a proposed interpretation of diet with what else is known about the organism, including such other lines of evidence, for example, as micro- and mesowear (chap. 8) and contextual information (e.g., paleobotanic record, sedimentary environment; chap. 9).

Diversity of Buccal Apparatuses in Vertebrates

In Radinsky's (1987) hypothetical ancestral marine vertebrate (referred to in chap. 5) that fed by organic particle filtration, the largest organ in the body was the pharynx, which functioned mainly in feeding and respiration (fig. 7.1). Water carrying mud with organic matter in suspension would have entered the oral cavity by suction resulting from expansion of the pharynx and opening of the mouth. Contraction of the pharynx, combined with closing of the mouth, would have decreased its volume, thereby increasing pressure and forcing water to pass out through the gill openings. During the process, ingested mud and organic

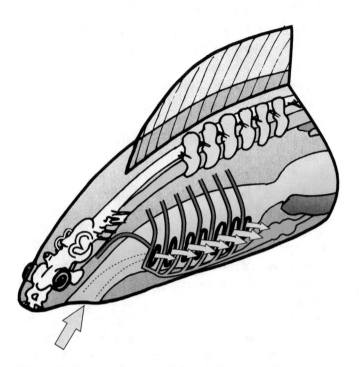

7.1. Detail of the cephalic region of a hypothetical ancestral vertebrate; note the direction of water currents (*green arrows*) passing through the branchial basket.

matter would have been retained in the pharynx and directed to the gut for digestion and nutrient absorption.

The development of jaws represents the most significant evolutionary innovation in vertebrate history (Romer 1962), as jaws allowed access to otherwise unattainable food sources and favored the development of larger forms. Jaws moved by muscles and armed with teeth make it possible to grasp objects firmly. However, the action of jaws is not limited to the procurement and reduction of novel prey items; jaws make possible a wide range of faculties and biological roles through object manipulation. They could thus take on the role of a multifunctional tool for digging, moving elements to build nests, restraining a mate during courtship, and supporting and mobilizing offspring, among other possibilities (fig. 7.2).

Jawed vertebrates constitute a monophyletic group, Gnathostomata, that includes most extant and extinct vertebrates. Its name (from the Greek *gnathos* for jaw and *stoma* for mouth) refers to its most outstanding feature, the presence of endoskeletal jaw elements, as well as a diversity of exoskeletal elements, such as teeth and dermal bones, for grasping, piercing, scratching, grinding, or cutting. Extant vertebrates are grouped into two larger clades, Chondrichthyes (sharks, rays, and chimaeras) on the one hand, and Osteichthyes (bony fish, both ray-finned and lobe-finned, and tetrapods) on the other. Also, two other major groups of extinct aquatic gnathostomes are recognized (Placodermi and Acanthodii), although our ideas on

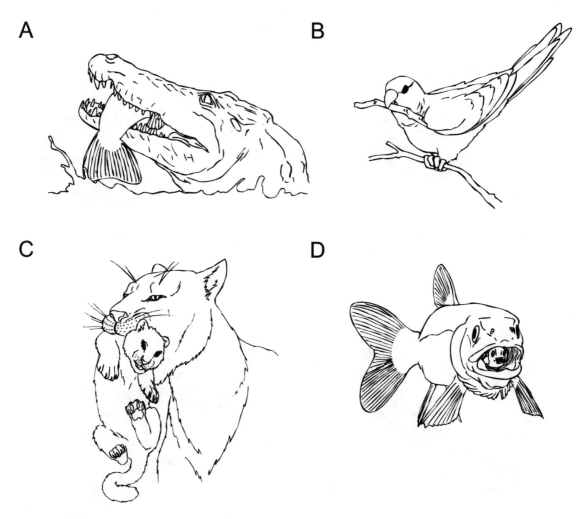

7.2. Biological roles of jaws. *A*, capture and killing of prey by a crocodile (procurement and manipulation of the feeding item). *B*, manipulation of surrounding objects by a parrot (nest building). *C* and *D*, offspring relocation and protection (lion and goby).

their monophyly and relationships are being reconsidered (see, e.g., De Iuliis and Pulerà 2019).

In the aquatic environment, jaw opening is accomplished mainly by the action of the hypobranchial musculature (the force of gravity is negligible in this regard). The difference in pressure established between the oropharyngeal cavity and the surrounding water (resulting from the abrupt increase in the cavity's volume during rapid opening of the mouth) generates a suctional component used, to a greater or lesser degree, by most fish for feeding, especially in cases where there is a considerable size discrepancy between predator and prey. In this sense, the presence of a jaw suspension mechanism that allows mobility of the upper jaw (hyostyly) allows the jaws to move forward (and in some cases also downward, as in skates and rays) quite independently of the cranium, favoring volume increase of the cavity and the consequent generation of suction. In sharks and other active predatory fish, jaw protrusion plays a significant role in prey biting. The

system of articulated elements that form the protrusible mouth of many bony fish, as well as the associated musculature, constitute an extremely complex structure of considerable functional plasticity (fig. 7.3).

Feeding by suction is not possible in the terrestrial environment due to the low density of air (water is 900 times more dense and 80 times more viscous than air), and so in tetrapods, jaws are especially prepared for grasping prey forcefully. Jaw suspension is of the autostylic type (the upper jaw is fused or firmly linked to the cranium) and the gill apparatus, once its original functions (filtering and gas exchange) had become redundant, was repurposed into new structures (e.g., glottis, larynx, and hyoid apparatus) and functions. Mouth opening is performed by depressor muscles, assisted by gravity, other than the hypobranchial musculature. Salivary glands are novel developments that function to lubricate and chemically process food items while they are in the oral cavity. In some clades (ophidians par excellence), salivary glands are modified into poison

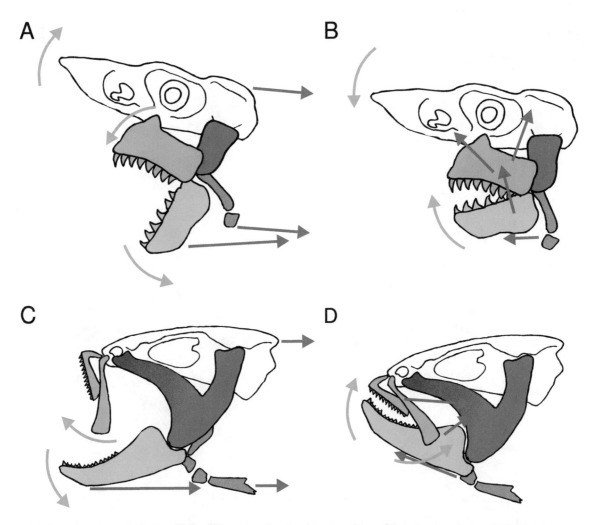

7.3. Jaw mechanics in vertebrates with hyostylic suspension. *A* and *B*, movements of the splanchnocranium during opening and closing of the mouth in an idealized neoselachian shark. *C* and *D*, movements of the splanchnocranium during opening and closing of the mouth in an idealized teleost fish. Upper jaw, *blue*; lower jaw, *green*; hyomandibular element of the hyoid arch, *red* (in teleost fused with other elements, *gray*, forming the suspensorium); remaining elements of hyoid arch, *orange. Blue arrows* indicate approximate direction of motion; *red arrows* indicate lines of action of the involved musculature. Modified from Hildebrand and Goslow (2001), Liem et al. (2001), and Kardong (2018).

glands. In tetrapods, the tongue also appears, a muscular organ developed on the floor of the oropharyngeal cavity and supported by the hyoid system. The tongue plays important roles in food intake, intraoral transport, and swallowing (fig. 7.4).

Among tetrapods, the evolution of primitive amniotes coincided with the diversification of insects as a food source, and this implied the development of more complex jaw mechanisms. Cranial kinesis, referring to intracranial mobility, developed in several amniote lineages (Hildebrand and Goslow 2001). Cranial kinesis involves several mechanisms of relative movement between parts of the skull (in addition to that between the upper and lower jaws) that allow the dissipation of loadings resulting from food intake and for the possibility of ingesting larger prey. Such a mechanism may be formed by, for example, four mobile units in many squamates, where the quadrate is

movably articulated with the neurocranium (streptostyly) and the degree of cranial kinesis increases based on the presence of other joints in the upper jaw and cranium. The complexity of this system reaches its maximum expression in snakes, especially endoglyphous snakes (elapids, including cobras and corals, and viperids, including rattlesnakes and American lanceheads) with eight movable units (fig. 7.5). In these, the skeletal, muscular, digestive, and integumentary systems have adaptations for facilitating prey ingestion.

Among diapsids that have an akinetic skull (such as chelonians, and crocodiles among archosaurians and tuataras among lepidosaurians), the quadrate is fixed firmly to the cranium (monimostyly). In these animals, unlike in the kinetic skulls of squamates and birds, cranial robustness resists mechanical stresses of feeding. The skull is also reinforced by the appearance of a secondary palate

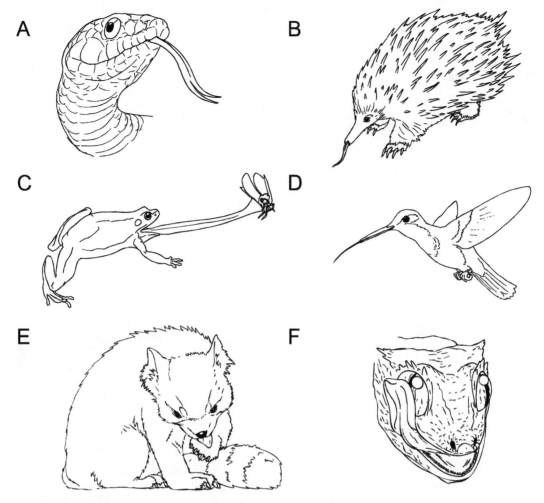

7.4. Biological roles of tongues. *A*, as a tactile chemosensitive organ (most tetrapods, here represented by an ophidian). *B, C,* and *D*, as a selection and food intake organ: insects in an echidna (*B*) and a toad (*C*), and nectar in a hummingbird (*D*). *E* and *F*, as a grooming organ in the red panda (*E*) and a gecko (*F*).

7.5. Cranial kinesis in an idealized viperid ophidian during opening of the mouth. The system of articulated bony elements produces erection of the venom fangs for biting. Modified from Liem et al. (2001) and Kardong (2018).

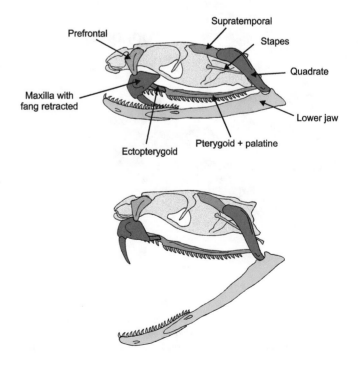

that largely separates the oral and nasal cavities. This structure is incipient in some chelonians and is maximally developed in crocodilians, where the premaxillae, maxillae, palatines, and pterygoids meet at the palatal midline.

Whereas the lower jaw of bony fishes is generally formed by an endochondral element (articular) and several dermal bones (e.g., dentary, splenial, angular, suprangular, prearticular, and coronoids), these elements can be variably absent or fused to varying degrees among different tetrapod groups. Chelonians possess a keratinized beak covering their edentulous jaws. Modern tuataras also lack teeth, but have serrated jaw margins that fulfill an analogous role until they wear by use. Snakes have teeth implanted in shallow pits (acrodonty), unlike the

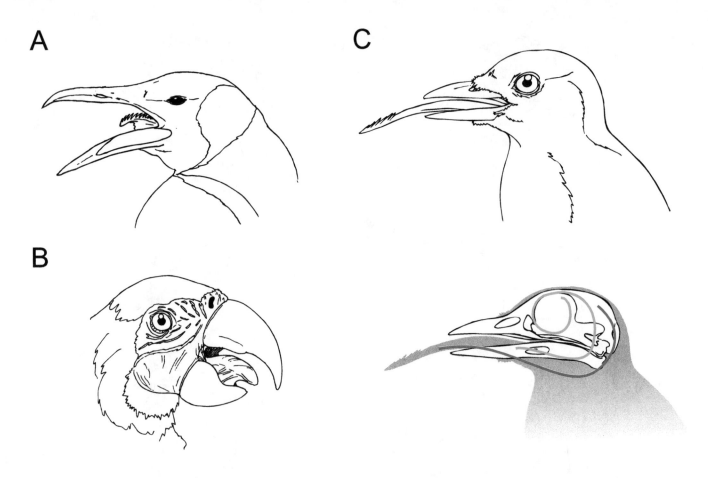

7.6. Tongue specializations in birds. *A*, pointed spiky tongue of an emperor penguin (Spheniscidae) for eating crustaceans. *B*, muscular tongue of a macaw (Psittacidae) for positioning hard-shelled fruits within the oral cavity. *C*, dart-shaped tongue of a woodpecker (Picidae) used in foraging, for example, for insects; *below* the *lower* image indicates the position of the hyoid arch with the tongue retracted (*blue*) and protruded (*red*). Modified from Liem et al. (2001) and Kardong (2018).

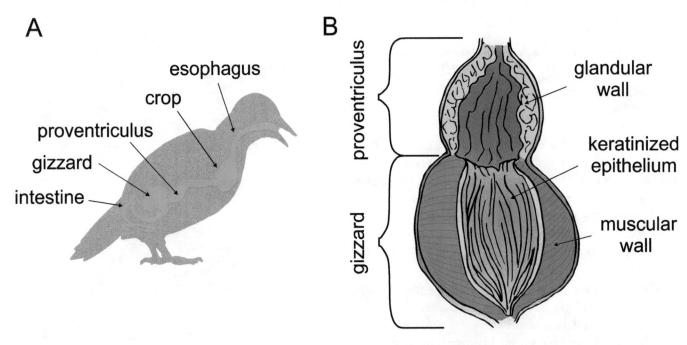

7.7. Specializations of the digestive system in birds. *A*, schematic diagram of an idealized granivorous bird; note the presence of a crop, a specialization of the esophagus in many birds, where food is stored and humidified. *B*, stomach specializations: a glandular portion (proventriculus) where chemical digestion of food is carried out, followed by a muscular portion (gizzard) for crushing food.

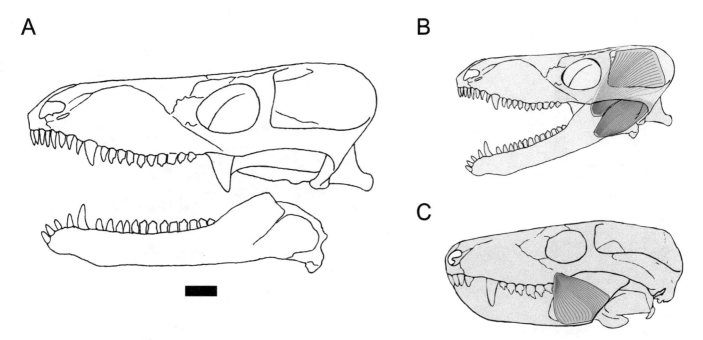

7.8. Cynodonts. *A, Procynosuchus* sp. The anterior teeth have a simple, conical form, whereas the cheek teeth have more complex crowns (scale = 1 cm). *B,* reconstruction of the mandibular adductor musculature (temporal and masseteric muscles) of *Procynosuchus* sp. *C, Thrinaxodon* sp., a more derived cynodont, with development of an additional masseteric component (superficial masseter muscle). Modified from Kemp (1979), Abdala and Damiani (2004), and Lautenschlager et al. (2017).

lateral implantation (pleurodonty) present in the rest of the squamates. Crocodilians and most groups of extinct archosaurs have teeth implanted in deep alveoli or sockets (thecodonty; see appendix).

In contrast to the relatively simple mandibular musculature of extant amphibians, in reptiles this musculature is markedly more complex. The adductor muscles are the most important; they arise from the bones of the dermal skull roof that form the braincase and the quadrate bone, and they insert on the mandible and are innervated by the trigeminal nerve. The reptile tongue is more complex than that of amphibians as well, with glands and a well-developed musculature.

Characteristics of modern birds include the absence of teeth and the presence of a keratinized beak of markedly varied morphology that covers the jaws. Cranial kinesis also developed during the adaptive radiation of the group, allowing a degree of independent movement of the beak with respect to the braincase. There are also specializations of the tongue for food capture (e.g., in woodpeckers and hummingbirds; fig. 7.6). As in crocodilians, the stomach of birds includes a muscular component, the gizzard, which reduces the size of food particles by crushing, and it is especially developed in plant eaters with a diet rich in fibers (fig. 7.7).

Several anatomical features of mammals developed among Permian to Triassic therapsids, although their original functions were not necessarily the same as those in extant mammals. Among nonmammalian cynodonts (fig. 7.8), the postcanine teeth had accessory cusps accompanying a main cusp, with wear facets that indicate a capacity for shearing and cutting, which would have allowed these animals to reduce food to smaller portions before swallowing. Differentiation of the adductor musculature resulted in a masseteric complex that may have been associated with the evolution of more precise occlusion and intricate mastication, a functional trait that further developed in early mammals and was a critical prerequisite for the ecological diversification of therians (Grossnickle 2017; Grossnickle et al. 2022). Although nonmammalian cynodonts retained an articulation between quadrate and articular, these bones were reduced and possibly contributed to sound conduction. Increased relative size, particularly in depth, of the dentary would improve the ability to resist load placed on the mandible during mastication, either by the presence of stronger muscles or by repetitive strokes.

In mammals, the skull has a secondary palate and the lower jaw is formed on each side only by the dentary (fig. 7.9). The bony palate (hard palate) is constituted by the premaxillae, maxillae, and palatines, and the soft palate, comprised mainly of muscular tissue and supported by the pterygoids, extends the separation of the oral and nasal cavities posteriorly.

The mouth is limited by lips and muscular cheeks, evolutionary novelties of therians (Theria): marsupials (Metatheria) and placentals (Eutheria). These soft-tissue

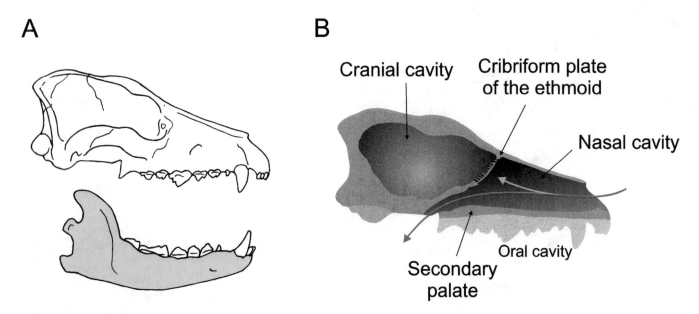

7.9. Idealized skull of a dog. *A*, the dentary is shaded in *blue*; the mandible is formed on each side by this single bony element. *B*, schematic sagittal section showing the secondary palate separating the oral cavity from the nasal cavity, allowing air (*red arrow*) to pass far back without passing through the oral cavity; the *blue arrow* indicates the flow of air directed to olfactory structures.

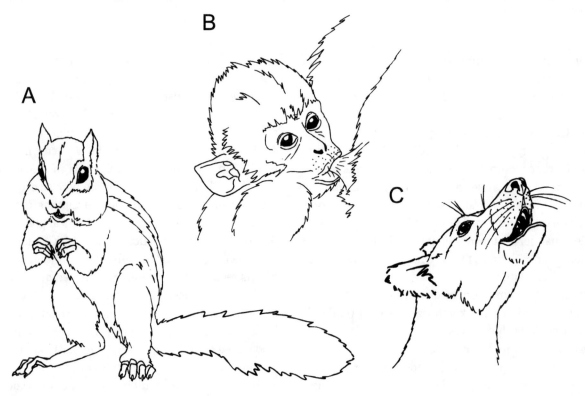

7.10. Lips and cheeks in mammals. *A*, role of cheeks (pouches) for temporary food storage in a squirrel. *B*, muscular cheeks and lips allow a newborn chimpanzee to suckle during breastfeeding. *C*, cheeks and lips participate in the modulation of sounds produced by the glottis in a howling domestic dog.

structures, together with nasal structures, form the muzzle and are vital for food intake (including lactation), mastication, and sound production (fig. 7.10). In some rodents, the cheeks have pouch-like extensions for storing food. The tongue, quite mobile, rich in taste buds, and generally keratinized, is involved in manipulating food in the oral cavity, positioning it between the upper and lower teeth for processing and farther back in the oral cavity for deglutition.

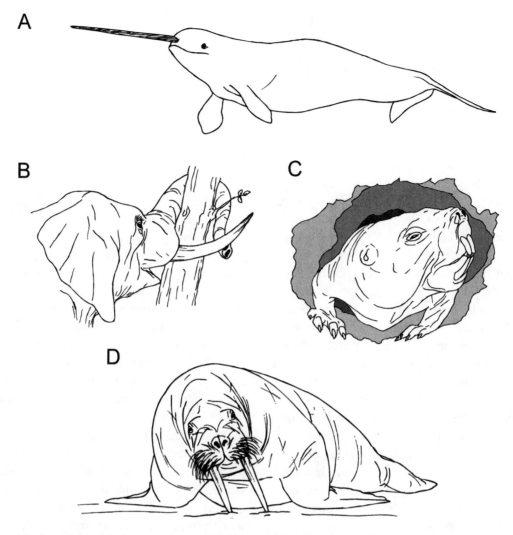

7.11. Specialized teeth of mammals. *A*, narwhal (Monodontidae) with a hypertrophied upper canine. *B*, elephant pushing a tree trunk with its tusks (hypertrophied incisors). *C*, mole rat (Bathyergidae) with hypertrophied incisors for digging and as defensive weapons. *D*, hypertrophied canines in a walrus (Odobenidae), used for bracing onto and for making holes in an ice shelf.

Teeth in mammals are in general more complex than in the rest of gnathostomes (see appendix). They are essential for food intake and mastication, though these are not their only functions. For instance, pigs (Artiodactyla, Suidae) use their enlarged canines for digging in soil in search of food (roots, tubers), walruses (Carnivora, Pinnipedia) use their enormous canines as icepicks for anchoring on the ice shelf in emerging from the water, and elephants (Proboscidea) use their tusks (massively enlarged incisors) for attack and defense or for manipulating branches drawn toward the mouth by the trunk (fig. 7.11).

Mammals also evolved toothless masticatory apparatuses in association with specialized feeding habits, in both terrestrial and aquatic environments. In terrestrial environments, myrmecophagy has been associated with the loss of the ability to chew in anteaters and pangolins (Naples 1999; Davit-Béal et al. 2009; Vizcaíno 2009). In water, some toothed cetaceans (i.e., odontocetes; also see chap. 8 for baleen whales or mysticetes) make use of suction in food intake. For example, the pilot whale (*Globicephala melas*) has been observed ingesting food without the use of its teeth (Werth 2000), and suction feeding is considered a main feeding strategy in the sperm whale (*Physeter macrocephalus*; see Werth 2004). The depression and retraction of a large, piston-like tongue generates negative intraoral pressures for prey capture and ingestion. These examples are considered further in chapter 8.

8 Feeding: Analysis of the Cephalic Feeding System

Feeding, the incorporation of food for energy and the production and repair of tissues, generally occurs by stages that include either or both mechanical and chemical breakdown, termed *digestion*. Once the food item been brought through the mouth into the oral cavity (ingestion), it is mainly subjected to physical processes (mechanical digestion) for reducing it into smaller portions that facilitate its passage to the pharynx (deglutition, the act of swallowing) and then the intestinal tract, where it is mainly exposed to several chemical processes (chemical digestion). Mechanical and chemical digestion (especially the latter) begun in the oral cavity are completed in the intestinal tract, reducing the food into its complex molecular components. These are further broken down into simple molecular components (nutrients) that pass from the digestive system to circulatory and lymphatic capillaries (absorption) and from there to the rest of the body. Finally, nonabsorbed materials are expelled from the digestive tract (egestion).

The oral cavity is involved in the selection and intake of food (the acts of separating the food item from its environment and its ingestion), and its predeglutition processing (including mastication) and deglutition. Therefore, the morphology of the components of the oral cavity is reflected by the features of food that pass through it and provides information about feeding aspects of extinct organisms. Further, though more indirectly, teeth can reveal particular aspects of digestive physiology, specifically of the processing in the intestinal tract. As deglutition almost exclusively involves soft tissues, the sections below deal only with the first two stages.

Selection and Food Intake

As noted in chapter 7, vertebrates in an aquatic environment can carry out ingestion by employing suction. Opening of the mouth in sharks involves the epaxial musculature lifting the head and the hypobranchial musculature pulling the mandible backward (Wilga et al. 2000). Most sharks can protrude the upper jaw while simultaneously enlarging the volume of the oral cavity and pharynx, producing suction that pulls prey through the mouth and into the oral cavity (see chap. 7, fig. 7.3A, 7.3B).

In derived osteichthyans, jaw shortening and decoupling of the upper jaw from the opercular skeletal elements, together with profound modifications of ligaments and muscles, allow a rapid expansion of the oral cavity along with jaw protrusion, enhancing suction (see chap. 7, fig. 7.3C, 7.3D). A great deal of morphological diversity is developed from this basic osteichthyan arrangement (fig. 8.1), from forms that maximize power (such as many coral reef fishes that extract food from hard substrates) to those that maximize speed in jaw closing (predators, mostly of other fish).

Among tetrapods, aquatic forms of amphibians, including the tadpoles of anurans, reptiles (e.g., some turtles), and mammals (e.g., some baleen whales, such as the gray whale, *Eschrichtius robustus*; Goldbogen et al. 2017) can also make use of suction in food intake (fig. 8.2A, 8.2B), with the tongue and hyoid apparatus playing a role. However, and as mentioned in the previous chapter, suction is impossible outside of water. In the remaining tetrapods (except for those mentioned in the preceding sentence), food intake is carried out by structures such as the tongue, teeth, beaks, lips, and, more rarely—as in, for example, some primates and rodents—by the forelimbs. In this section, we will deal primarily with the tongue—and its skeletal support, the hyoid apparatus—as an important organ in food ingestion (figs. 8.2C, 8.2D, 8.3).

The tongue is a muscular organ located on the floor of the oropharyngeal cavity. Although the term *tongue* is applied, for example, to the raised mucosa forming the floor of the oral cavity in such vertebrates as the shark, this structure is not a "true" or muscular tongue, and as such is often termed a *primary tongue*. A true tongue is composed of and controlled by voluntary musculature that is supported mainly by elements of the hyoid apparatus, although portions of other branchial arches may contribute, depending on the vertebrate; in some instances, such as in the mudpuppy, the hyoid apparatus is composed of elements of the hyoid arch and the first three branchial arches (see De Iuliis and Pulerà 2019). More commonly,

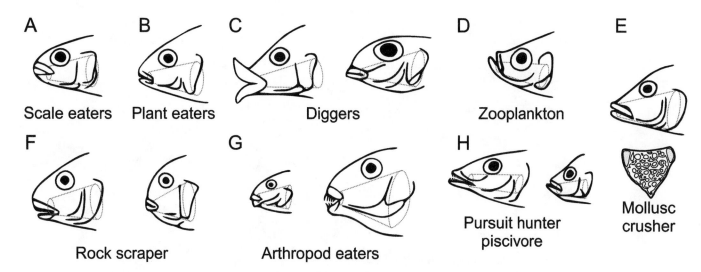

8.1. Morphological and functional diversity of the oral cavity (indicated in *dotted red lines*) of cichlid fishes of Lake Tanganyika, Africa. *A, Perissodus. B, Limnotilapia. C, Lobochilotes* and *Xenotilapia. D, Haplotaxodon. E, Lamprologus; lower* image illustrates its lower pharyngeal dental plate in dorsal view. *F, Petrochromis* and *Simochromis. G, Eretmodus* and *Tanganicodus. H, Bathybates* and *Lamprologus.* Modified from Liem (1991a).

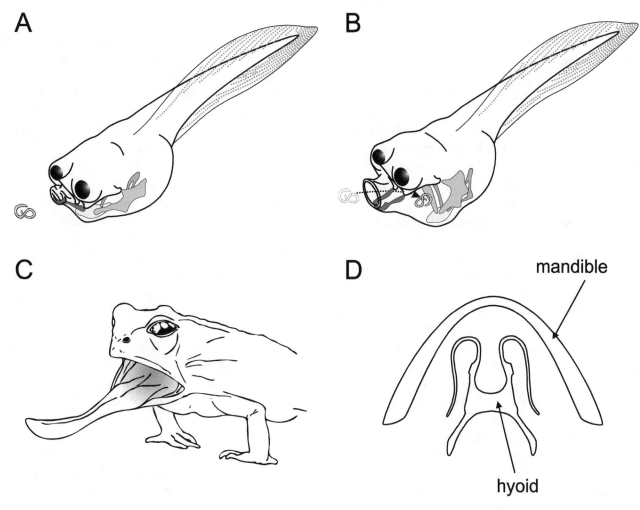

8.2. Dynamics of the mouth system in anurans. *A* and *B*, in tadpoles protrusion of the mouth and movement of hyobranchial elements result in rapid expansion of the oropharyngeal cavity, which decreases the pressure within the cavity and produces suction. Meckel's cartilage (mandible) is indicated in *red*, the ceratohyoids are indicated in *yellow*, the copula (the basibranchial cartilage) in *green*, and the ceratobranchials in *blue. C*, during feeding in adults, the mouth opens while the sticky tongue, which is attached to the anterior floor of the oral cavity, is flung forward to capture prey. *D*, the hyoid apparatus, shown in ventral view, is derived from hyoid arch and other larval branchial elements. *A* and *B* modified from Deban and Olson (2002); *D* modified from Schwenk (2000).

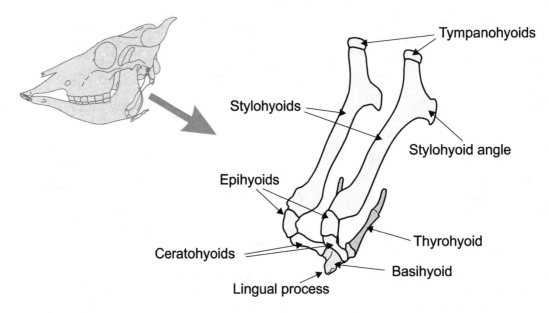

8.3. The hyoid apparatus of mammals, as exemplified by that of the cow (*Bos taurus*), in anterolateral view, with *inset* image indicating its position relative to the skull. Modified from Barone (1976).

however, there is a fairly standard set of elements forming this apparatus in nearly all vertebrates. In general, it consists of a median ventral, unpaired element, the basihyoid, which provides the main support for the tongue and is linked to the otic region of the skull by a chain of paired lateral elements—which, in particular, may differ depending on the group (figs. 8.2, 8.3). In tetrapods, although a basic pattern may be recognized, the musculature of the hyoid apparatus and of the tongue (derived from the hypobranchial musculature of fishes) is complex, and its nomenclature and homology vary among groups depending on authors' interpretations. One group of muscles (of the mandibular and hyoid arches) is arranged to link the apparatus with the medial surface, on each side, of the lower jaw, and forms the floor of the oral cavity. A second group extends between the hyoid apparatus and sternum. Finally, some muscles link the hyoid arch with the base of the skull or extend between the various elements of the apparatus. Animals with a more mobile tongue have a series of muscles that extend from the hyoid apparatus to the tongue base. Animals that feed by means of "licking" (e.g., birds and nectarivorous mammals), birds that extract invertebrates from bark, and anteaters have an extremely well-developed tongue and hyoid apparatus.

Nearly all adult amphibians (pipid frogs being an exception, e.g., see Carreño and Nishikawa 2010), reptiles, birds, and mammals possess a tongue (Iwasaki 2002). In adult anurans, the tongue allows the capture and ingestion of insects and other invertebrates that mainly comprise their diet. The presence of abundant mucus facilitates capture, and the tongue is extensible in many taxa. The tongue of

turtles and some crocodiles can be protruded but is not extensible. In squamates, especially in snakes, the tongue is much more developed, and that of chameleons is as long as the body (when extended) and sticky, enabling them to capture insects. Whereas the tongue of agamids and iguanids is short, that of the remaining squamates is long, narrow, bifid, quite mobile, and retractable into a sheath. In these vertebrates, the tongue plays a crucial role in trapping molecules from the air and transferring them to Jacobson's (or vomeronasal) organ, an olfactory chemoreceptor located in the palate in many vertebrates. In birds (which lack a Jacobson's organ), the tongue is commonly covered by a heavily keratinized mucosal epithelium, and a prominent bony component of the hyoid apparatus is embedded in it. The tongue and hyoid apparatus of these vertebrates display a wide variety of morphology and features, complementing the diversity of beaks associated with different feeding habits. Birds generally lack intrinsic lingual musculature (muscles that enable control of the tongue itself), parrots being an exception. Woodpeckers and hummingbirds have long tongues that protrude rapidly and are covered with sticky mucus for trapping insects or nectar, although some woodpeckers, such as the pileated woodpecker, have a shorter tongue with a spear-like tip and bearing backward-facing barbs. Certain parrots have a keratinized brushlike tip for absorbing nectar by capillary action; birds that feed on fish have spinelike papillae that help catch, position, and consume prey; birds of prey have a keratinized epithelium for scraping the meat of their prey; and the tongue of ducks and filter-feeding birds bears barbed or hairlike edges that let

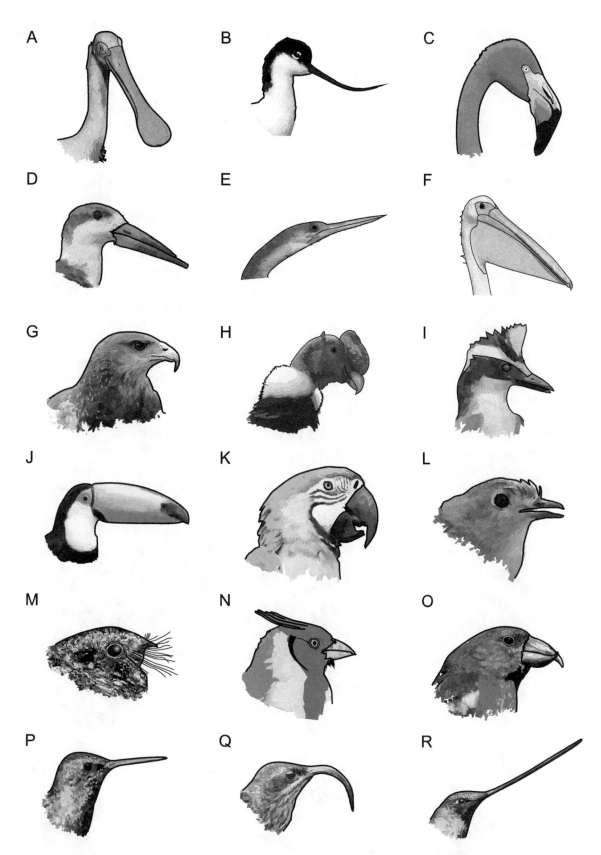

8.4. Diversity of bird beaks. *A*, roseate spoonbill (*Platalea ajaja*). *B*, pied avocet (*Recurvirostra avosetta*). *C*, American flamingo (*Phoenicopterus ruber*). *D*, black skimmer (*Rynchops niger*). *E*, anhinga or darter (*Anhinga anhinga*). *F*, great white pelican (*Pelecanus onocrotalus*). *G*, black-chested buzzard-eagle (*Geranoaetus melanoleucus*). *H*, Andean condor (*Vultur gryphus*). *I*, great kiskadee (*Pitangus sulphuratus*). *J*, toucan (*Ramphastos toco*). *K*, blue-and-yellow macaw (*Ara ararauna*). *L*, European robin (*Erithacus rubecula*). *M*, band-winged nightjar (*Systellura longirostris*). *N*, red-crested cardinal (*Paroaria coronata*). *O*, Hispaniolan crossbill (*Loxia megaplaga*). *P*, glittering-bellied emerald hummingbird (*Chlorostilbon aureoventris*). *Q*, white-tipped sicklebill (*Eutoxeres aquila*). *R*, sword-billed hummingbird (*Ensifera ensifera*).

water pass, thus straining out food particles. In mammals, the tongue participates not only in food intake, but also in food manipulation within the oral cavity during mastication and deglutition, as well as in sound production. In many ungulates, the tongue, in concert with the incisors, is used for grasping and pulling out shoots or bunches of grass. Giraffes (*Giraffa giraffa*) use their tongue for taking leaves from trees, and the giant anteater (*Myrmecophaga tridactyla*) and other ant- and termite-feeding specialists have a long tongue covered with viscous mucus to which their prey adhere. The masticatory musculature is reduced in anteaters, but the hyoid apparatus and its musculature are prominently developed, as are the salivary glands (e.g., Naples 1999; Ferreira-Cardoso et al. 2020).

In reptiles generally, teeth or a cornified beak fulfill the task of seizing food items. In birds the beak exhibits different morphologies that are correlated with different feeding habits, and displays significant morphological plasticity (fig. 8.4). For example, some birds search through mud and water and have a long, flattened, and spatulate beak that sweeps through water (spoonbill; fig. 8.4*A*) or a long, slender beak (avocets; fig. 8.4*B*) that can probe through mud and loose sediment. Flamingos are filter feeders that have a peculiarly shaped beak used to feed with the head upside down (fig. 8.4*C*). Fish-eating birds display a variety of beak shapes: some have a long beak that facilitates catching their prey (skimmers; fig. 8.4*D*), others have a slender, sharply pointed beak for "spearing" fish (anhinga; fig. 8.4*E*), and pelicans scoop up fish with an expandable, pouched beak (fig. 8.4*F*). Meat-eating birds generally have a hooked beak suited to tearing meat from the bones of prey (eagles, vultures; fig. 8.4*G*, 8.4*H*). Omnivores may have straight and very stout beaks (great kiskadee; fig. 8.4*I*). Birds that primarily eat large fruits have a large, long, and stout beak, such as toucans (fig. 8.4*J*). Parrots, which feed on a variety of hard or soft items, have a strong, curved beak that produces high bite forces (fig. 8.4*K*). Insect-eating birds usually have a small, sharp beak to snare insects swiftly and deftly (robins, nightjars; fig. 8.4*L*, 8.4*M*). Those that crack nuts and seeds have strong, thick, and cone-shaped beaks (cardinals and finches; fig. 8.4*N*, 8.4*O*). Hummingbirds reach into flowers to sip nectar with a long, straw-like beak (fig. 8.4*P*, 8.4*Q*, 8.4*R*). Although there is a relationship between feeding habits and beak form (see Olsen 2017), an analysis by Navalon et al. (2019) indicated that the considerable variation in beak morphology is largely unexplained, which suggests that nonfeeding functions of the beak influence its evolution (van Wassenbergh and Baeckens 2019).

Many mammals use their anterior teeth for taking food. In addition, the mouth and oral cavity are limited by lips and cheeks that form the muzzle. Powered by facial

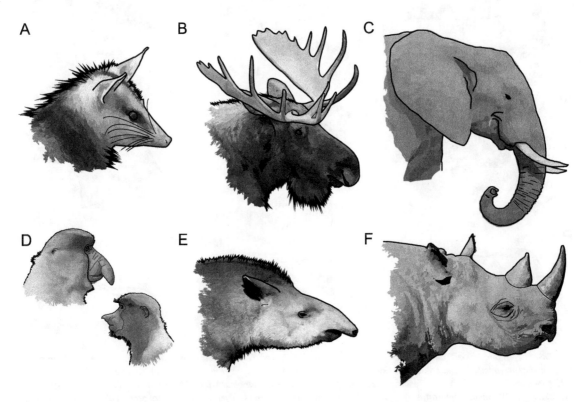

8.5. Diversity in muzzle morphology among mammals. *A*, white-eared opossum (*Didelphis albiventris*). *B*, moose (*Alces alces*). *C*, African elephant (*Loxodonta africana*). *D*, proboscis monkey (male and female; *Nasalis larvatus*). *E*, South American tapir (*Tapirus terrestris*). *F*, black rhinoceros (*Diceros bicornis*).

musculature, muzzles are essential for suckling in infants and in food selection and intake in many adults (fig. 8.5). An extreme example is the elephant's trunk. Another, more subtle, example is furnished by the black rhinoceros, *Diceros bicornis*. Its thick, conical upper lip forms a prehensile structure used for selecting the plants or plant parts (e.g., leaves, shoots) on which it feeds. The nasal cartilages and muzzle musculature have a very close relationship with the skull bones. Facial bones around the nasal cartilages bear discernible entheses (marks, grooves, and contours) that allow inferring, to some extent, the form and size of the nasal cartilages and the development of many of the muzzle muscles that attach on them. Overall, apart from the anterior teeth, the bony morphology of the anterior region of the face and of the hyoid apparatus provide useful information for reconstructing the soft tissues that are relevant for food selection and intake in tetrapods.

Applications in paleobiology

The masticatory apparatus of Pleistocene giant ground sloths (*Megatherium, Scelidotherium, Mylodon, Glossotherium*, and *Lestodon*) has been studied from morphofunctional and biomechanical perspectives (Bargo 2001; Bargo and Vizcaíno 2008). To infer their possible food habits and define their niches in a paleoecological context, Bargo et al. (2006) considered the form of their muzzle, which is quite relevant in these mammals given the varied morphology and edentulous condition of the muzzle, to understand the different styles of food intake. For this purpose, the authors reconstructed the nasal cartilages and facial musculature based on the available information from closely related extant taxa (sloths and anteaters), together with that on groups of artiodactyl and perissodactyl herbivores as functional analogs (fig. 8.6). In the

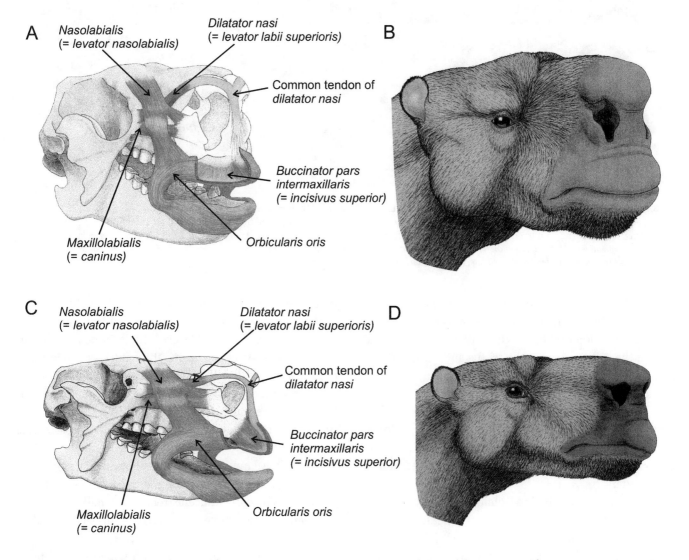

8.6. Reconstruction of nasal cartilages and facial musculature in two Pleistocene ground sloths and their respective life reconstructions by N. Toledo. *A* and *B*, *Glossotherium robustum*, a wide-muzzled bulk feeder. *C* and *D*, *Scelidotherium leptocephalum*, a narrow-muzzled selective feeder (modified from Bargo et al. 2006; see also fig. 8.7).

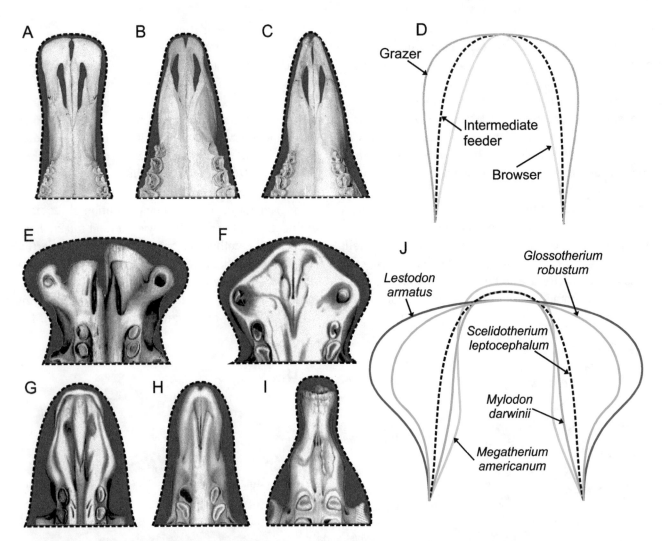

8.7. Feeding habits inferred from the contour of the muzzle. Extant ungulates: *A*, common wildebeest, *Connochaetes taurinus*, grazer; *B*, impala, *Aepyceros melampus*, mixed feeder; *C*, gerenuk, *Litocranius walleri*, browser (modified from Solounias and Moelleken 1993). *D*, contours of *A–C* carried out by digitization of premaxillary form. Pleistocene ground sloths: *E*, *Lestodon armatus*; *F*, *Glossotherium robustum*; *G*, *Mylodon darwinii*; *H*, *Scelidotherium leptocephalum*; *I*, *Megatherium americanum* (modified from Bargo et al. 2006). *J*, muzzle contours of ground sloths: *blue*, bulk feeder; *dashed line*, mixed feeder; *green*, selective feeder.

case of *Megatherium*, some specimens preserved a calcified nasal septum, which contributes to the reliability of the reconstruction of the muzzle. With these reconstructions of the soft tissues as a basis, the authors conducted a morphofunctional analysis of the muzzle. By means of an index that reflects the relative muzzle width (RMW) and the analysis of its contour (premaxillae + maxillae), they classified the sloths in two functional groups: forms with narrow muzzles, which selected their food (selective feeders), and forms with wide muzzles, which fed in bulk (bulk feeders; fig. 8.7).

With these data and those obtained from the reconstruction of the facial musculature (particularly, the lip retractor musculature), Bargo et al. (2006) proposed that narrow-muzzled sloths (*Megatherium, Scelidotherium*, and *Mylodon*) had a thick, conical prehensile lip that helped

them select food (plant parts), whereas wide-muzzled sloths (*Glossotherium* and *Lestodon*) had a quadrangular and nonprehensile lip and were bulk feeders, probably on grass (fig. 8.6). As body size and the muzzle characteristics of extant sloths do not cover the diversity of Pleistocene sloths, their interpretation relied on extant functional analogs, the wide-muzzled white rhinoceros *Ceratotherium simum* and the narrow-muzzled black rhinoceros *Diceros bicornis*.

The relative muzzle width also allowed distinguishing two main feeding forms among glyptodonts (Vizcaíno et al. 2011b). The early Miocene, relatively small (body mass between 80 and 100 kg), and narrow-muzzled propalaehoplophorids would have fed selectively, whereas large (body mass more than 1 tonne) and wide-muzzled post-Miocene forms would have been bulk feeders. With its

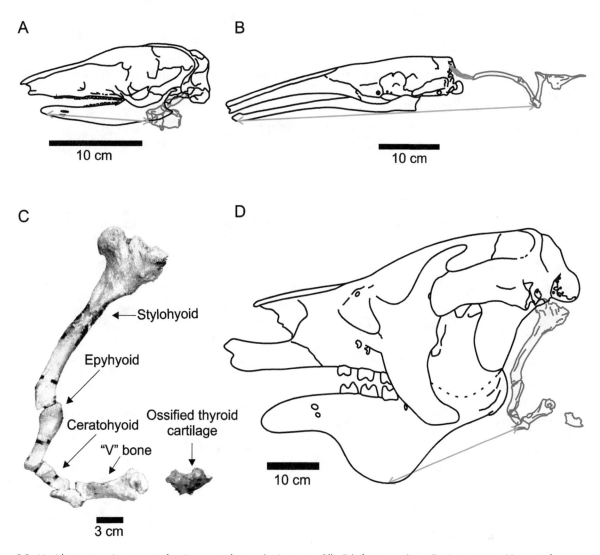

A

B

C

- Stylohyoid
- Epyhyoid
- Ceratohyoid
- "V" bone
- Ossified thyroid cartilage

3 cm

D

10 cm

8.8. Hyoid apparatus in extant and extinct xenarthrans. *A,* giant armadillo *Priodontes maximus. B,* giant anteater *Myrmecophaga tridactyla. C,* hyoid elements of *Megatherium americanum* (MNHN PAM 297; Muséum national d'histoire naturelle, Paris) in lateral view. *D,* skull of *M. americanum* with the hyoid apparatus in position (modified from Pérez et al. 2010). The *blue arrow* indicates the direction of contraction of the geniohyoid muscle, which arises from the inferior mentonian spine of the mandible, near the symphysis, and inserts on the base of the hyoid apparatus (basihyoid). Its function is to pull the tongue and hyoid arch forward.

narrow muzzle, however, the Pleistocene *Glyptodon* appears to have been an exception, implying a reversion to selective feeding.

Pérez et al. (2010) carried out an analysis of functional morphology of the hyoid apparatus of several fossil Xenarthra, including sloths exhibiting several morphotypes and body sizes, armadillos, and glyptodonts, for which the apparatus is nearly completely preserved. They performed detailed descriptions and muscular reconstructions based on the related extant forms and on other mammals (fig. 8.8). Action lines, orientations, and relative lengths of some of the main muscles linked to tongue mobility were inferred. Pérez et al. (2010) concluded that in some taxa, the anteroposterior mobility of the hyoid apparatus

was restricted, precluding considerable protrusion of the tongue. For instance, they proposed that those giant sloths (e.g., *Glossotherium*) with a short preorbital region, wide muzzle, and short lips had a considerably mobile tongue, whereas those (e.g., *Scelidotherium* and *Megatherium*; fig. 8.8) with a longer preorbital region, narrow muzzle, and longer, conical, and prehensile lips had less capacity for tongue protrusion. This analysis casts doubt on the classical reconstructions of *Megatherium* represented as feeding by means of a long, protrusible tongue. As additional evidence, Pérez et al. (2010) described a specimen of this taxon with a hyoid apparatus that had broken and healed during the animal's life, which suggests that tongue mobility was not an essential factor for feeding.

Predeglutition Processing of Food

Vertebrates that do not swallow their food items whole perform predeglutition processing of food in which the items are reduced to smaller sizes to provide increased surface area for chemical digestion.

Among fish, durophagous chondrichthyans (those feeding mostly on invertebrates with shells) have flat teeth, and the mechanism of crushing prey differs from that for biting and transport by suction (Wilga and Motta 2000). This crushing mechanism is carried out by lengthening the activity duration of the mandibular adductor muscle and modifying the bite kinematics by the addition of the second phase of mandibular closing. The diversity in chondrichthyans' dental morphology has been used for predicting their feeding style. Long and thin (i.e., piercing) teeth would allow a firm grasp on the prey, sharp and wider (i.e., slicing) teeth would permit cutting and

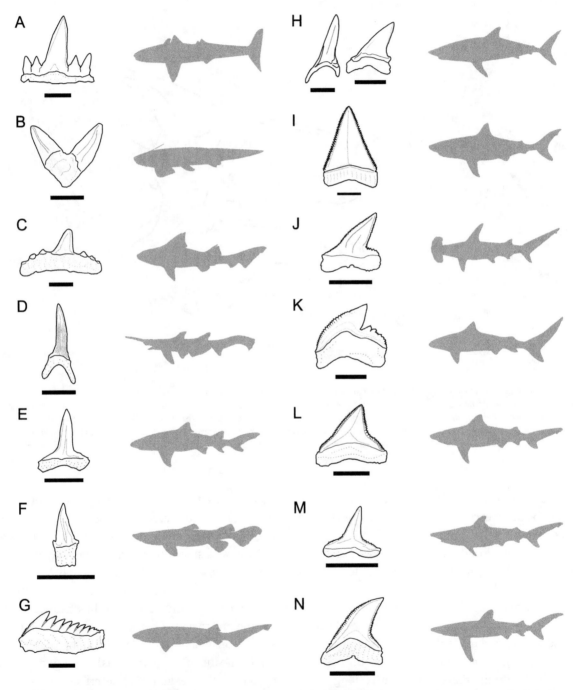

8.9. Dental diversity in extinct (†) and extant selachians. A, †*Cladodus occidentalis*. B, †*Xenacanthus compressus*. C, †*Hybodus sp.* D, †*Scapanorhynchus lewisi*. E, *Negaprion brevirostris*. F, *Scymnodon ringens*. G, *Hexanchus griseus*. H, *Isurus oxyrinchus* (*left*, anterior tooth; *right*, lateral tooth). I, *Carcharodon carcharias*. J, *Sphyrna mokarran*. K, *Galeocerdo cuvier*. L, *Carcharhinus leucas*. M, *Carcharhinus limbatus*. N, *Prionace glauca* (modified from Whitenack and Motta 2010). Scale bar = 1 cm; body silhouettes are not to scale.

slicing tissues by lateral shaking of prey, and those bearing midtooth apicobasal grooves may serve to prevent the tooth from becoming anchored to the prey (see Hildebrand and Goslow 2001; Whitenack and Motta 2010). However, Whitenack and Motta (2010) cautioned that these attributed functions are rather qualitative and have not been demonstrated biomechanically. Although these latter authors identified some correlation between teeth with a fairly sharp morphology (fig. 8.9) and their function and efficiency, they concluded that their results provide limited support for using elasmobranch tooth morphology as a reliable indicator of biological role.

In birds, cranial kinesis may have enabled a general improvement in food manipulation by the jaws; for example, in those, such as parrots, specialized for opening seeds and extracting their content. However, many birds swallow the food whole (or nearly so), and crushing mechanical digestion is performed by the gizzard. These birds often ingest stones, termed *gastroliths*, which are housed in the gizzard and contribute to the reduction of food; their action may be thought of as analogous to millstones. The recovery of similar stones among the remains of extinct reptiles such as dinosaurs and plesiosaurs allows proposing a similar mechanism for these vertebrates.

Mastication is characteristic of mammals and has had a profound influence on skull design, with the great variety in mammalian skull morphology reflecting different trophic specializations. Mastication is characterized by the need for considerable precision in occlusion between the upper and lower teeth, which allows increased efficacy in the reduction of food items. Multiple tooth replacement (polyphyodonty and oligophyodonty) is the general condition in other vertebrates, but in mammals is reduced to two (diphyodonty and hemiphyodonty) due to requirements of food processing. Precise occlusion is not feasible in growing jaws, and therefore eruption of the molars is not completed until the jaws reach their definitive (i.e., adult) size. The solution for feeding in this situation of lacking cheek teeth appears to be correlated with the evolution of lactation. Other features associated with mastication are the evolution of a secondary palate that helps reinforce the skull, prominently developed adductor musculature, and considerable diversity in jaw morphology.

The process of mastication is conditioned by the physical properties of food (Lucas 2004), because the effectiveness of the forces acting on food when the teeth come into contact with it depends on the characteristics of the food (fig. 8.10). These include its external physical attributes

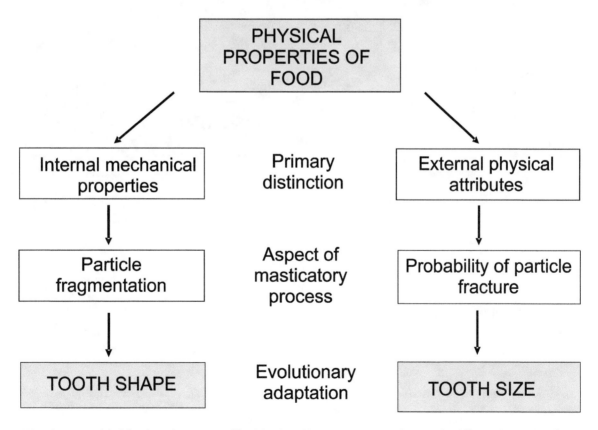

8.10. A binary model of the physical properties of food developed by Lucas (2004), emphasizing the difference between surface attributes (*right* column) and properties that act to hinder the generation of increased surface area by reduction of food into smaller particles (*left* column).

(particle contour and size, total volume, and properties of the particle surface including roughness, softness, and adherence) and internal mechanical properties (toughness or resistance of food to being reduced, which exposes increased surface area that can be subjected to chemical digestion), which include Young's modulus, strength, and toughness. The rate at which food particles are reduced in size depends on several variables, both anatomical and physiological. Conceptually, however, this rate can be considered as the result of two events: the likelihood that a particle comes into contact with a tooth and the fragmentation produced when a particle is contacted by a tooth. The probabilities of contacting and fragmenting a food particle with a tooth are greater if the tooth is larger (tooth-size increase). By contrast, the strength effect that the tooth exercises on that particle depends on the configuration of the occlusal surface of the tooth (tooth shape).

According to the internal mechanical properties of food, the mechanical principles of tooth design involved in mammal mastication can be summarized in three basic patterns (fig. 8.11):

1. A system of blades for cutting tough, nonfibrous (i.e., soft) but chewy, food (i.e., muscle and skin), as in the carnassial tooth of carnivores.
2. A series of low and sharp ridges (enamel crests and dentine grooves) that act as a grinding machine for tough, fibrous food (i.e., grass), as occurs in rodent and ungulate molars.
3. A mortar-and-pestle arrangement suitable for crushing hard and brittle (e.g., nuts) or turgid (e.g., fruit pulp) food, as occurs in the bunodont molars of an omnivore.

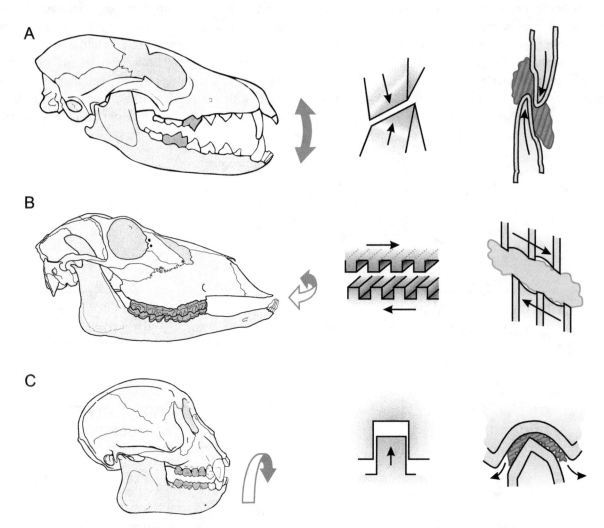

8.11. Chewing in mammals; mechanical principles of tooth design. *A*, skull of a carnivore (gray pampean fox, *Lycalopex gimnocercus*) showing the position of carnassial teeth (*shaded in gray*) that act as scissorlike cutting blades by means of orthal or vertical movements. *B*, skull of a herbivorous ungulate (pampas deer, *Ozotoceros bezoarticus*) showing the position of grinding teeth (*shaded*); the corrugated surface (with cutting ridges) grinds fibrous food by means of lateral movements. *C*, skull of an omnivorous primate (capuchin monkey, *Cebus* sp.) showing the position of the compression teeth (*shaded*) that reduce hard foods, acting like a mortar and pestle. Modified from Kardong (2018).

Masticatory Movements

Mastication in mammals (except in some highly specialized forms) is characterized by the following features:

1. Food reduction on one side of the jaws at a time (working side; the opposite side is the balancing side).
2. Presence of transverse (= lateromedial) movements of the mandible against the upper jaws (minimal in carnivores and maximal in some herbivores), which typically involve translational and lateral and axial (yaw and roll, respectively) rotational components (see Maynard Smith and Savage 1959; Grossnickle 2017, 2020b; Bhullar et al. 2019).
3. Upper and lower molars with complex occlusal surfaces and cusps (tribosphenic molar; see appendix) that occlude precisely.

In mammals that have upper and lower jaws of different transverse widths, especially in the region of the molars (anisognathus), mastication is unilateral. In some groups of mammals (rodents and humans) in which the transverse diameter of the upper and lower jaws is the same (isognathus), bilateral mastication takes place; that is, food reduction can occur on both sides at the same time.

Mastication requires proper positioning of the lower teeth with respect to the upper ones, though this position may vary as food is incorporated and processed. Several mammals have a mixed mastication (e.g., primates, including humans, and pigs), because the craniomandibular articulation is fairly mobile and allows the mandible to undergo a combination of vertical, lateral, and anteroposterior movements. Many rodents exhibit mainly propalinal movements: the glenoid fossa, composed of an anteroposteriorly oriented groove, allows the mandibles to move forward and backward; the teeth, with their lophs oriented transversely (thus perpendicular) to the movement of the jaws, act to grind plant matter. Apparently, no vertebrate exhibits truly propalinal (both fore and aft) mastication; technically, rodents and elephants chew with a proal (anterior) power stroke (see below), and some extinct lineages (e.g., dicynodonts, multituberculates) had a palinal (posterior) power stroke (von Koenigswald et al. 2013; Grossnickle et al. 2022). In rodents, the incisors and molars do not occlude at the same time. Forward displacement of the mandible allows the upper and lower incisors to come into contact for gnawing, whereas the backward position of the jaws allows the molars to engage in grinding. The anterior

gnawing system is separated from the posterior grinding system by an edentulous region, termed a *diastema*.

Other herbivores exhibit mainly transverse mandibular movements, including yaw and roll (as mentioned above), with a nearly flat glenoid fossa and varied dentitions adapted for ingesting and grinding fibrous and tough food (e.g., grasses, leaves, shoots, roots). Most herbivores crush and grind food with their cheek teeth, which includes premolar and molar teeth, whereas their anterior teeth, the incisors, are specialized for cutting, gnawing, or pulling food items. Generally, a diastema, as noted for rodents, is present, separating the anterior teeth from the cheek teeth. The premolars and molars of herbivores, designed for grinding or crushing, tend to resemble each other (i.e., the premolars in particular tend to become molarized) and are often hypsodont (see appendix), which, given that teeth undergo wear, extends their functional lifetime. Their occlusal surface is wide and relatively flat but not smooth, a feature that enhances grinding efficiency. The enamel develops folds between the dentine, creating a series of crests and valleys. The folding pattern is specific for each species and varies broadly among them; indeed, it may vary throughout the lifetime of the individual as a tooth experiences wear. The valleys can be filled with cementum (e.g., horses, elephants) or composed only of dentine (e.g., deer, guanacos). As the enamel is harder than dentine and cementum, it is more resistant to wear and thus stands in relief above these other substances as the tooth wears, creating a battery of vertical sheets effective in grinding. Other herbivores, particularly sloths among xenarthrans, have a homodont dentition, with teeth that are reduced in number, lack enamel, and exhibit continuous growth (euhypsodont). In xenarthrans, dentine is differentiated as durodentine (the functional analog of enamel, forming crests) and vasodentine (a softer form that functions as the valleys).

The masticatory apparatus of specialized carnivorous mammals has features that confer movements that are nearly diametrically opposed to those just described for herbivores. The glenoid fossa is transversely concave, in some cases bordered by anterior and posterior ridges, so that movement of the mandibular articular process is restricted to rotation in the transverse axis. These animals have orthal mastication—that is, much limited to vertical movements (thus predominantly pitch)—well suited to cutting/tearing meat and crushing bones. For the former function, they make use of secodont (large, bladelike) teeth, such as the carnassial pair formed by the fourth upper premolar and the first lower molar in Canidae and Felidae, with pointed cusps and sharp, well-developed ridges for cutting (see appendix). Marine mammals that feed on fish and squid have conical teeth. Carnivora also includes herbivorous (e.g., panda, *Ailuropoda melanoleuca*) and omnivorous (e.g., raccoons, Procyonidae, and bears, Ursidae) forms, which have molarized premolars and well-developed molars.

In terrestrial vertebrates generally, movements of the mandible comprise fairly complex masticatory cycles. For descriptive purposes, the cycle in mammals is divided into opening and closing phases, and each of these can, in turn, be subdivided based on differences in mandible velocity. A generalized masticatory cycle consists of four phases: (1) slow opening, (2) fast opening, (3) closing stroke, and (4) power stroke. In most mammals, in frontal view, the mandible describes an orbit in which the direction of the initial opening is vertical, but once the closing phase begins, the mandible moves laterally toward one of the sides (working side) so that the food particles can be processed between the teeth (fig. 8.12). The general scheme of all cycles that have been analyzed is comparable, but the duration and amplitude are highly variable among mammals, as well as in any particular mammal depending on the food and other factors (e.g., think of crushing ice and chewing gum). Then, whether the food is cut, ground, or some combination of the two depends on the cheek teeth.

In mammals, the slow opening phase occurs mainly by contraction of the digastric muscle and, during this phase, the hyoid apparatus is moved forward. In the following phase (fast opening), several hyoid muscles participate in actively drawing the mandible downward. In the next phase (closing stroke), the mandible rises quickly by contraction of the masseter and pterygoid muscles. The final phase (power stroke) is characterized by strong contractions of the masseter, pterygoid, and temporal muscles and posterior movement of the hyoid. During this phase involving greatest strength, food is reduced to small particles, or crushed.

Two types of cycles have been described in a masticatory sequence in extant mammals such as opossums, shrews, squirrels, some monkeys, cats, pigs, and humans:

1. *Puncture-crushing cycle*: occurs during the first phase of mastication; the teeth do not occlude while the food is being broken and reduced (tooth-food-tooth contact); there is no cutting or crushing. The cusp tips hook food and puncture it to break and reduce it, without the cutting crests coming into contact. This results mainly in wear of the cusp tips. The prevailing direction of the movement is vertical, with transverse movement being limited.

2. *Chewing cycle s.s.*: occurs when food has the proper consistency and particle size. Food is reduced by means of a rhythmical series of chewing cycles in which the teeth progressively approach each other until reaching full occlusion. Wear is also produced by direct tooth-tooth contact.

There are two examples of variations of this pattern that have been widely studied. The pig has a puncture-crushing cycle with two stages, due to the hard nature of its food. Two chewing cycles follow, during which teeth rarely reach complete occlusion. Rats and rabbits do not perform the food puncture-crushing cycle, probably due to the efficiency of the incisors for gnawing (cutting) food until it is the proper size for chewing.

In paleobiological studies, the most common approach to the study of movements and chewing cycles is by means of the analysis of wear facets and striations (Greaves 1973; Rensberger 1973; Costa and Greaves 1981). There are two types of wear areas: those produced by tooth-tooth contact, resulting in small scratches or striations that indicate the orientation, but not the direction, of movement; and those produced only by food, thus involving tooth-food-tooth contact, the evidence for which is the lack of striations.

The direction of mandibular movement was clarified by Greaves (1973) for the selenodont teeth of cervids and bovids, in which the direction of mandibular movement is

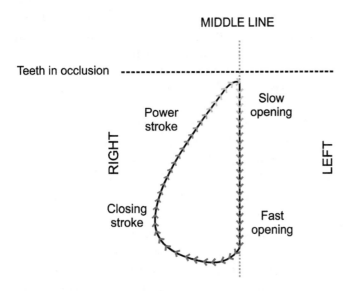

8.12. Generalized masticatory cycle in mammals. The cycle is subdivided into four phases: slow opening; fast opening; fast closing or closing stroke, during which vertical movement aligns the upper and lower molars for the following phase; and slow closing or power stroke, marked by the beginning of tooth-food-tooth contact. It is this last phase that results in food reduction. Based on Hiiemae (1978); Hiiemae and Crompton (1985).

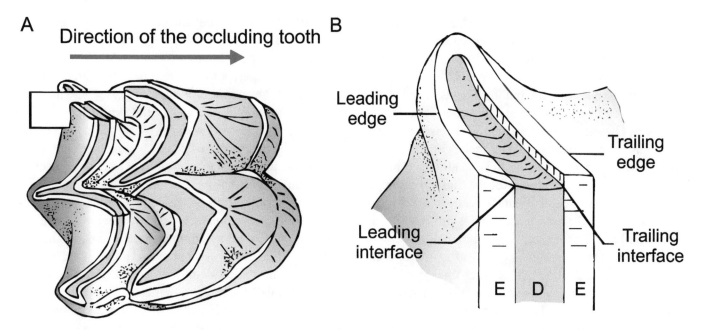

8.13. *A*, occlusal view of an upper molar of a selendont artiodactyl. The molar remains relatively stationary with respect to the lower molar with which it occludes. The *red arrow* indicates the direction (from lateral to medial) of mandibular (and thus lower molar) movement across the occlusal surface of the upper molar. *B*, transverse section through the ectoloph of the upper molar (see *A* for position of the section) showing detail of a wear facet and the interfaces between enamel and dentine. *E* = enamel; *D* = dentine. Modified from Greaves (1973).

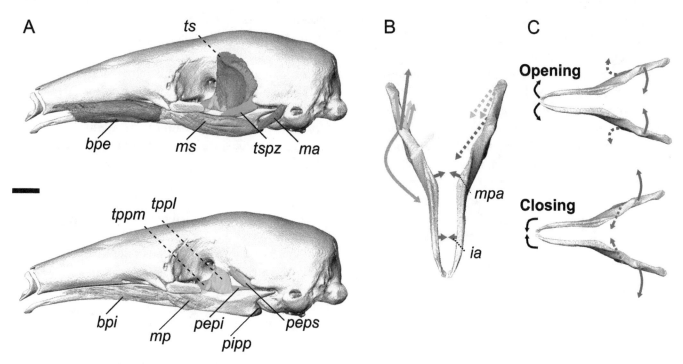

8.14. *A*, facial and masticatory musculature of *Tamandua tetradactyla* in lateral views; *lower* image illustrates the deeper musculature covered by the superficial musculature illustrated in the *upper* image. *bpe, m. bucinatorius pars externa; bpi, m. bucinatorius pars interna; ms, m. massetericus superficialis; mp, m. massetericus profundus; ts, m. temporalis superficialis; tspz, m. temporalis superficialis pars zygomatica; tppm, m. temporalis profundus pars medialis; tppl, m. temporalis profundus pars lateralis; ma, m. mandibulo auriculis; pepi, m. pterigoideus externus pars inferior; peps, m. pterigoideus externus pars superior; pipp, m. pterigoideus internus pars posterior.* Scale bar = 10 mm. *B*, dorsal view of the mandible of *T. tetradactyla* with the mediolateral component of the lines of action of the masticatory muscles (*color code of the lines* of action corresponds to that used for the muscles). *mpa, m. mylohyoideus pars anterior; ia, m. intermandibularis anterior.* *C*, mandibular dynamics. Schematic illustration of the mediolateral rotational movement of the dentaries during mandibular opening and closing. *Dotted lines* represent the lines of action of muscles completely or partially hidden in lateral view. Modified from Ferreira-Cardoso et al. (2020).

from outside to inside (lateromedially) and perpendicular to the enamel ridges. During this movement, wear facets are formed in which the leading edge of the occluding upper and lower teeth (that is, when they initially meet in occlusion) is sharp and the trailing edge is rounded (fig. 8.13). Even clearer evidence for the direction of movement is provided by wear facets at the interfaces between enamel and dentine of ectolophs of upper molars and in lingual lophids of lower molars. As the molars move in occlusion, even wear is produced between the dentine and enamel at the leading enamel-dentine interface, due to the fact that dentine is softer than the enamel but is protected by the enamel, which precedes it during mandibular movement. By contrast, at the trailing enamel-dentine interface, the dentine is worn below the level of the enamel, resulting in the formation of a discontinuity or step between dentine and enamel. This is probably due to the food accumulation in this interface; that is, food particles are retained by the tougher tissue forming a discontinuity. A similar pattern is present between the durodentine and vasodentine of xenarthrans. Consequently, by observing where these even and discontinuous interfaces are located, movement of the upper and lower jaws relative to each other during the power stroke can be discerned.

As mentioned in chapter 7, the myrmecophagous anteaters and pangolins have lost the ability to chew, and this is reflected by several aspects of the masticatory apparatus of these mammals, such as reduced musculature (fig. 8.14A). For example, anteaters do not have a digastric muscle, the main mandibular abductor in other mammals, implying loss of the typical mandibular adduction/abduction cycle and development of a new feeding strategy based on protrusion and retraction movements of an elongated and sticky tongue (Ferreira-Cardoso et al. 2020 and references therein). Contractions of the masseter complex and the superficial temporal, in association with a loosely articulated mandibular symphysis, permit rolling of the dentaries, which results in opening of the mouth and increased volume of the oral cavity during tongue protrusion. The internal pterygoid muscle reverses the roll of the dentaries to close the mouth (fig. 8.14B, 8.14C).

Mandibular Musculature

The mandibular musculature of gnathostomes can be functionally divided into two groups: abductor or depressor musculature (which opens the jaws) and adductor musculature (which closes the jaws). Whereas the musculature that opens the jaws differs among groups (hypobranchial group in fish, mandibular depressor in tetrapods in general, and digastric in mammals), the adductor musculature is consistent and homologous in all gnathostomes. In fish, it consists of a single muscular component that becomes more complex in tetrapods, differentiating into several muscles that have received different names according to taxonomic group and authors. These muscles are often anchored on bones of the skull (e.g., maxilla, jugal, pterygoid) and lower jaw (e.g., coronoid, gonial, dentary) by means of visible and well-defined entheses. Mammals have three main groups of adductor musculature: temporal muscle (origin: temporal fossa and sagittal crest; insertion: coronoid process of the dentary), masseter muscle (origin: zygomatic arch, and sometimes from the orbit or maxilla; insertion: masseteric fossa and angular process of the dentary), and pterygoid muscle (origin: pterygoid, on the underside of the skull behind the palate; insertion: medial part of the angular process). These muscular masses are divided into two functional components: one formed by the temporal and deep masseter muscles, which are responsible for the vertical mandibular movements, and the other formed by the superficial masseter and pterygoid muscles, which are responsible for a combination of elevation, lateral, and forward and backward movements. As noted above, the digastric muscle is the main abductor muscle (mandibular depressor).

The reconstruction of the mandibular musculature in extinct organisms is usually relatively straightforward and, in many cases, is a necessary step for carrying out functional studies (e.g., Vizcaíno et al. 1998; De Iuliis et al. 2000; Bargo 2001; Vizcaíno and De Iuliis 2003; Bargo and Vizcaíno 2008; Desojo and Vizcaíno 2009; Sakamoto 2010; Cassini and Vizcaíno 2012; Herrera 2012; Nabavizadeh 2020).

As previously mentioned in this book, the reconstruction of soft tissue is carried out by analyzing specimens qualitatively and relying on anatomical knowledge of extant forms as thoroughly as possible, mainly by means of dissections. This allows the reconstruction of adductor muscles (closing) and abductor or depressor muscles (opening) of the mandible, the contraction and relaxation of which allow mandibular movement.

Biomechanics of Mandibular Movement

From a biomechanical perspective, the system formed by the mandible and its associated musculature together with the pivot at the craniomandibular articulation is considered a mechanism of levers that transmits forces from one place to another in the system (chap. 2). The input force (F_i) is generated and applied to the lever by the masticatory musculature, whereas the output force (F_o) is applied by the teeth on food (fig. 8.15).

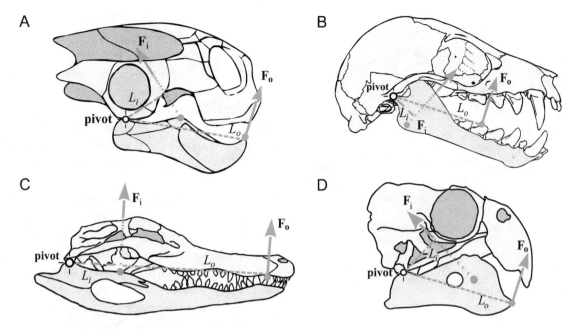

8.15. Third-class lever system exemplified in tetrapod mandibles, with the moment arm of components of the adductor musculature indicated. *A*, African mud turtle (*Pelusios castaneus*), external mandibular adductor (modified from Lemell et al. 2000). *B*, flying fox (*Pteropus lylei*), masseter muscle (adapted from Giannini et al. 2006). *C*, alligator (*Alligator* sp.), pseudotemporal (internal mandibular adductor; modified from Desojo and Vizcaíno 2009). *D*, Pliocene parrot (*Nandayus vorohuensis*), superficial external mandibular adductor (following Carril et al. 2014). Abbreviations: F_i, input force; F_o, output force; L_i, input lever arm; L_o, output lever arm.

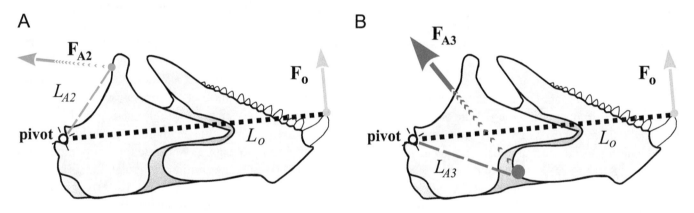

8.16. System of levers of the lower mandible of the tripletail wrasse (*Cheilinus trilobatus*; Perciformes). *A*, lever mechanism of the A2 muscle, which inserts onto the posterior edge of the coronoid process. *B*, lever mechanism of the A3 muscle, which inserts onto the medial surface of the dentary. F_{A2} and F_{A3} input forces of muscles A2 and A3 respectively; F_o, output force; L_{A2} and L_{A3}, input lever arm of muscles; L_o, output lever arm. Modified from Westneat (2003).

The type of lever and the mechanical advantages of the system depend on the group under study. For instance, in bony fish the pivot is at the quadrate/articular articulation, and the jaw-closing muscle, the adductor mandibulae, has three subdivisions (A1, A2, and A3). A1 participates in protruding the upper jaw, whereas the input force for closing the jaw is provided by A2 and A3 (Westneat 2003; fig. 8.16).

In chelonians and lepidosauromorphs, the pivot also occurs at the quadrate/articular joint (as in all nonmammal gnathostomes, except for the cartilaginous fishes, in which these skeletal elements are not recognized as such), and the input force is provided mainly by the mandibular adductor, which produces its effect through the transilien cartilage (an element that works as a trochlea to increase muscular efficiency), though the pterygoid muscle also participates (Herrel et al. 1999; Lemell et al. 2000). The mandibular adductor is subdivided into external, internal, and posterior parts. These, in turn, can be considered individual muscles that act together, as in turtles (Lemell et al. 2000), or each may be subdivided into several bellies, as in squamates (Haas 1973; fig. 8.17); such differences

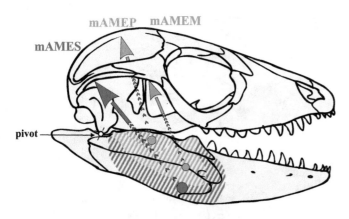

8.17. System of levers of the external mandibular adductor muscle in the giant tegu lizard (*Salvator merianae*). The muscle is divided into three portions: superficial (*m. adductor mandibulae externus superficialis*; mAMES), middle (*m. adductor mandibulae externus medialis*; mAMEM), and deep (*m. adductor mandibulae externus profundus*; mAMEP). Modified from Gröning et al. (2013).

require that biomechanical aspects such as lines of action of the musculature and their moment arms be taken into consideration. In crocodiles, the pterygoid bones that laterally flange the palate can form extra craniomandibular joints, buttressed against the medial side of lower jaws, that function to help resist bending loads induced by the horizontal component of forces produced by the powerful adductor muscles (van Drongelen and Dullermeijer 1982; Walmsley et al. 2013).

In mammals, the pivot occurs at the squamosodentary articulation (given that in mammals the quadrate and articular are incorporated into the middle ear as the incus and malleus, respectively) and the muscles are, as just noted, the temporal, masseter, and pterygoid (Maynard Smith and Savage 1959; Vizcaíno and Bargo 1998; Cassini and Vizcaíno 2012). Although there is considerable structural diversity among mammalian skulls, several commonalities may be identified. In nearly all, for example, the glenoid cavity lies above the level of the upper tooth row; conversely, however, the articular condyle of the dentary may lie at different positions relative to the level of the lower tooth row (Greaves 1980). In the literature, the conformation of this latter feature is often emphasized in explaining the clearly differentiated masticatory models exemplified by mammalian carnivores and herbivores, with the mandibular condyle lying slightly above in the former and well above in the latter (but see Grossnickle 2020a); and these differences have generally been viewed as related to improving the mechanical advantage of the musculature. Greaves (1974) discussed the geometric relationships of the two parts of the joint and the upper and lower tooth rows in these masticatory apparatuses to provide clarification on their functioning.

Greaves (1974) considered a generalized skull, based on that of a raccoon (*Procyon*), in which the lower tooth row is on the same level as the jaw joint (fig. 8.18*A*), and the same skull (i.e., with the joint at the same glenoid cavity height, *n*, above the tooth row) but with the horizontal ramus of the dentary rotated so as to reposition the lower tooth row below the level of the joint (fig. 8.18*B*). In doing so, he demonstrated that repositioning the lower tooth row (by altering the relative positions of the ascending and horizontal rami to each other) with respect to the jaw joint does not alter the lines of action and moment arms of the adductor musculature, thus establishing that differences in the level of the mandibular condyle with respect to the tooth row are not related, in contrast to the common supposition, to improving the leverage (mechanical advantage) of the apparatus.

An important functional outcome of the differences in the two masticatory mechanisms may be appreciated by considering figure 8.18*C* and 8.18*D*, based on an idealized herbivore skull (*Ovis*) but depicting variations on the condition described for carnivorans; that is, with the mandibular condyle level with the lower tooth row ($m = 0$) or slightly above it ($m > 0$) but less than the height of the glenoid cavity above the upper tooth row ($n > m$). In either of these scenarios $n \neq m$, so the distances of the upper and of the lower tooth rows with respect to the position of the craniomandibular joint differ. It is this critical feature of the carnivoran design that allows the teeth of the lower jaws to approach those of the upper jaws in a manner analogous to scissor blades, thus making it effective in slicing food. This is the general condition that holds in the carnivoran lineage, as illustrated in the pampean fox (fig. 8.18*G*), where the craniomandibular joint is a vertical distance above the upper tooth row (*n*) and the lower tooth row (*m*), but a shorter distance from the former than the latter; that is, $m < n$. As noted earlier, the craniomandibular articulation of carnivorans functions as a hinge, an arrangement that restricts anteroposterior and side-to-side movements of the lower jaw.

By contrast, the generalized herbivore condition, represented schematically (based on *Ovis*) in simplified (fig. 8.18*E*) and more realistic (fig. 8.18*F*) depictions, requires that the distances between the craniomandibular joint (i.e., of the glenoid fossa and mandibular condyle) and the upper tooth row and the lower tooth row are equal. Figure 8.18*E* illustrates the condition in which the joint is level with the upper tooth row and the lower tooth row ($n = m = 0$), resulting in simultaneous occlusion of the upper and lower tooth rows along their entire length. This design is akin to wire cutters, in which the cutting edges of the blades, both at the same level as pivot, meet along

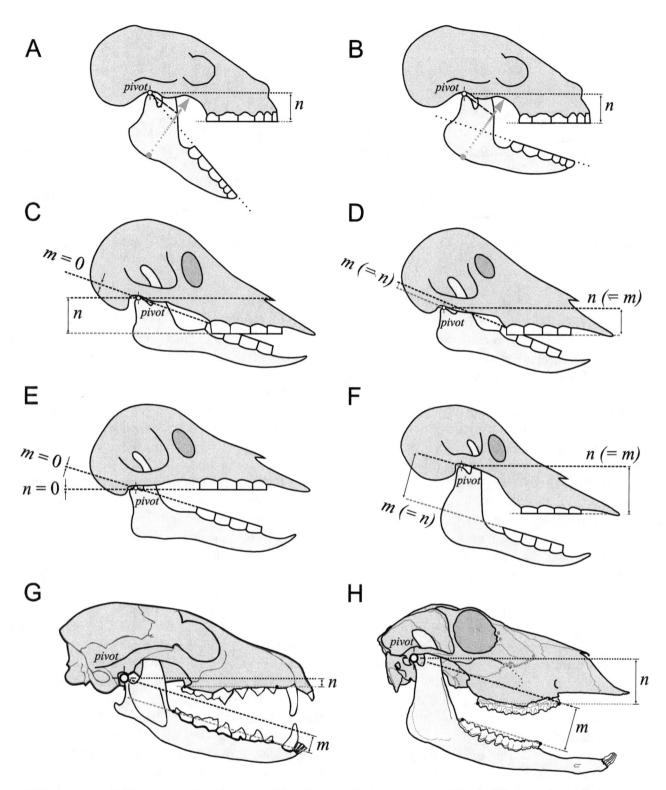

8.18. Functional implications of craniomandibular joint (CMJ) position. *A*, schematic generalized skull with the lower tooth aligned with the CMJ. *B*, the lower tooth row is unaligned with the CMJ; the lever arms of the masticatory musculature (*blue* vectors) are unaffected. *C* and *D*, occlusion of the upper and lower tooth rows acting analogously to scissor blades, as a result of unequal distances between the CMJ and the cutting edges of the upper and of the lower tooth rows. *E* and *F*, simultaneous occlusion of the upper and lower tooth rows when these distances are equal (*A–F* modified from Greaves 1974). *G*, example of a generalized carnivoran in which distances *n* and *m* are different and a scissorlike occlusion is produced between the carnassial complex (fourth upper premolar and first lower molar), and *H*, of a herbivorous ungulate, in which *n* and *m* are equal and occlusion of the upper and lower tooth rows is thus simultaneous. Abbreviations: *m* = vertical distance between the upper tooth row and glenoid cavity; *n* = vertical distance between the lower tooth row and mandibular condyle.

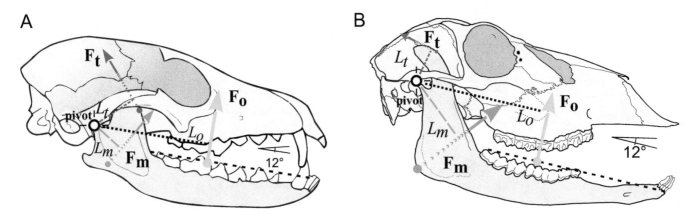

8.19. Mandibular mechanics in mammalian carnivores and herbivorous ungulates. *A*, grey pampean fox (*Lycalopex gymnocercus*). *B*, pampas deer (*Ozotoceros bezoarticus*). Abbreviations: F_m, input force of the masseter; F_t, input force of the temporal; F_o, output force; L_m, input lever arm of masseter; L_t, input lever arm of temporal; L_o, output lever arm. Angle of jaw opening = 12°.

their entire length. In figure 8.18*F*, the vertical distances between the joint and the upper tooth row are greater than zero, but they are equal ($n = m \neq 0$). As in the case depicted in figure 8.18*E*, the upper and lower cheek teeth on one side of the head occlude nearly simultaneously, and biting strength is distributed more evenly along the cheek teeth, as occurs in the pampas deer *Ozotoceros* (fig. 8.18*H*). This design is appropriate for herbivores that process great quantities of food. The mandibular condyle and glenoid fossa are relatively flat, allowing considerable freedom of movement in all directions, which is advantageous for shearing and grinding plant material.

In carnivorans, the temporal muscle is well developed and constitutes more than half of the adductor mass; it arises mainly from the temporal fossa and inserts onto the coronoid process of the dentary, with a long lever or moment arm for the temporal muscle, which results in a greater mechanical advantage than that for the masseter muscle (fig. 8.19*A*). The vertical component of the temporal's line of action provides a strong bite, whereas the horizontal component pulls the mandible backward to maintain the articulation's integrity. This is very important, as it compensates for the great strength produced at the anterior part of the jaws when the canines are used to pull on prey. The masseter and the pterygoid, given that they are smaller, contribute to a lesser degree to bite strength. In felids, the relative strength at the tip of the muzzle is greater than in canids, owing to differences in the lever arms of the masseter and temporal. The former have a design that enhances output force, whereas in the latter the jaws are optimized for a higher closing velocity, the differences reflecting the two classical models for prey capture practiced by these carnivorans (see Christiansen and Adolfssen 2005 and references cited therein).

In contrast to the condition just described for carnivorans, in ungulates the masseter and pterygoid are more developed and form the largest part of the adductor musculature mass. They insert onto a more expanded mandibular angular process, and their longer lever arms confer a greater mechanical advantage over the temporal (fig. 8.19*B*). In addition to vertical components for mandibular adduction (pitch), their contraction produces combinations of lateral translation, roll, and yaw, which together with a higher position of the craniomandibular joint and simultaneous occlusion of the cheek teeth allow the mandible to move from one side to the other for grinding food.

Despite the undeniable didactic value of considering the mandible as a lever system with the pivot at the squamosodentary articulation for a first approach to the mammalian masticatory biomechanics, detailed analyses of masticatory dynamics, including consideration of the structural skull and dental diversity among mammals (see appendix), reveal considerable functional complexity. Although such details are beyond the scope of this book, it is worth pointing out several aspects that may lead to a more comprehensive understanding of mammalian mastication. As already mentioned, there are significant yaw and roll components in chewing cycles in many species (fig. 8.20; some carnivorans being an exception) that imply the existence of vertical and horizontal rotational axes. Also, likely due to yaw, significant anteroposterior translation of the mandible occurs in herbivores (see fig. 6 in Maynard Smith and Savage 1959; Crompton et al. 2006). Further, the condyle's movement describes an arch that often positions the mandible's axis of rotation well below the jaw joint (Iriarte-Diaz et al. 2017).

The considerable diversity in jaw-joint elevation among mammals deserves further consideration. For example,

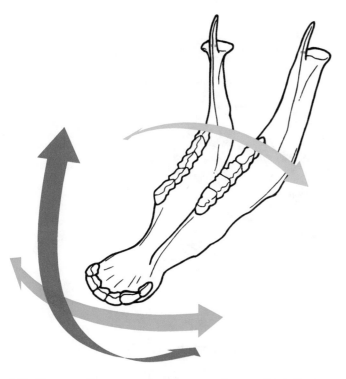

8.20. Diagrammatic representation of a mammalian mandible indicating its main rotational movements: *red*, pitch; *blue*, yaw; *green*, roll.

as the jaw joint typically lies well above the level of the cheek-tooth rows in herbivores such as ungulates and lagomorphs, an elevated jaw joint is generally viewed as the stereotypical morphology allowing simultaneous occlusion of the cheek teeth. However, whereas this pattern is also observed among many herbivorous (and omnivorous) rodents (e.g., beaver, agouti, viscacha rat), other rodents with similar dietary preferences tend to have a jaw joint that is at nearly the same level as the cheek-tooth row (e.g., capybara, some squirrels), and they can chew with multiple cheek teeth occluding simultaneously. Some carnivorans with an herbivorous/omnivorous diet (kinkajou, bears) also have a jaw joint at the same approximate elevation as the lower molar row, and they too can chew with multiple molars occluding simultaneously, whereas the scissorlike analogy for tooth occlusion often applied to the more carnivorous carnivorans may be particularly appropriate only for large carnassial teeth (as in, e.g., felids) rather than for the tooth rows as a whole. The raised jaw joint of herbivores such as ungulates and lagomorphs may be related to additional functional considerations, such as gape (Herring and Herring 1974) and/or increased yaw (Grossnickle 2017, 2020a). Grossnickle (2020a) considered the relative depth of the angular process from the jaw joint as a much stronger dietary correlate than jaw-joint elevation.

MASTICATORY MOVEMENTS

Bargo et al. (2009) conducted a morphofunctional study of the feeding system of the Miocene sloth *Eucholoeops* (Megatherioidea, Megalonychidae) and applied their results to the feeding system of other contemporary megatherioid sloths (e.g., *Hapalops*). The analysis on the tooth wear facets allowed them to infer the occlusal movements and, therefore, to propose the type of food that this sloth was able to process and its probable diet. The authors established a dental nomenclature, represented by cusps (upper teeth) and cuspids (lower teeth), analogous to that used in other mammals and necessary for describing occlusal patterns and masticatory movements (fig. 8.21; see also Pujos et al. 2012). The analysis and mapping of the wear facets allowed reconstructing two different mandibular movements during the closing (or power) stroke. The first, denoted movement A, is equivalent to that previously noted in this chapter for mammals having a tribosphenic molar configuration or a pattern derived from it; it is a mainly orthal (vertical) movement that results in piercing, tearing or shearing, and cutting of food. The second, denoted movement B, is a different and nonrelated movement of the working side of the mandible, characterized by an upward movement in a distal direction and slightly lingual, the main result of which is a cutting action.

The analysis of wear facets, together with the morphology of the craniomandibular articulation, the presence of a fused symphysis and of a well-developed temporal muscle, allowed Bargo et al. (2009) to conclude that the orthal component prevailed during closing, with smooth medial or anteroposterior movements when the teeth are in occlusion. Further, they proposed that *Eucholoeops* and probably most Miocene megatherioid sloths may have consumed mainly leaves, which they processed by cutting and shearing rather than by the grinding action that occurs in ungulates.

BIOMECHANICS

Vizcaíno et al. (1998) proposed a geometrical model that allows calculation of the average lever arms of each modeled muscle. Conceptually, the model consists of calculating the moment arms without the necessity of hypothesizing the direction of the muscle's line of action, which is, indeed, generally ambiguous. To this end, the areas of origin and insertion of the main masticatory muscles (i.e., temporal and masseter) under evaluation should first be discerned as accurately as possible, and the skull and mandible must be articulated with the teeth in occlusion

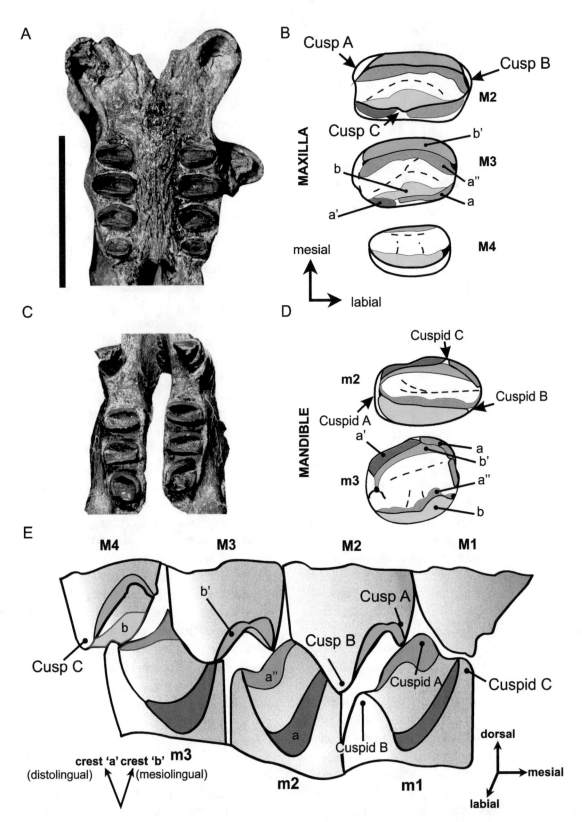

8.21. Analysis of wear facets of the Miocene megalonychid sloth *Eucholoeops ingens* (MPM-PV 3401; Museo Regional Provincial Padre M. J. Molina, Río Gallegos, Argentina). *A*, occlusal view of the palatal region, showing the upper tooth rows. *B*, mapping of wear surfaces (in *different colors*) of right M2–M4 in occlusal view. *C*, occlusal view of the lower tooth rows. *D*, mapping of wear surfaces of right m2 and m3 in occlusal view. *E*, lateral view of right tooth rows in partial occlusion, illustrating how teeth are positioned when occluding. M, upper molariform; m, lower molariform; a, a', a'', b, and b', wear facets. Modified from Bargo et al. (2009).

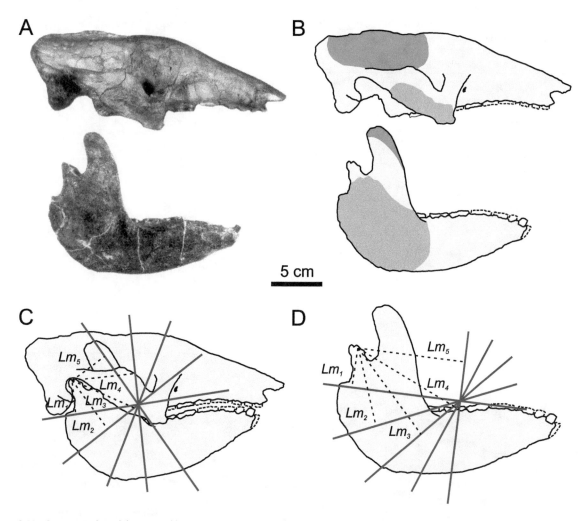

8.22. Geometrical model proposed by Vizcaíno et al. (1998) for calculating moment arms of the masticatory musculature. *A*, skull and mandible of the cingulate pampathere *Vassallia maxima* (FMNH P14424; Field Museum of Natural History, Chicago). *B*, reconstruction of origin and insertion areas of the masseter (*blue*) and temporal (*red*) muscles. *C*, lever arms (Mm$_{1-5}$; *dashed lines*) perpendicular to pivot calculated from the center of the masseter's origin. *D*, as for *C*, but from the most anterior part of the masseter's origin (skull excluded for clarity). Modified from Vizcaíno et al. (1998).

(if it is not possible to do so manually with the specimens, this step may be conducted digitally; see below). Then, a line is drawn from the most anterior part of the muscle's origin to the most posterior part of its insertion, and vice versa. Next, the angle formed by these lines is subdivided into five equidistant angles, and lines, representing hypothetical lines of action, are drawn along these angles; from these lines of action, the lever arms for each muscle are subsequently calculated as the perpendicular line to the pivot (fig. 8.22). These five obtained values are averaged for each muscular unit under study (table 8.1). The method enables direct comparison of values obtained for species ranging from phytophagous to zoophagous habits, and, as noted just above, obviates the need for determining a priori a muscle's line of action. This method has been applied to different groups of extinct Xenarthra (Vizcaíno et al. 1998; De Iuliis et al. 2000; Bargo 2001; Vizcaíno and De Iuliis 2003; Bargo and Vizcaíno 2008).

In a study of the mandibular mechanics of aetosaurs, Desojo and Vizcaíno (2009) broadened the application of this model to archosaurian reptiles (fig. 8.23). The masticatory musculature of archosaurs is generally more complexly subdivided than that of mammals. The muscular reconstruction was further complicated due to the uncertainty resulting from the insertion of several muscles via aponeuroses, rather than more nearly directly on bones. For this reason, the authors had to make several assumptions, based on dissections of extant crocodiles, for modeling the lever (or moment) arms of several of the muscles (see fig. 8.23 for details). The authors found that the skull design of aetosaurs favored strength over fast jaw closing. They also found that representatives from the Northern Hemisphere may have had a more powerful bite, whereas Southern Hemisphere aetosaurs had faster jaw closing, as the former may have had more capacity for processing food in the oral cavity. Considering other morphological

Table 8.1. Moment arms (M1 to M5) of the masseter muscle (Mm) calculated for its most posterior, middle, and anterior points of insertion in several extant and extinct cingulate xenarthrans

Genus	Angle	Sub-angle	M1	M2	M3	M4	M5	Σ	Mean
			Mm (posterior)						
Vassallia	105	26.25	17	26	30	26	18	117	23.24
Holmesina	120	30	15	22	22	17	8	84	16.8
Euphractus	75	18.75	8	6	9	5	1	29	5.8
Dasypus	90	22.5	7	10	12	12	9	50	10
			Mm (middle)						
Vassallia	120	30	20	35	41	35	22	153	30.6
Holmesina	120	30	18	34	41	35	22	150	30
Euphractus	90	22.5	26	30	31	28	20	135	27
Dasypus	105	26.25	8	17	21	22	18	86	17.2
			Mm (anterior)						
Vassallia	90	22.5	20	39	52	58	54	223	44.6
Holmesina	105	26.25	17	30	42	53	48	190	38
Euphractus	82	20.5	29	39	44	44	32	188	37.6
Dasypus	105	26.25	9	21	28	29	24	111	22.2

Note: Modified from Vizcaíno et al. (1998).

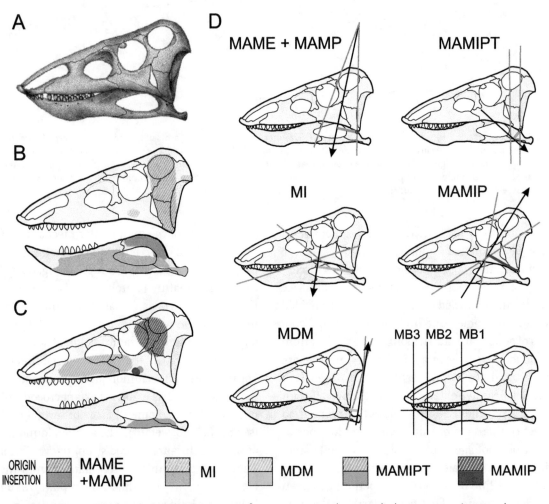

8.23. Geometrical model for calculating moment arms of masticatory musculature applied to aetosaurs. *A,* restored skull of *Neoaetosauroides engaeus. B* and *C,* reconstruction of origin and insertion areas of the main mandibular muscles. *D,* moment arms and lines of action of *m. adductor mandibulae externus* (MAME) and *m. adductor mandibulae posterior* (MAMP) together, *m .adductor mandibulae internus pseudotemporalis* (MAMIP), *m. intramandibularis* (MI), *m. adductor mandibulae internus pterygoideus* (MAMIPT), and *m. depressor mandibulae* (MDM); and moment arms of the posterior, middle, and anterior bite points (MB1, MB2, and MB3). Modified from Desojo and Vizcaíno (2009).

differences, for example that teeth of the Southern Hemisphere representatives are not serrated and lack wear facets (features present in Northern Hemisphere aetosaurs), Desojo and Vizcaíno (2009) concluded that *Neoaetosauroides* had a relatively soft, nonabrasive diet (e.g., soft leaves and/or insect larvae). It is likely that *Neoaetosauroides* represented a trend toward insectivorous feeding habits, exploiting a food source that was very widespread in continental environments throughout the Triassic.

Using a similar though somewhat less sophisticated approach to investigate another group of herbivorous archosaurs, Nabavizadeh (2016) estimated relative bite forces for several ornithischian dinosaurs. This researcher reviewed and reconstructed jaw adductor muscles, inferred muscle attachment sites, and calculated, in two dimensions, the mechanical advantage of these muscles from skulls in lateral view at several positions along the jaws (fig. 8.24). Nabavizadeh (2016) found that larger ceratopsians and

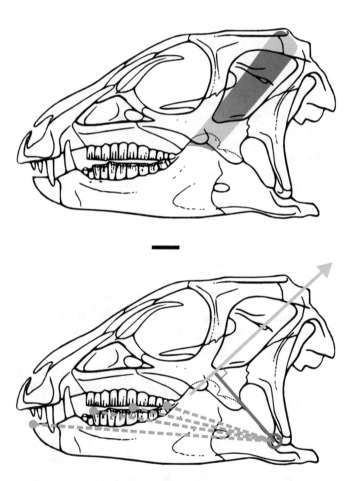

8.24. Jaw adductor mechanics in *Heterodontosaurus*. *Above*, schematic reconstruction of the deep external mandibular adductor muscle. *Below*, lower jaw modeled as a lever, with the pivot at the craniomandibular joint (*red circle*), the action line of the deep external mandibular adductor muscle (*green arrow*), its lever arm (*red line*), and different output lever arms (*dotted blue lines*) at four different occlusal sites (*blue circles*). Scale bar = 1 cm. Modified from Norman et al. (2011) and Nabavizadeh (2016).

hadrosaurs exhibited the highest relative bite forces, especially for distal tooth occlusion, whereas stegosaurians, ankylosaurs, basal ornithopods, iguanodontians, and pachycephalosaurians showed lower values overall, with smaller ceratopsians and iguanodontians showing slightly higher relative bite forces for predentary occlusion. Ceratopsians and hadrosaurs, which were among the main larger megaherbivores during the Late Cretaceous, convergently developed higher mechanical advantages and large dental batteries to process tough vegetation such as conifers and angiosperms.

Using a morphogeometric approach, Cassini and Vizcaíno (2012) extended the two-dimensional geometrical model method proposed by Vizcaíno et al. (1998) to a three-dimensional analysis by retaining the simplicity of the model but incorporating a methodological step of digital articulation between the skull and mandible, which makes it more versatile for working with fossil organisms (fig. 8.25). The authors validated the model based on a sample of extant mammals composed of herbivorous, omnivorous, and carnivorous species, which generated results aligned with those expected for the masticatory mechanics of these sample groups. For instance, when comparing the lever arm length of the two main masticatory muscles, they observed that herbivores exhibit a lever arm for the masseter that is up to twice as long as that for the temporal (table 8.2). In carnivores, conversely, the lever arm of the temporal is equal to or up to 1.3 times longer than that of masseter. Once the model was validated, Cassini and Vizcaíno (2012) applied it to a group of early Miocene ungulates from Patagonia (see chap. 9), revealing that only the typotherian *Pachyrukhos* corresponded to the lever arm ratios of extant herbivores (i.e., masseter-temporal ratio 2:1). In the remaining Miocene ungulates, the relationships between these two large muscles were similar, with the minor difference that in astrapotheres and litopterns the masseter prevailed, whereas in notoungulates the temporal prevailed. When considering both muscles, notoungulates (except for *Pachyrukhos*) exhibited an increased mechanical advantage for biting strength compared with astrapotheres and litopterns. These results led Cassini and Vizcaíno (2012) to conclude that notoungulates may have had a capacity for processing a diet rich in rather hard plant materials (e.g., grasses or even bark) compared with litopterns, which probably consumed softer plants (e.g., dicots). These conclusions lend support to the hypothesis of Billet et al. (2009) that increased hypsodonty in notoungulates was a response to increased masticatory effort.

Ercoli et al. (2020) analyzed the masticatory mechanics of the latest pachyrukhines *Paedotherium* and *Tremacyllus* (late Miocene–late Pliocene). The tooth morphology

A

B

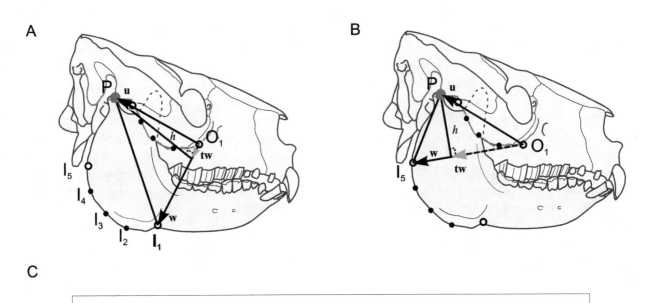

C

$$\vec{u} = \overline{O_iP} \qquad \vec{w} = \overline{O_iI_j} \qquad \vec{tw} = \text{proy}\,\frac{\vec{u}}{\vec{w}} = \frac{\vec{w}\cdot\vec{u}}{\vec{w}\cdot\vec{w}}\,\vec{w} \qquad h = \left\| \vec{u} - \vec{tw} \right\|$$

8.25. Calculation of lever arms in 3-D based on Cassini and Vizcaíno (2012). *A–B*, skull and mandible of *Adinotherium* (Notoungulata, Toxodontidae) in right lateral view showing landmarks indicating the first (*A*) and the final step (*B*) for calculating lever arms of the masseter. O_1, origin of masseter; P, pivot (craniomandibular joint; *red circle*); $I_{1\,to\,5}$, the most anterior to most posterior insertion of the masseter on the angular process; *open circles*, landmarks that indicate the most extreme points of muscular origin/insertion; *black dots*, relaxed semilandmarks representing intermediate positions along origin/insertion sites, based on muscle scars. *C*, definitions of the vectors and formulas used for calculation of *h*, the length of the lever arm of input forces; **u**, vector O_1P; **w**, vector O_1I_j; **tw**, vector of the orthogonal projection of **u** on **w** (*blue arrows*); *h*, indicated above as the length of the lever arm, is also the perpendicular distance between the pivot (P) and vector **w** perpendicular to the latter (resulting from calculating the norm of the subtraction of vectors **u** and **tw**), and may thus also be interpreted as the lever arm for the hypothesized line of action. *C* as in Cassini and Vizcaíno (2012).

and reconstructed muscular configuration (in particular the superficial masseter, which approximates the sciurid condition) in these pachyrukhines indicate an important mediolateral component during chewing, the predominance of crushing over grinding, and an anteroposterior component for the coupling action of stronger gnawing incisors. These actions are more compatible with hard and brittle or turgid fruit consumption than specialized folivorous and, in particular, grazing habits.

Teeth and Digestive Physiology

We have already mentioned that the probability of a tooth striking and fragmenting a food particle is higher if the tooth is larger; that is, it has a larger occlusal surface. In the absence of other adaptations, an animal that requires more extensive oral processing of food will have teeth with a larger occlusal surface than an animal of the same size that requires less oral processing. This allows the possibility, at least for mammals, of inferring aspects of digestive physiology in the intestinal tract from elements of the oral cavity.

Using length of the cheek-teeth (premolars and molars) series as a proxy for occlusal surface area (OSA), several authors (Janis 1990, 1995; Janis and Constable 1993; Mendoza et al. 2002) pointed out that among both placental (ungulates) and marsupial (kangaroos) phytophages, OSA is larger in grazers than in browsers. For instance, monogastric ungulates such as perissodactyls (i.e., those with a simple or undivided stomach) have longer rows of molarized premolars and larger, more quadrangular molars than ruminant artiodactyls, differences that may reflect feeding strategies linked to the distinct digestive physiologies of monogastric and ruminant mammals (Janis 1988, 1995; Janis and Constable 1993). Thus, in comparison with cows, which ruminate, horses have larger OSA and spend more time chewing, especially when consuming higher fiber content food.

Vizcaíno et al. (2006a) quantified OSA more directly by calculating area as a two-dimensional representation

Table 8.2. Descriptive statistics [mean ± standard deviation] of lever arms (in mm) for the input forces generated by the masticatory muscles and the resulting output (bite) forces produced at different positions along the mandible in Miocene ungulates from Patagonia (astrapotheres, litopterns, and notoungulates) and the extant mammals used to validate the model (carnivorans, artiodactyls, and perissodactyls)

Taxa	M. masseter	M. temporalis	Infradentale	First premolar	Premolar / Molar	Last molar
Astrapotheria						
Astrapotherium	88.537 ± 15.438	84.300 ± 16.770	432.322 ± 56.354	252.075 ± 41.708	224.656 ± 46.193	113.744 ± 6.564
Litopterna						
Diadiaphorus	38.194 ± 4.485	37.550 ± 5.765	187.325 ± 6.825	166.822 ± 5.130	106.461 ± 2.841	55.220 ± 5.731
Tetramerorhinus	34.772	31.210	150.901	128.956	83.915	47.412
Theosodon	40.847 ± 2.852	40.684 ± 8.309	235.001 ± 13.953	197.941 ± 9.337	124.959 ± 10.270	57.961 ± 10.132
Notoungulata						
Adinotherium	45.042 ± 4.411	54.734 ± 6.031	186.611 ± 17.754	153.085 ± 18.214	110.847 ± 15.121	63.723 ± 3.309
Nesodon	90.108 ± 9.095	98.517 ± 20.025	348.412 ± 26.708	297.935 ± 24.819	237.035 ± 25.395	123.473 ± 15.904
Protypotherium	16.485 ± 2.007	20.155 ± 4.917	75.640 ± 8.154	62.094 ± 6.112	45.553 ± 7.598	25.160 ± 3.021
Interatherium	13.177 ± 2.866	16.111 ± 1.230	59.113 ± 5.197	44.439 ± 4.595	32.950 ± 4.742	20.125 ± 2.292
Pachyrukhos	16.118	8.242	60.621	46.070	35.357	20.609
Carnivora						
Chrysocyon	30.306	39.422	172.513	132.494	79.629	58.958
Puma	44.975	39.043	138.990	102.222	64.605	62.767
Artiodactyla						
Hippocamelus	49.589	28.375	232.277	138.933	101.294	56.609
Camelus	70.995	76.494	378.543	289.896	184.646	88.146
Lama	65.219	42.028	242.890	144.349	125.494	64.346
Sus	50.148	37.943	297.257	193.707	143.735	80.026
Perissodactyla						
Equus	125.506	65.746	450.010	321.065	224.809	140.132
Tapirus	75.864	61.265	285.885	193.457	122.558	64.897

Note: Modified from Cassini and Vizcaíno (2012).

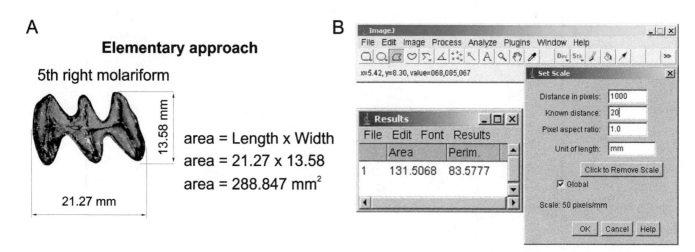

8.26. Two methods for the calculation of OSA (occlusal surface area) of a tooth of a Pleistocene glyptodont. *A*, elementary approach by multiplying maximal length and width of a tooth. *B*, use of orthogonal projection of occlusal surface by Vizcaíno et al. (2006a).

of the occlusal surfaces of the upper cheek teeth, based on digital images of these teeth (fig. 8.26). This method, though, assumes that a two-dimensional quantification, based on the occlusal periphery of a tooth row, is a reliable approximation of a three-dimensional surface, bearing in mind that the occlusal surface of each tooth is not flat, but variably folded to produce crests and valleys. Thus, Vizcaíno et al. (2011b) evaluated the approach by considering OSA as the total area of the cheek teeth, this time taking into account the folded nature of the occlusal surface. To

Table 8.3. Allometric relationship between occlusal surface area (OSA) and body mass (BM)

Taxa	Size range (kg)	No. of spp.	R^2	Intercept BM = 1 kg	Slope	Residuals	Isometry
Mammalia	.084–4,637	54	.958	1.470	.701	E > X	pos
Xenarthra	1.175–3,780	24	.931	1.334	.620	C > F	iso
Cingulata	1.175–1,061	14	.978	1.336	.681		iso
Dasypodidae	1.175–1,637	5	.940	1.336	.683		iso
Glyptodontidae	73.40–1,061	8	.955	1.182	.729		iso
Folivora	1.937–3,780	10	.918	1.274	.599		iso
"Epitheria"	.084–4,637	47	.967	1.585	.664	P > A	iso
Rodentia	.084–50.07	8	.924	1.567	.630		iso
Artiodactyla	12.17–3,729	18	.937	1.640	.602		neg
Ruminantia	12.01–450.6	12	.955	1.610	.603		iso

Note: Results of the simple linear regressions of each group. The most common parameters, R^2 value and the estimators of regression coefficients, are indicated in bold. The results of the residual analysis are included when the pair of comparisons is statistically significant ($p < 0.01$). E, "Epitheria"; X, Xenarthra; P, Perissodactyla; A, Artiodactyla; C, Cingulata; F, Folivora. Modified from Vizcaíno et al. (2006a).

this end, these authors digitized the contour and occlusal crests and valleys of each upper tooth, and then converted the obtained data points to a grid surface to calculate the total tooth row OSA.

Vizcaíno et al. (2011b) then compared these more precise estimates of OSA values with those obtained by Vizcaíno et al. (2006a) and found that the absolute value (modulus) of the mean difference obtained was only 38 mm² (i.e., less than 5%). Therefore, the results from Vizcaíno et al. (2011b) are comparable with those from Vizcaíno et al. (2006a), obviating the need for recalculating

values from the extant sample. Vizcaíno et al. (2011b) evaluated the allometric relationship between OSA and body mass, calculating the regression lines by the ordinary least squares (OLS) method, with the logarithm of OSA as the dependent variable and the logarithm of body mass as the independent variable. They tested, statistically, the bias of the observed slope with respect to that predicted by isometry and that predicted by Kleiber's law (0.66 and 0.75, respectively; table 8.3) and found that it does not differ significantly from isometry but is different from that expected based on Kleiber's law.

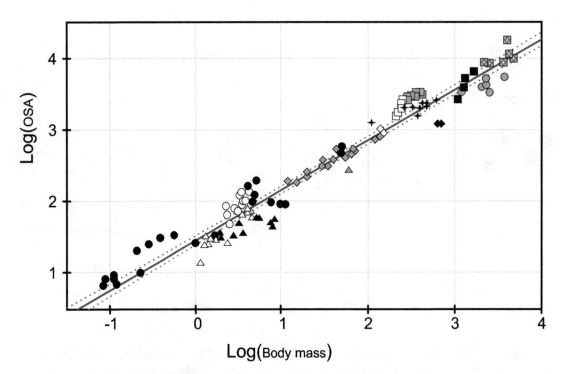

8.27. Regression of dental occlusal surface area (OSA) against body mass (log-transformed values) in extant mammals (*n* = 125). Symbols: △, Dasypodidae; ▲, Bradypodidae; ▲, Tubulidentata; ●, Caviomorpha; ○, Hyracoidea; □, Tapiridae; ▣, Equidae; ■, Rhinocerotidae; ◆, Cervidae; ◇, Bovidae; ◆, Giraffidae; ◐, Hippopotamidae; ✦, Camelidae; ▦, Elephantidae. *Dotted red lines* encompass the 95% confidence interval.

An important observation of Vizcaíno et al. (2006a) is that, among ungulates, artiodactyls (particularly ruminant artiodactyls) are distributed below the regression line with smaller OSA values, whereas the monogastric (i.e., nonruminant) forms lie above the regression line, indicating that their OSA values are larger than expected for their body size (fig. 8.27). This is in agreement with the conclusions of Janis (1988, 1995), noted above. Further, Vizcaíno et al. (2006a) found that among perissodactyls, equids, which are mainly grazers and live in more open habitats, have relatively larger OSAs than tapirs, which are browsers and live in more closed habitats. Similarly, among artiodactyls, bovids and camelids, which are mixed feeders or grazers, have larger OSAs than browsers such as cervids and giraffids. This suggests that the available OSA for chewing in ungulates is also related to the nutritional value of food; browsers consume succulent leaves, while grazers ingest forage, which is both of lower nutritional value and more abrasive on teeth. This link between food and OSA is reinforced by the findings of Texera (1974), who reported that, among ruminants, subadult individuals of the cervid *Hippocamelus* spend more time chewing than adults, a behavior that Vizcaíno et al. (2006a) attributed to the smaller OSAs of subadults, in which the third molar has not yet erupted. An exception is the hippopotamus, the teeth of which are unusually small for such a large grazing mammal, a condition probably related to its low metabolic rate (Owen-Smith 1988).

As Vizcaíno et al. (2006a) were interested in analyzing fossil faunas with abundant xenarthrans, they dealt especially with this group. They found that both fossil and extant xenarthrans have smaller OSAs available for processing food than other eutherians of similar size. This may be related to the low basal metabolic rate of extant xenarthrans, which is between 40% and 60% of the rate expected for a placental mammal of their body mass according to Kleiber's law (McNab 1985). This implies that xenarthrans have lower energetic requirements than other therians and, therefore, for a specific type of food, require less food than other therians of similar body mass.

Applications in paleobiology

Vizcaíno et al. (2006a, 2011b) studied the relationships among molariform OSA, body mass, diet, and other biological factors in several extinct xenarthrans. Apart from having smaller OSAs than other therians, these authors found that, among xenarthrans, Pleistocene mylodontid sloths (most of them with a body mass over 1 tonne) exhibited much smaller OSAs (fig. 8.28). This was interpreted as an indication of poor efficiency in food processing in

the oral cavity, which was probably compensated by high fermentation in the digestive tract (Vizcaíno et al. 2006a). Surprisingly, the Pleistocene megatheriid *Megatherium americanum* had the expected OSA for a mammal of its size (i.e., comparable to an elephant), much larger than that of mylodontids. If a parallel with ungulates is presumed, mylodontids may have been foregut fermenters, whereas *M. americanum* may have been a hindgut fermenter, which, in turn, implies diets of lower and higher nutritional qualities, respectively (Alexander 1996). However, the absence of a caecum in extant xenarthrans casts doubt on *M. americanum* as a hindgut fermenter (although the sample size of extant sloths, in particular, is exceedingly small). In this case, the pronounced difference in OSA values between *M. americanum* and mylodontids may then be interpreted to suggest the former had fewer and/or smaller stomach chambers (which are, by contrast, present in extant sloths) than mylodontids. Although stomach anatomy cannot be compared among fossil taxa except in some exceptional cases, it is evident that *M. americanum* was better suited for processing food in the oral cavity than were mylodontids. This is in agreement with the morphological and biomechanical evidence widely discussed by Bargo (2001).

Vizcaíno et al. (2006a) reported that, with the exception of *M. americanum*, both fossil and extant cingulates (armored xenarthrans) have higher OSA values than sloths, and this suggests greater food processing in the oral cavity. Vizcaíno et al. (2011b) found deviations in the correlation between OSA and body mass, especially in post-Miocene glyptodonts and particularly among the large Pleistocene forms with an adult body mass of more than 1,000 kg. Whereas specimens of *Glyptodon* fell on, or very near, the glyptodont regression line, which indicates that they have the expected OSA for glyptodonts of their size, those for *Panochthus* and *Doedicurus* lay above and well below the regression line, indicating higher and lower OSA values than expected based on their body mass, respectively (fig. 8.28). This suggests that these two genera followed divergent evolutionary paths.

Ecomorphological Approaches

As noted in chapter 2, ecomorphological approaches converge in quantifying variation in form and in evaluating the factors that explain variation, focusing on historical (ontogenetic, phylogenetic) and environmental (ecological) aspects. A known reference framework (a sample of extant animals) is required so that its patterns can then be applied to extinct organisms. We cannot summarize all the literature on ecomorphological approaches to the

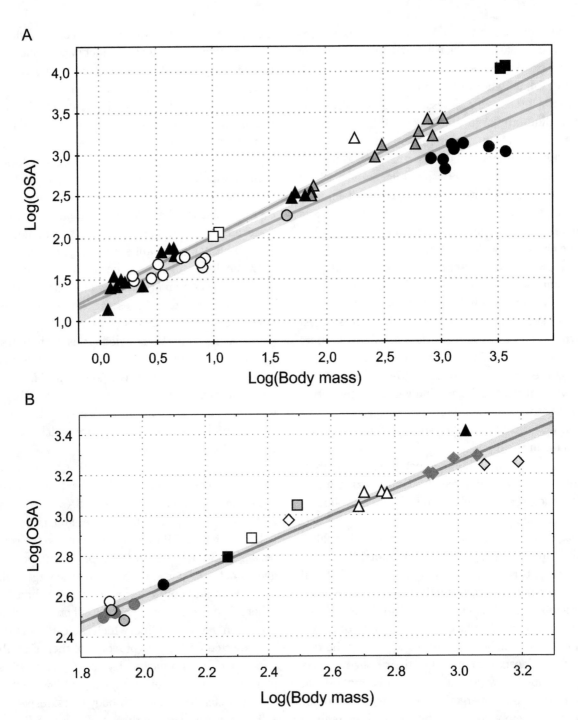

8.28. Regression of dental occlusal area (OSA) against body mass in xenarthrans (log-transformed values). *A*, extant and extinct xenarthrans, excluding glyptodonts. In *green*, the regression line for Cingulata. In *blue*, the regression line for Folivora. Symbols: ▲, dasypodid cingulates; ◮, pampatheriid cingulates; △, glyptodontid cingulates; ○, extant sloths; ◓, *Eucholoeops* (Miocene sloth); □, *Hapalops* (Miocene sloth); ■, *Megatherium* (Pleistocene sloth); ●, mylodontid sloths (Pleistocene). Modified from Vizcaíno et al. (2006a). *B*, Glyptodonts. *Red line*, regression. Symbols: ○, *Asterostemma* (Miocene); ◓, *Cochlops* (Miocene); ●, *Eucinepeltus* (Miocene); ◓, *Propalaehoplophorus* (Miocene); □, *Hoplophractus* (Mio-Pliocene); ▣, *Neuryurus* (early Pliocene); ■, *Eosclerocalyptus* (Mio-Pliocene); △, *Neosclerocalyptus* (Pleistocene); ▲, *Panochthus* (Pleistocene); ◇, *Urotherium* (Pleistocene); ◈, *Doedicurus* (Pleistocene); ◆, *Glyptodon* (Pleistocene). Modified from Vizcaíno et al. (2011b). *Shaded* areas around regression lines = 95% confidence interval.

study of extant vertebrate communities, as it is quite extensive; here we provide only a sample of articles relevant either to the study of certain groups or to the discipline in general (e.g., Janis 1990 on ungulates; van Valkenburgh and Wayne 1994 on carnivorous mammals; Klingenberg and Ekau 1996 on fish; Herrel et al. 1999 on squamates).

Artiodactyl and perissodactyl ungulates have been frequently studied from an ecomorphological perspective (among others, Gordon and Illius 1988; Janis and Ehrhardt 1988; Solounias and Dawson-Saunders 1988; Janis 1990, 2008; Spencer 1995; Pérez-Barbería and Gordon 2001; Mendoza et al. 2005; Mendoza and Palmqvist 2006, 2008; Fraser and Theodor 2011). Some of the earliest attempts at quantitative analyses, such as those involving Janis and Solounias, took as their starting point an article by Boué (1970), who analyzed qualitatively the morphology of incisors in ungulates with the aim of characterizing browsing species (central incisors wider than lateral incisors) and grazing species (all incisors subequal in width). Although, as noted in the previous chapter, the meanings of the terms *browser* and *grazer* are ambiguous, in most ecomorphological works feeding categories have been defined as a function of the proportion of food items consumed (e.g., grasses or Gramineae). In Janis's initial forays into this subject (Janis and Ehrhardt 1988; Janis 1990), three categories were recognized: grazers (more than 90% of Gramineae in their diet), browsers (less than 10% of Gramineae in their diet), and mixed feeders (between 10% and 90% of Gramineae in their diet). Even though researchers have been studying ungulates ecomorphologically for several decades and have found morphological differences—that is, that there are patterns of variation among the different trophic groups of ungulates (e.g., Mendoza 2005; Clauss et al. 2008)—it is still a fertile field

for the application of new methodological and statistical techniques.

Applications in Paleobiology

Cassini (2013) used geometric morphometrics to infer the use of habitat and diet in Santacrucian Age (early Miocene of Patagonia; see chap. 9) South American native ungulates. This author analyzed a wide sample of skulls and mandibles of extant mammals, including artiodactyl and perissodactyl ungulates, hyraxes (Hyracoidea), and kangaroos (Diprotodontia, Macropodidae), and evaluated phylogenetic effect through generalized estimating equations (GEE) on the regression residuals and principal component analysis (PCA) scores. Cassini (2013) found that skull morphology exhibited evidence for mainly phylogenetic influence (or signal), rather than for ecological patterns. Conversely, he detected patterns of mandibular shape variation explained by a complex interaction between habitat use and diet composition, in addition to some variation of shape due to common ancestry; that is, skull morphology seemed more constrained by phylogeny than did the mandibular morphology. This author also assessed the effect of size by means of regressions of shape against centroid size (fig. 8.29), empirically demonstrating the influence of size on the type of diet, as had been suggested

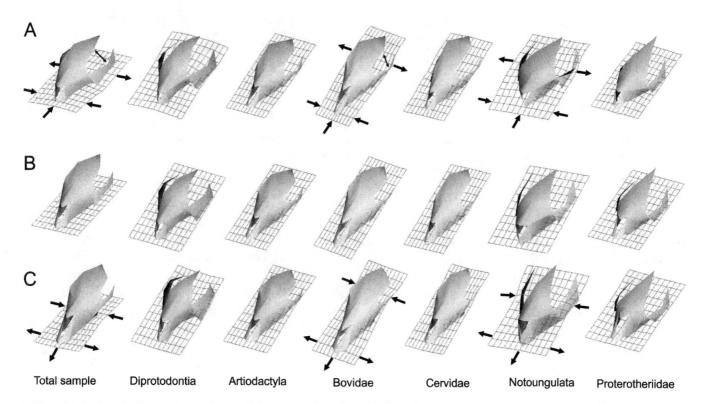

8.29. Deformation grids of regressions of the cranial shape versus logarithm to base 10 of centroid size for Santacrucian native South American ungulates (Miocene of Patagonia, Argentina). *A*, small forms. *B*, consensus. *C*, large forms. The *arrows* indicate main directions of changes in shape. Reproduced from Cassini (2013).

A   C

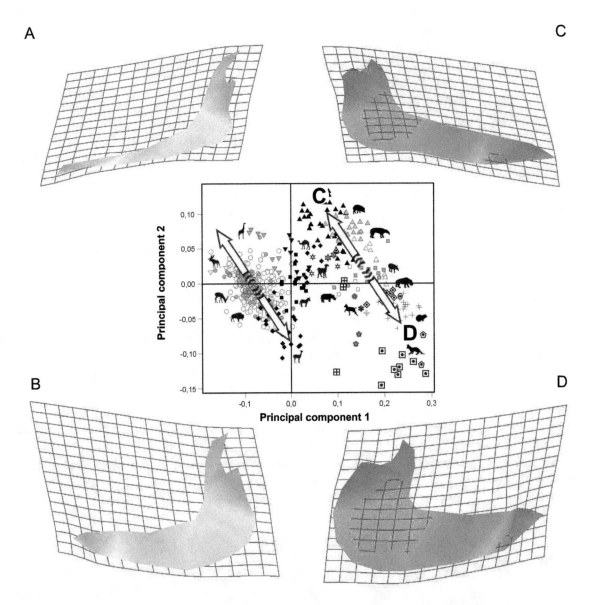

8.30. Deformation grids in the morphospace depicted by PC1 and PC2 of PCA of mandibles for the complete sample. A, to *Litocranius*. B, to *Vicugna*. C, to *Hylochoerus*. D, to *Procavia*. Extant mammals: +, Hyracoidea; ■, Equidae; ▨, Tapiridae; ☐, Rhinocerotidae; △, Hippopotamidae; ▲, Suidae; ▲, Tayassuidae; ▽, Tragulidae; ▼, Antilocapridae; ▽, Giraffidae; ◆, Camelidae; ○, Bovidae; ◉, Cervidae. Santacrucian ungulates: Astrapotheria: ✿, *Astrapotherium*; Litopterna: +, *Anisolophus*; ⊞, *Diadiaphorus*; ✳, *Tetramerorhinus*; ✳, *Thoatherium*; ✭, *Theosodon*; Notoungulata: ▣, *Interatherium*; ◉, *Protypotherium*; ▼, *Hegetotherium*; ◉, *Adinotherium*; ◈, *Nesodon*. Reproduced from Cassini (2013).

B   D

in classical studies (e.g., Jarman 1974). Like De Esteban-Trivigno (2011), Cassini (2013) found that morphological variation of the mandible related to different types of diet was, in turn, conditioned by the presence of hypertrophied canines and incisors (fig. 8.30).

Prevosti and Palmqvist (2001) studied the trophic relationships of the Pleistocene canid *Theriodictis platensis* from the Pampean Region of Argentina, for which hypercarnivorous habits had been previously proposed by van Valkenburgh (1991). Prevosti and Palmqvist (2001) estimated its body mass as 37 kg using linear measurements and multiple regression equations adjusted for phylogeny

in canids and body masses available in the literature. They then analyzed craniomandibular and dental dimensions with multivariate statistics, PCA and discriminant analysis (DA). The results of their PCA analysis indicated that extant hypercarnivorous species (i.e., those in which meat accounted for more than 70% of the diet), occupying the morphospace of positive values of PC1 and PC2, were well separated from omnivorous species (fig. 8.31). The omnivorous species (occupying the negative morphospace of PC2) were characterized by a longer talonid of the lower carnassial tooth and an elongated rostrum, whereas the hypercarnivorous species exhibited large canines and

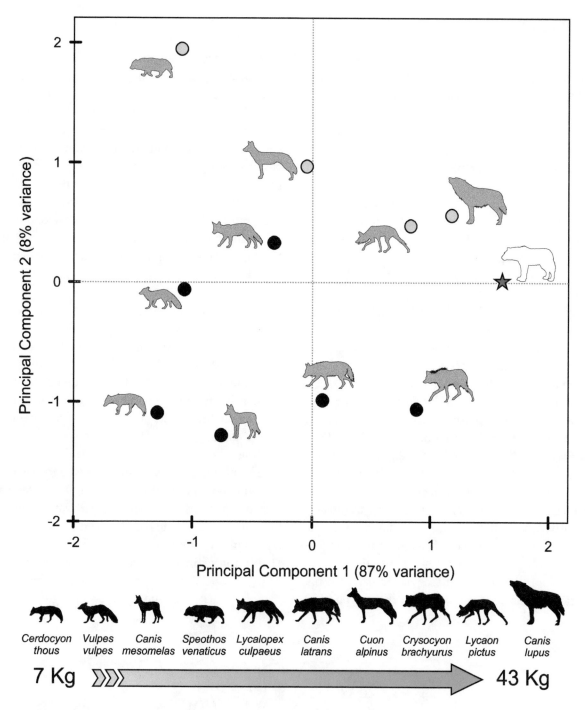

8.31. Morphospace depicted by PC1 and PC2 of the PCA obtained from craniodental variables of extant canid species (*gray* silhouettes) and the Pleistocene species *Theriodictis platensis* (*white* silhouette). ●, omnivorous canids; ◯, hypercarnivorous canids; ⭐, *T. platensis*. The observed body-mass gradient, from 7 kg (crab-eating fox) to 43 kg (wolf), correlates positively with the PC1. Modified from Prevosti and Palmqvist (2001).

a lower carnassial tooth with a relatively large trigonid compared with the talonid. *Theriodictis platensis* lies closer to the hypercarnivorous species than to the omnivorous species. The results of the DA, which distinguishes between these two dietary groups in extant canids, placed *T. platensis* among the hypercarnivores. Thus the authors, based on both PCA and DA results, concluded that the

diet of *T. platensis* included more than 70% meat; that is, it was indeed a hypercarnivore. They suggested as potential prey artiodactyls such as cervids, tayasuids, and camelids; perissodactyls such as equids (of the genera *Hippidion* and *Equus*); mesotherids (native ungulates); rodents; and large armadillos.

Dental Microwear and Mesowear Analyses

There are other tools for the formulation of hypotheses about feeding that complement the form-function approach in vertebrate paleobiology, such as dental micro- and mesowear analyses. Although they have been applied mainly to extant and fossil ungulates, their use has been extended to other groups of mammals (e.g., xenarthrans). Both methods evaluate physical features produced on enamel (or dentine in xenarthrans) by tooth-tooth (attrition) and tooth-food (abrasion) interaction during chewing. Based on the micro- and mesowear patterns present in extant mammals, hypotheses on the diets of extinct mammals can be proposed.

Microwear analysis focuses on the study of microscopic features (by electron microscopy), such as striations, scratches, and pits, produced by food on enamel (Solounias et al. 1994; Solounias and Semprebon 2002). The information obtained reflects the last food ingested by an individual and thus is not necessarily representative of the animal's regular feeding habits.

Mesowear analysis, proposed by Fortelius and Solounias (2000), was originally based on features present on the second upper molars (M2s) of ungulates and later (Kaiser and Fortelius 2003) expanded for equids to include the last four upper teeth, P4–M3. Mesowear analysis has proved to be a robust tool, providing consistent results for the three conventional feeding categories for ungulates: browser, mixed, and grazer. It involves analysis, with the unaided eye or under low magnification, of features (see below) on the buccal cusps of teeth—that is, those forming the ectoloph (on the labial side of cheek teeth; fig. 8.32), because it has been demonstrated that the physical properties of the food influence the development of these cusps. Unlike microwear, the food effects that produce mesowear are cumulative, and thus mesowear is a better indicator (following the lactation stage) of an animal's usual diet. Abrasive foods, such as grass, tend to round and wear down cusps, whereas softer plants tend to keep them sharp and prominent. The variables used in mesowear analysis include occlusal relief (high or low) and cusp contour (sharp, rounded, or blunt). Ackermans (2020) provided a useful and comprehensive review of published mesowear analyses conducted over the past two decades. This author's work summarized the development of techniques in this field of research and provided a template for the application of each technique to a particular research question.

Applications in Paleobiology

In their initial work on mesowear analysis, Fortelius and Solounias (2000) used a sample of 2,200 individuals from 64 extant ungulate species to evaluate the stability of mesowear during ontogeny. Their analysis correctly identified the feeding habits of a succession of Serengeti (African savanna) grazers. These authors then extended this analysis as a pilot application to several extinct equids and bovids, noting that it was superior to microwear analysis in two cases where the diet of fossil ungulates had been previously studied by microwear and other conventional analyses. According to Fortelius and Solounias (2000), the advantages of mesowear analysis for fossil taxa include: (1) placing extinct species within the dietary spectrum of extant species (e.g., extinct species X had a diet like extant species A, B, and C) or assigning fossil species to

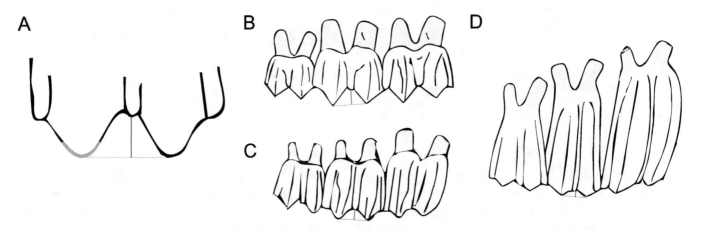

8.32. Features analyzed in mesowear studies. *A*, height, indicated by a *red line*, of occlusal relief (high or low with respect to cusp length) and cusp contour (indicated by a *blue line*: sharp, rounded, or blunt). Examples in ungulates: *B*, white-tailed deer (*Odocoileus virginianus*), brachydont teeth, with high relief and sharp cusps; *C*, barasingha (*Rucervus duvaucelii*), mesodont teeth, with high relief and rounded cusps; *D*, goat (*Capra hircus*), hypsodont teeth, with high relief and sharp cusps. Modified from Fortelius and Solounias (2000).

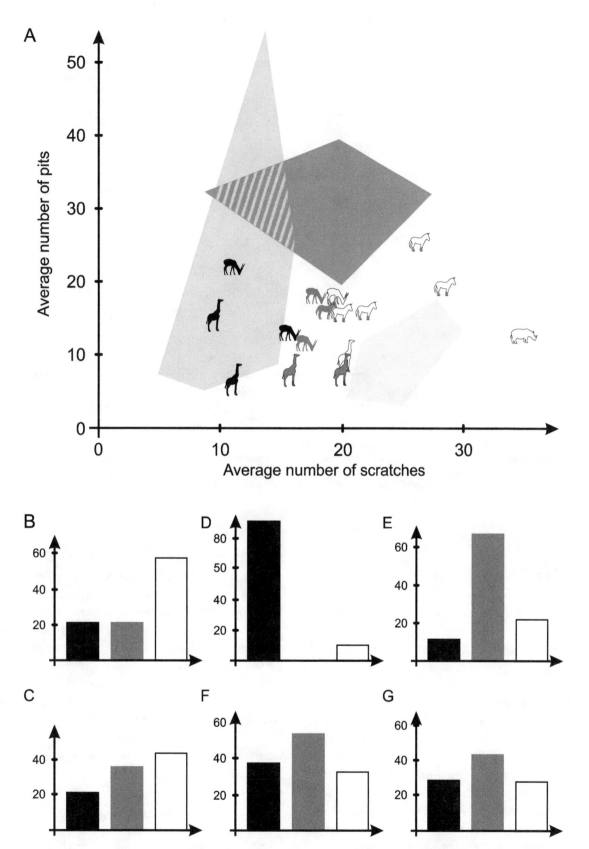

8.33. *A*, dispersion diagram of the average number of pits versus the average number of scratches detected at 35× magnification in extant ungulates and fossil representatives of Equidae, Rhinocerotidae, Giraffidae, and Antilopini (browsers in *black*, mixed feeders in *gray*, and grazers in *white*) of the late Miocene from Pikermi and Samos (Greece). Polygons indicate the distribution of extant taxa: browsers of leaves (*green*), frugivorous (*orange*), grazers (*yellow*). *B–G*, comparison of percentages of browsing (*black*), mixed-feeding (*gray*), and grazing (*white*) taxa in some extant biomes and in the ungulate fauna from Pikermi and Samos. *B*, African savanna. *C*, African woodland. *D*, African forest. *E*, Indian woodland. *F*, Samos. *G*, Pikermi (excluding *Hipparion*). Modified from Solounias et al. (2010).

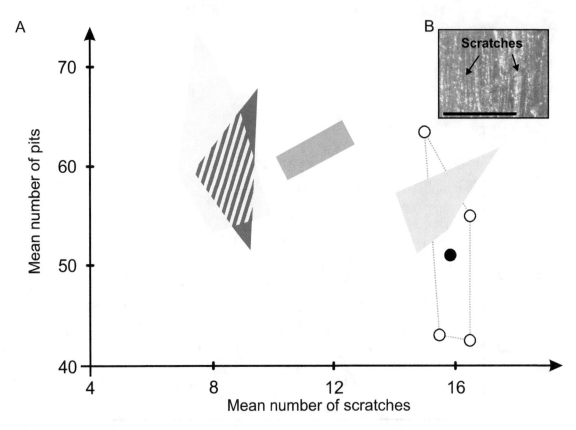

8.34. Dispersion diagram of the average number of pits versus average number of scratches in the Pleistocene ground sloth *Nothrotheriops shastensis* in relation to dietary ecomorphospaces of extant xenarthrans. *Yellow*, insectivores; *red*, carnivore-omnivores; *orange*, frugivore-folivores; *green*, folivores; *white circles*, specimens of *N. shastensis*; *black circle*, average for this species. *B*, low-magnification image of microwear patterns observed on the dentine of a specimen of *N. shastensis* (scale bar = 0.50 mm). Modified from Green (2009a).

food categories of extant species, and (2) direct comparison among fossil species in terms of the physical properties of food (e.g., extinct species X had a diet that caused less general wear but more abrasion than that of extinct species Y).

Solounias et al. (2010) used microwear analysis in a study of 809 extant ungulates and 561 late-Miocene ungulates from Pikermi and Samos (Greece) to infer food adaptations of the ungulate fauna of the Pikermian Biome. They concluded that the feeding habits of the extinct taxa were more consistent with those from a forest habitat than with a savanna habitat. They also found that the species richness and diversity of ungulates of the Pikermi and Samos faunas were greater than those of extant ungulates from savannas, forests, and African woodlands, and that their diets were more similar to those of ungulates from Indian and, to some extent, African forests (fig. 8.33).

Xenarthran teeth lack enamel, and thus tooth-wear analyses must be conducted on dentine (orthodentine). As this substance is more ductile than enamel, wear patterns produced may differ from those produced on enamel, thus

requiring investigation of patterns present in extant xenarthrans. Green (2009b) therefore carried out microwear studies in extant xenarthrans (armadillos and sloths) based on a sample of 255 teeth belonging to 20 species. Discriminant analysis allowed distinguishing between arboreal folivorous and frugivorous-folivorous sloths, and between fossorial omnivorous-carnivorous and insectivorous armadillos. Green (2009a) then applied these results to an analysis of the North American Pleistocene sloth *Nothrotheriops shastensis* to elucidate its possible diet. This author found that the microwear patterns of this species were most closely comparable to those of extant folivorous sloths (fig. 8.34).

Dental Topography

Dental topographic techniques, such as relief index (Thiery et al. 2019), Dirichlet normal energy (Bunn et al. 2011), and orientation patch count (OPC; Berthaume et al. 2020), are increasingly employed for inferring diets in extinct mammals. Such analyses employ geographic information systems (GIS) technology to study functional aspects of tooth shape by modeling dental features as analogous to geographic features, such as, for example, representing tooth cusps as mountains and fissures as valleys. The resulting data are used to relate tooth form to function in extant mammals, infer diets from extinct species, and consider how tooth shape and function change with the wear that occurs over a lifetime. These techniques involve the collection of three-dimensional point cloud data from an occlusal surface that is used to interpolate a virtual surface for analysis. The tools used to collect the initial point clouds depend on the resolution required for a particular sample and research question and include electromagnetic digitizers (Zuccotti et al. 1998), touch probe scanners (King et al. 2005; Ungar 2006), laser scanners (Ungar and M'Kirera 2003; Bunn and Ungar 2009), and confocal microscopes (Jernvall and Selanne 1999). Following the collection of point cloud data, a surface is modeled and analyzed using GIS software. Detailed explanations of these methodologies may be found in Klukkert et al. (2012), Evans (2013), Pineda-Munoz et al. (2017), Evans and Pineda-Munoz (2018), and Berthaume et al. (2020).

Among the advantages of dental topographic techniques are that they are useful in overcoming several limitations of more traditional methods. Among these is tooth wear, which in mammals causes, nearly universally, changes in shape that hinder quantitative characterization of functionally relevant aspects of tooth shape because many landmarks used in common measurements become obliterated. Dental topographic analysis, however, permits the measurement and comparison of occlusal tooth surface features without the use of landmarks. Another limitation is the scarcity of unworn teeth, especially in the paleontological record, which renders comparison among individuals difficult. Because dental topographic analysis allows characterization of a surface without the use of landmarks, it permits analysis and comparison of variably worn teeth. The inclusion of worn specimens in analyses results in a dramatic increase in sample size, particularly for species that wear their teeth quickly. Although these methods seem to be suitably effective for groups with relatively complex crown morphology, their usefulness for clades with simpler crowns, such as sloths and aardvarks, remains to be tested.

We may cite, by way of example of its application to paleobiology, the work of Wilson et al. (2012) on multituberculates, a globally distributed mammalian clade that originated by the Middle Jurassic epoch and became extinct in the late Eocene. These mammals had a distinctive dentition consisting of procumbent incisors, bladelike premolars, molars with longitudinal rows of cusps (fig. 8.35), and a predominantly posteriorly directed (palinal) chewing motion. In exploring multituberculate ecomorphological diversity through time, Wilson et al. (2012) quantified dental complexity in 41 genera using GIS analyses of three-dimensional crown surfaces of

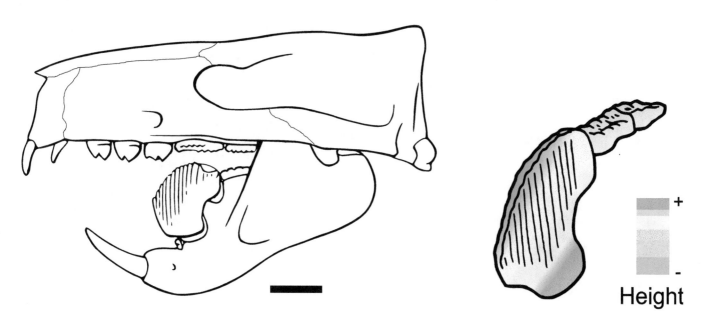

8.35. Dental occlusal surface topography in the multituberculate *Ptilodus*. On the *left*, skull and mandible (modified from Krause 1982, scale approximately 5 mm); note the prominent premolar, the large, compressed, bladelike, and serrated plagiaulacoid tooth. On the *right*, topographic 3-D *shading* of the plagiaulacoid tooth. Modified from Wilson et al. (2012).

the lower cheek teeth. These authors calculated OPC, a measure of dental complexity, as the number of discrete surfaces on the cheek-tooth row as distinguished by differences in orientation. The Late Jurassic through to the Early Cretaceous multituberculates exhibited low and tightly constrained dental complexity, with OPC values corresponding to carnivory and the low end of animal-dominated omnivory (e.g., eating both insects and fruits) among extant mammals. In the early Late Cretaceous, multituberculates had a slightly higher mean OPC and low morphological disparity. In the latest Cretaceous, mean OPC rose and peaked, and disparity increased markedly as well. Of the 17 taxa for this time interval, five exhibited OPC values greater than 160 and two greater than 200, corresponding to values for plant-dominated omnivory and herbivory among extant mammals, respectively. In the early Paleocene, multituberculates maintained high OPC (mean OPC, 138) and disparity peaked. The early Paleocene *Taeniolabis* from North America had the highest OPC value (348) not only among multituberculates but also among extant herbivorous rodents and carnivorans. *Ectypodus*, the only known genus from the middle to late Eocene, had a low OPC (109), corresponding to the high end of the range for extant mammalian carnivores. The degree of detail in dental morphology permitted by dental topographic techniques in inferring the varying diets among multituberculates over a long period of geological time allowed Wilson et al. (2012) to determine that these multituberculates were able to take advantage of new ecological opportunities in the Mesozoic and that at least some of these opportunities persisted through the Cretaceous–Paleogene mass extinction.

Paleoecology

Many paleontologists tend to regard paleoecology as the discipline that deals with fossils as a means for reconstructing the environments in which extinct organisms lived, or are hypothesized to have lived, a view that is rather closer to geology than biology. By contrast, in *Deciphering Earth History*, Gastaldo et al. (1996) offered a more encompassing vision. They considered paleoecology within the context of ecological parameters, defining it as the investigation of individuals, populations, and communities of past organisms, as well as their interactions with environmental changes and dynamic responses to them. These authors considered modern ecosystems (a geographically defined physical area together with its biological component) as a continuum of the ecosystems that have existed throughout geological history and a product of the trials and tribulations to which past ecosystems were subjected. Thus, the paleoecological perspective constitutes a broader view of ecology that considers how organisms (individuals, populations, or communities) have responded to biotic and abiotic factors over long time intervals.

Environmental factors that influence an organism can be abiotic (nonliving) and biotic (those involving other living entities). The abiotic influences can be chemical or physical and include water, light (particularly for photosynthesis, but also for the synthesis of other compounds in organisms), temperature, atmospheric gases, chemical substances in solution in air and water currents, gravity, pressure, and the nature of the medium in or surface on which plants and animals live and the latter move. Organisms are capable of surviving within a tolerance range for each of these parameters. The biotic factors that influence the survival of any organism include its relationships with other organisms and behavioral and functional adaptations. Understanding the history of these past associations allows us to better comprehend how the present was formed and, eventually, to prepare us for the future.

In keeping with the analogies between ecology and paleoecology, paleoecological studies focus on single species (paleoautoecology) or many species (paleosynecology) and consider organismic responses to the environment, including morphological and physiological adaptations

that help an organism meet the minimal requirements for survival, behavioral features of the organism acquired for exploiting its environment more efficiently, or the environmental impacts on an individual. As organisms do not usually function as isolated entities, but as a part of a larger population of individuals that interact, paleoecological studies of populations fall under the aegis of paleoautoecology. Some aspects of fossil populations are difficult to assess due to the fragmentary nature of the fossil record and to time-averaging. The latter reflects the loss of temporal resolution; that is, the processes that make events appear to have occurred synchronously in the geological record, even though they happened at different times (Kowaleswski 1996). Paleoautoecological studies have focused on the structure and evolution of past populations, rather than on the characters of those populations, that are useful for interpreting paleoenvironments. These population-level attributes provide information that shows the organismic adaptations for survival within abiotic (physical) and biotic (biological) limitations imposed during their geologic history.

Paleosynecological studies attempt to assess the broader view that ecosystems (total biota interacting with its abiotic environment) and biomes (interaction of regional climates with biotic and abiotic components that results in a readily distinguishable unit) imply. Approaches for researching each one of these ecological units are similar, but the research scale differs. Paleosynecological research focuses on describing, understanding, and interpreting the organisms in the context of other coexisting organisms (biocenosis). Thus, paleosynecology has to do with communities of the past: associations of coexisting preserved organisms (thanatocenosis) within a determined area (generally considered as the environment that the organisms had in common). Communities are distinguished using several basic criteria: (1) characteristics of taxonomic composition and structure that separate a particular assemblage from others at the scale of investigation, (2) internal homogeneity, (3) readily definable geographic boundaries, and (4) persistence through time and geographic distribution.

It is worth clarifying here the unit area of application in paleoecology. In some circumstances, as is the case in this book, it refers to the paleoecology of a geological formation. From the strictly geological point of view, a formation is a body or unit of rock with certain characteristics that exists in the present time. The environments that supplied and deposited the sediments and organic remains are a historical component of a formation but are not formally involved in defining a formation in a geological sense. For this reason, expressions such as "the paleoecology of the XYZ Formation" are open to scrutiny. By contrast, paleoecological approaches often assume that formations are acceptable *proxies* (representatives) of the physical environments that existed in an area during a particular period of time, including their historical aspects and their past biota, even though these aspects are not formally recognized geologically as components of a formation as it exists in its present form.

The interpretation of paleoecological data, in the conceptual framework defined here, requires practical knowledge of biology and initially proceeds by adherence to the concepts of parsimony, uniformitarianism, and analogy. Parsimony, or simplicity, is a central principle of every scientific investigation that prefers selection of the most parsimonious (simplest) explanation for the data; in other words, faced with a choice between two (or more) alternatives (hypotheses), the one requiring fewer ad hoc assumptions for explaining the observations should be selected. Uniformitarianism is based on the understanding that materials, conditions, and rates of processes have remained relatively constant through time (are largely invariant). This concept has played a very important role in the interpretation of paleoecological data. Analogy (or actuopaleontology) involves the application of features of modern organisms to extinct organisms. This principle can be applied to individuals (form and function), the community structure (species diversity, organizational and trophic structure), and the population dynamics (response to environmental factors independent of time).

Certainly, the present can, and must, be used as a guide for the past, but only when we are reasonably certain that the extinct forms are very similar morphologically, physiologically, and phylogenetically to the extant forms with which we compare them. However, as we have seen in chapters 1 and 2 (see also the next section, "Models to Investigate the Past"), many extinct organisms differ considerably from their close extant relatives; some even exhibit a general morphology not represented nowadays, and it is for this reason that we say that such organisms lack modern analogs. An example among mammals is that of sloths (Xenarthra, Folivora), which we will consider in

further detail. The few extant species are arboreal animals weighing less than 10 kg and adapted to moving and hanging from branches with their back facing the ground. By contrast, the great diversity of extinct taxa includes many that surpass by up to three orders of magnitude the body mass of modern sloths. Further, their limbs are more similar to those of mammals of other lineages, such as anteaters (Xenarthra, Vermilingua), wombats (Metatheria, Vombatidae), and pangolins (Pholidota, Manidae), although their masticatory systems and diets are quite different. Consequently, inferences made on their paleobiology cannot be based on a simple uniformitarian reference to the biology of their close living relatives. In these cases, if the use of the modern forms as elements of comparison is not duly contextualized, it often produces dissatisfactory and confusing reconstructions. In these cases, paleontologists can turn to other tools, such as using nonbiological mechanical analogs and accepting arguments not based on homology or parsimony criteria (i.e., ad hoc) or that are less well supported but retain heuristic value. Following Witmer (1995), morphological evidence is often strong enough so that attractive hypotheses, though not strictly supported by parsimony or homology, can be reasonably accepted. In a broad example, using a bear instead of an extant sloth as an analog for a giant Pleistocene sloth may require fewer ad hoc hypotheses from a functional point of view, even though it requires more ad hoc hypotheses from a phylogenetic point of view (convergence, in this case).

Models to Investigate the Past

Uniformitarianism is generally defined as the constancy of physical variables and processes through Earth history. Together with its derivative actualism (as it applies here, the use of extant species to model the structure-function and infer the behavior of extinct species), these concepts have been and continue to be among the most robust paradigms of the Earth sciences (Kay et al. 2021). Uniformitarianism is often familiarly expressed as "the present is the key to the past." In paleontology, paleoclimatic and paleoenvironmental conditions are commonly inferred based on the distribution of extant taxa that are closely related to the extinct taxa for which the paleobiology is to be inferred, as has been done by many authors (e.g., Tauber 1997; Croft 2001; Vizcaíno et al. 2010; Kay et al. 2012b, 2021) for the Santacrucian fauna.

However, whereas this approach is generally appropriate, its restricted and uncritical application, termed *naive actualism* by Vizcaíno et al. (2017), may generate hypotheses that are poorly supported and have little or no heuristic value; indeed, in many cases this condemns the past to have been just like the present (see Vizcaíno et al. 2023). A clear example is furnished by the paleoclimatic inferences that would be drawn based only on the current distribution of the jaguar *Felis onca* and the Pleistocene records of this species from Patagonia and the pampas region of Argentina. Whereas this species today lives exclusively in tropical closed environments, in historical times it had a much wider distribution, including arid environments. Another example, noted by Kay et al. (2012b), would be assignment of the climatic conditions of modern Andean-Patagonian forests to those of the early Miocene based on the present distribution of microbiotheriid marsupials. Yet another, noted by Kay et al. (2021), would be the unwarranted inference that the paleoenvironment of the Santacrucian, based on the presence of the caenolestid paucituberculatan *Stilotherium dissimile*, resembled the high-elevation cold and wet climate inhabited by extant caenolestid marsupials. The reconstruction of paleoenvironments and paleoclimates necessitates comprehensive analyses that take into account the many interweaving threads of biotic and abiotic evidence, as the intrinsic and extrinsic factors that establish the current distribution of taxa do not necessarily reflect their maximal ranges of environmental and climatic tolerances (Vizcaíno et al. 2017).

Analogously, a species' morphology and behavior are determined by a complex interplay of phylogeny and adaptation. As such, it cannot be assumed that an extinct animal's morphology is solely a function of the niche and habitat occupied by its extant relatives. This is particularly true for many extinct xenarthrans, a group abundant in Santacrucian times and indeed throughout most of the South American Cenozoic. Large sloths and glyptodonts differ morphologically from their living sloth and armadillo relatives to such a degree as to readily suggest markedly different modes of life; they have no modern analogs, and "naive" application of an actualistic approach may produce nonsensical reconstructions (Vizcaíno 2014; Vizcaíno et al. 2018). This does not invalidate actualism, but for such cases, a wider comparative net must be cast to include other, more distantly related mammals and biomechanical approaches that address form-function relationships (Vizcaíno et al. 2018).

"The present is the key to the past" is not the sole paradigm among the historical branches of the Earth sciences; indeed, several of them have also proposed that "the past is the key to the present" and even "the future." For example, in ichnology the expression *reverse uniformitarianism* has been applied to situations where structures that were first described and interpreted from the fossil record were later identified as also existing in the present and produced by modern organisms (Frey and Seilacher 1980). And a major frontier in geological research, initiated in the 1970s, involves predicting future geologic trends or events based on study of past and present events and processes (Doe 1983).

It may, further, be the case that the past can also be a key to the past. Vizcaíno et al. (2004) articulated this concept in suggesting that fossil forms could serve as biomechanical analogs for fossils belonging to different lineages or lacking living representatives. These authors applied this concept using the armadillo *Peltephilus* from the Miocene of Patagonia as a model for interpreting the enigmatic Paleocene *Ernanodon antelios* from Asia. The biomechanical study of the masticatory apparatus of *Peltephilus* by Vizcaíno and Fariña (1997) revealed a pattern not recorded among extant mammals but similar to that of *Ernanodon*. Vizcaíno (2014) proposed such an approach, with fossil forms acting as analogs for extinct forms that are distantly related or lack living representatives, as a robust investigative tool for analyzing evolutionary paleobiology. This approach may be referred to as fossilism (Kay et al. 2021) and may be extended to encompass consideration of paleosynecology, allowing the possibility for well-understood extinct faunas and biotas becoming models for investigating and generating hypotheses for other extinct faunas. A more detailed example of this approach, based on the Miocene Santacrucian fauna of South America, is considered further below.

Paleoecological Reconstructions

The application of the protocol for paleobiological studies, described in chapter 1, permits paleoautoecological reconstructions based mainly on biotic factors, including the three interrelated disciplines of functional morphology, biomechanics, and ecomorphology, which have been treated thoroughly in previous chapters. Another discipline that has grown recently is paleohistology (see below). Further, several methodological approaches permit incorporation of a wider range of abiotic factors that contribute to a paleosynecological understanding, thus allowing the development of more complete paleoecological reconstructions. Among them are taphonomy, ichnology, and isotopic paleontology, and, particularly in the context of our discussions, ecometrics, which provides a link for a paleoecological integration among form, function, and environment.

Paleohistology

The fossilization process can preserve not only the external form of hard tissues but also internal microscopic

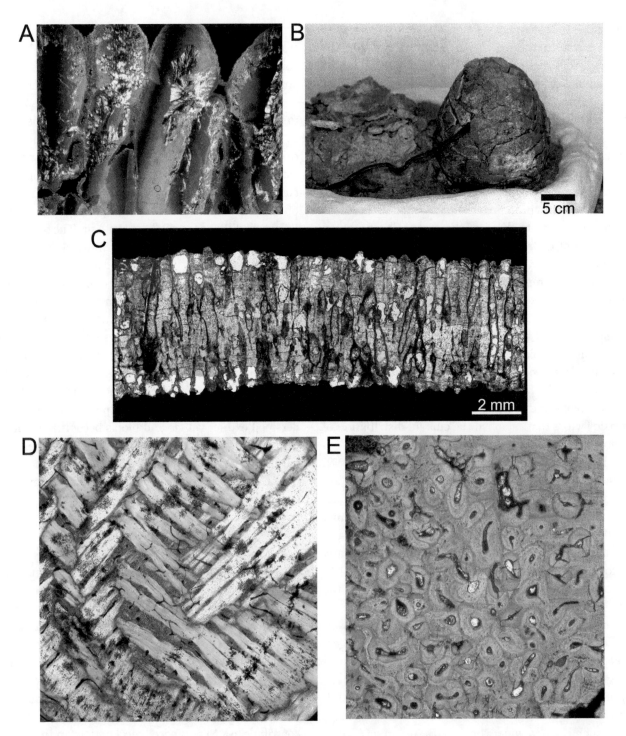

9.1. Paleohistology. *A*, light microscope image of a portion of a faveoloolithid eggshell, a type of spherical egg approximately 20 cm in diameter, assigned to sauropod dinosaurs (image courtesy of Mariela S. Fernández). *B*, egg of a sauropod dinosaur from the hydrothermal Cretaceous site in Sanagasta (La Rioja Province, Argentina), a locality preserving evidence of colonial nesting that allowed confirmation of both extended periods of parental care and a symbiotic reliance of the broods on the singular hydrothermal environment that favored incubation (Grellet-Tinner and Fiorelli 2010). *C*, transmission light microscope image of a section of an eggshell from the Sanagasta site (images *B* and *C* courtesy of Lucas Fiorelli). *D*, image of a sectioned osteoderm of the armored titanosaur dinosaur *Saltasaurus loricatus* from Argentina. *E*, histological section through a femur of a basal sauropodomorph dinosaur from the Triassic of Argentina showing dense Haversian bone tissue. Images *D* and *E* courtesy of Ignacio A. Cerda.

structure. Paleohistology examines, through thin sections, the fossilized remains of bone, ossified tendons, eggshells, and teeth (fig. 9.1), among others. Sectioning of such material combines methodologies of petrography, metallurgy, and bone histology. The standard process (Lamm 2007; Cerda et al. 2020) involves preliminary photography, measurement, drawing, and even molding of the specimen or sample. In this regard, sampling procedures involve irreversible modifications or even destruction of the specimen, and their implementation therefore requires careful consideration. The original sample is then embedded in resin under a vacuum, and thin sections (<1.0–2.0 mm thick) are cut from this block with a diamond-blade circular saw, a diamond-bladed drill, or a handsaw. It is recommended that the sectioned area be replaced with a cast to preserve the original morphology of the specimen. The section to be studied is embedded into epoxy resin, fastened with glue or the resin itself to a ground glass, and this assemblage is cut and polished to a thickness of 50–250 microns. The sections are examined under a light microscope to investigate histological features of the fossilized tissue, with plane-polarized light being the most common method to enhance contrast.

Paleohistology can provide information on the physiology, ontogeny and growth, pathology, functional anatomy, and behavior of extinct animals (Lamm 2007; Cerda et al. 2020). There are many examples of paleobiological inferences on dinosaurs based on paleohistological studies, including, for example: growth dynamics and physiological features in a population of individuals of several age ranges (Horner et al. 2000; Zhao et al. 2019); approximate age and developmental stage of an individual using growth lines combined with methods for retrocalculating remodeled bone (Horner and Padian 2004); early sexual maturity (Cerda and Chinsamy 2012); pathological reactions to injuries and diseases (Hanna 2002); an exoskeletal covering of the frontoparietal dome based on an abundance of Sharpey's fibers in skull bones (Goodwin and Horner 2004); absence of locomotion in juveniles and thus requirement of parental care based on the presence of heavily calcified cartilage in the ends of long bones of nesting juveniles (Horner et al. 2001); and water-vapor conductance in eggshells (Grellet-Tinner et al. 2012). Paleohistology has also been applied to Mesozoic marine reptiles; for example, in studies of ontogenetic variation in ichthyosaurs (Talevi et al. 2012), on relative osseous density among different species and its implication on the way of life and swimming mode of ichthyosaurs (Talevi and Fernández 2012), and the comparison between the level of osseous remodeling between different elements of the same skeleton, which provides an idea of

the use of a particular element or of its parts (Talevi and Fernández 2015).

The paleohistological study by Whitney and Pierce (2021) investigated the paleobiology of *Greererpeton*, an early tetrapod from the early Carboniferous of North America. These authors reasoned that the bony signature in the femora could best be explained as indicating a dispersal event, a common life history strategy among aquatically bound vertebrates, as opposed to being related to events such as polymorphism or migration. Paleohistological investigations have also been conducted on early synapsids. Besides permitting assessment of growth patterns and variation within or among closely related therapsid species (e.g., Botha and Angielczyk 2007; Huttenlocker and Botha-Brink 2013, 2014), analysis of growth patterns (which are associated with organismic fitness) has been employed by Botha-Brink and Angielczyk (2010) to investigate whether certain growth strategies conferred success on some groups during the end-Permian mass extinction, such as, for example, rapid growth in the dicynodont *Lystrosaurus* and its relatives.

Taphonomy

In 1940, aware that the greatest difficulty in paleontological reconstruction was the unconnected and incomplete nature of the remains and the causal factors leading to their preservation in rock, the Russian paleontologist Ivan Antonovich Efremov (1908–1972; fig. 9.2) developed the theoretical and practical basis for a new field of paleontological investigation, for which he proposed the term *taphonomy*, from the Greek taphos (τάφος) meaning burial and nomos (νόμος) meaning law. This field deals with the transition (in all of its details) of organismic remains from the biosphere to the lithosphere; that is, it is the study of a process that results in the complete separation of organisms (or their parts) from the biosphere and, once fossilized, incorporation into the lithosphere (Efremov 1940). This author stated that the indissoluble unity of biological-geological analysis is the key to the most important problems in paleontology, which could not be determined by then-conventional methods.

More broadly, taphonomy deals with the processes through which an organism becomes part of the fossil record and how these processes influence the conservation of the information in this record (Gastaldo et al. 1996). Recent discussions on the application of this term imply that taphonomy is typically framed in terms of the processes that turn one or more organisms into objects of interest to paleobiologists. These objects can be the microscopic and macroscopic remains of plants or animals, their condition

9.2. Ivan Antonovich Efremov (1908–1972). The Russian paleontologist, born near St. Petersburg, established taphonomy as a new branch of paleontology. A renowned writer, he wove his scientific works into science fiction novels. He directed several paleontological expeditions to the Gobi Desert to important Upper Cretaceous sites preserving dinosaur fossils. The expeditions provided inspiration for some of his novels. Image from the Borissiak Paleontological Institute of the Russian Academy of Sciences; courtesy of Sergey Rozhnov.

with respect to a great variety of variables, and their locations, orientations, distributions, and associations. According to Lyman (2010), taphonomists focus on the search for answers to questions such as: Are the remains different from when they formed part of an extant organism? If so, how different are they? What do the similarities and differences indicate about paleobiology and paleoecology?

Taphonomic processes include events that affect an organism at the end of its life, transference of the organism (or part of it) from the living world (biosphere) to the sedimentary record (lithosphere), and the physical and chemical interactions that affect the organism from the moment it becomes buried until it is recovered in the field. Before this can happen, however, the remains of an organism must successfully pass through three different and unrelated stages (fig. 9.3).

The first stage, necrobiosis, encompasses events that an organism undergoes during its death or a loss of one or more of its parts. Although we generally think of organisms as having to die before passing to the next stage, this is not necessarily the case. Most plants, for instance, do not need to do so to contribute one or several of their parts

to the fossil record. Many animals, too, can leave parts in the environment without dying, such as eggshells, scales, teeth, feathers, hair, antlers, and all or part of the tegument or exoskeleton in molting.

The second stage, biostratinomy, includes all the interactions that the remains of an organism undergo in the transition from the living to the inorganic world, ending in definitive burial. Apart from the preservation of the more obvious external and internal characteristics of the fossil, subtler details that record what happened to the organism (or a part of it) before fossilization are often preserved. The study of such details allows understanding of how death occurred or how part of an organ became detached (necrobiosis), the biological processes that may

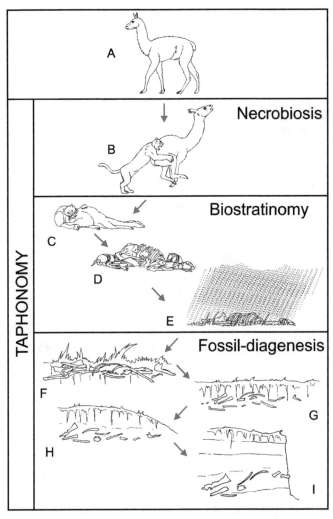

9.3. Stages of taphonomic processes, as exemplified in a guanaco. *A*, during necrobiosis, all or part of the organism becomes biogenic remains, in this case by means of predation. *B*, during biostratinomy, the biogenic remains undergo disarticulation, fragmentation, and other alterations as a result of the activities of predators (*C*), scavengers and decomposers (*D*), and exposure to weather (*E*). Fossil-diagenetic processes may begin with final burial (*F*) and cause the remains to suffer chemical and physical alterations that end with their recollection at the fossil site (*G–I*). Modified from Shipman (1981).

have modified the remains before burial (e.g., decomposition, disarticulation, fragmentation, chemical alteration, scavenging), agents of transportation (e.g., animals, water, wind), and the interval, termed *residence time*, that the organism remained in the environment before burial. Further, the preservation in recognizable form of organic remains requires the existence of specific physical and chemical conditions in the burial environment that minimize or deter biological (e.g., enzymatic and bacterial) and chemical (e.g., dissolution and recrystallization) processes that degrade or destroy the remains.

Burial can be autochthonous, with remains buried in the same habitat in which the organism lived; parautochthonous, with remains transported from the site of death though still buried within the same habitat; and allochthonous, with burial occurring in an environment located distantly from the habitat in which the organism lived. Thanatocenosis refers to an assemblage of remains comprising organisms that died in the same place, whereas taphocenosis refers to assemblages of organisms from different environments. Biofabric is the primary spatial orientation of the components from the biological origin of a sediment; it reflects dynamic conditions at the moment of deposition and allows carrying out paleoenvironmental reconstructions.

The third stage, diagenesis, may follow burial. Diagenesis involves all processes responsible for sediment lithification and chemical interactions with the water existing between clasts (called *pore water*). These generally long-term processes produce the changes that sedimentary deposits undergo before becoming rocks; they commonly include cementation, recrystallization, replacement, dissolution, and compaction. These processes, often long-lasting, affect the organic remains contained in sediments and constitute the main modelers of fossils. The fossilization processes (sometimes called *fossil-diagenesis* to distinguish them from general diagenesis) seem to be site specific with respect to deposition environments. Although no two

9.4. Predation marks in an Upper Cretaceous peirosaurid crocodyliform from Patagonia. Many of the preserved elements of this specimen, particularly its osteoderms and caudal vertebrae, show clear signs of injury and bite marks. The taphonomic and faunal context analyses indicate that the presumable predator of this 3 m long peirosaurid crocodile may have been a larger crocodile or a theropod dinosaur (Fiorelli 2010). *A*, nearly complete caudal region. *B*, detail of articulated tail that exhibits several bite marks (perforations) and other injuries (*white arrows*). Images courtesy of Lucas Fiorelli.

fossil assemblages are identical, particularly with regard to the manner in which they were formed, general patterns may be recognized. Understanding features of the taphonomic assemblage within an environmental context allows a more accurate interpretation of the fossil record.

An example of the biostratinomical analysis is the report of abundant bite marks and injuries in a crocodyliform peirosaurid of the Upper Cretaceous from northern Patagonia (Fiorelli 2010). This specimen presents injuries distributed across all its preserved parts, with the highest concentration of bite marks, perforations, and breakage in the caudal region. The author concluded that injuries represent bite marks made by a predator rather than the result of intraspecific combats (fig. 9.4). Another example is the report of predation by a Miocene crocodylian on the mylodontid sloth *Pseudoprepotherium* (Pujos and Salas-Gismondi 2020), based on the pattern and size of the pits and punctures preserved on the left tibia of the sloth.

The work of Montalvo et al. (2008, 2019) furnishes examples of taphonomic analysis. Montalvo et al. (2008) conducted a taphonomic analysis of an assemblage of vertebrate remains from the Cerro Azul Formation (late Miocene, La Pampa Province, Argentina). The fossil-bearing component of this formation was interpreted as a loessic deposit (sediment of aeolian origin) with two similar and poorly developed paleosols. The sample was collected in an area of 48,000 m² and contained 5,598 remains, which included microvertebrates and macromammals. Most remains, disarticulated and highly fragmented, ranged from small to very small in size. The authors inferred two taphonomic histories. The association of microvertebrates was considered the result of predation activities (biostratinomy); after a short period of preburial exposure, the remains were dispersed from their area of original deposition. With regard to the macromammals, they interpreted that there was a process of natural and gradual mortality (biostratinomy), followed by a lengthy period of exposure to weathering and dispersal by physical agents. Once buried, the remains of both groups underwent diagenetic processes as the sediments lithified within the bearing rock. In short, this faunal association would represent a condensed assemblage corresponding to two different periods; the microvertebrate accumulation took place over a relatively short time interval, whereas that of macromammals occurred over a longer period and coincided with the formation of both paleosols.

Montalvo et al. (2019) conducted a taphonomic analysis of the Santa Cruz Formation (early Miocene) of southern Patagonia. Their effort is notable for, among other reasons, being the first such analysis to investigate the continental vertebrate assemblage from a single stratigraphic level and/or restricted locality of this formation. The assemblage includes mainly mammals (143 specimens), three birds, and several coprolites of carnivorous mammals recovered from an area of 123,349 m². The taphonomic features and the sedimentological interpretation suggest that the skeletal remains were preserved in the floodplain of a fluvial system influenced by volcaniclastic input. Death of the mammals was interpreted as having occurred over a short time span, and the residence time of carcasses on the surface was variable. Although most skeletal elements were quickly buried, exposure was sufficiently long for their disarticulation and scattering and the action of other processes, such as trampling and weathering. The loss of skeletal elements was attributed to the action of water flows generated by rainfall and over-bank flooding, which transported those elements more susceptible to being moved. After burial, diagenetic processes (e.g., soil corrosion, infilling, impregnation, deformation) affected the specimens. Sedimentological and taphonomic evidence suggest the floodplain taphonomic mode sensu Behrensmeyer (1988) for this Santacrucian assemblage.

Ichnology

Ichnology, from the Greek *ichnos* (track, trace) is the study of the products, termed *traces*, of organismal behaviors (Bromley 1996; Hasiotis 2007). A trace is a physical, three-dimensional structure produced by the interaction of an organism with its environment and can thus provide information on an organism's behavior. Traces are generally classified as resting, locomotion, dwelling, feeding, escaping, grazing, farming, or multipurpose (Bromley 1996; Hasiotis 2002; Buatois and Mángano 2011). Although this classification was developed initially for marine invertebrates, it can readily be extended conceptually to vertebrates.

Ichnology comprises neoichnology, the study of modern producers of traces, their tracks, and behavior, and paleoichnology, the study of fossilized traces, which are usually referred to as trace fossils (fig. 9.5). This includes the description, interpretation, classification, and distribution of traces left behind by microbes, plants, invertebrates, and vertebrates and their distribution (Miller 2007). Modern and fossil traces can be used to interpret the biophysicochemical parameters of an environment, such as medium consistency, depositional energy, sedimentation velocity, nutrient availability, salinity, atmospheric oxygen concentration, sediment or pore water, and the level of water turbidity (Hasiotis 2007; Hasiotis et al. 2007).

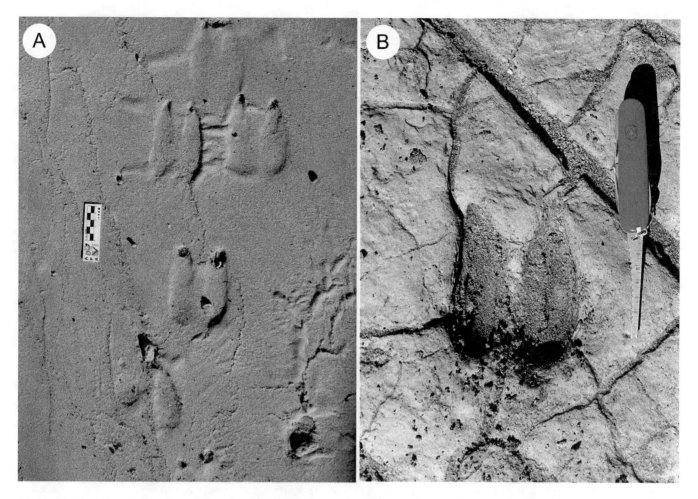

9.5. Neoichnology and paleoichnology. *A*, footprints of extant guanacos (*Lama guanicoe*). Photos taken in Talampaya National Park, La Rioja, Argentina; image courtesy of Verónica Krapovickas. *B*, fossil footprint of the large extinct camelid *Megalamaichnum tulipensis* preserved at Pehuen Có, Buenos Aires Province, Argentina (Aramayo and Manera de Bianco 1987); photo by the authors (SFV and MSB).

In a strict sense, the inferences resulting from morphofunctional approaches that we have applied to fossil vertebrates are unlikely to be verifiable, as it is impossible to observe behavior of fossilized organisms, but from time to time ichnology provides indirect evidence about that behavior. For instance, traces (such as trackways) assignable to taxa under study can offer a tool for supporting or developing functional hypotheses on substrate use, including gait styles and behaviors such as digging and scratching. They are also useful variables for calculating the size of the organisms that produced them.

In most cases, vertebrate traces and skeletal elements are not associated in the same stratigraphic units or facies. Thus, in the absence of the latter, traces may become the main source for recording the presence of the producer organism at those levels. Because tracks are usually preserved in situ in the facies where they were produced, they represent markedly reduced temporal and spatial margins of error for analyses of the taxonomic composition and faunal distribution, and a close relation with the environments where the track producers lived. Therefore, fossil tracks are useful tools for such analyses and for paleoenvironmental and paleocommunity interpretations.

Within an ichnological context, paleoecology can be approached from two perspectives. One is the reconstruction of environmental parameters, such as the consistency of substrate, climate, nutrient supply, topography (e.g., littoral, exposed fluvial bars), water saturation, current energy, salinity, oxygenation, sedimentation velocity, light, and temperature, among others. The other is to analyze the paleoautoecological aspects of different species that allow reconstruction of paleosynecology (Falcon-Lang et al. 2007; Krapovickas et al. 2009; Minter and Braddy 2009; Wilson et al. 2009; Kubo 2011). Although these approaches have not been commonly applied in analyses of South American ichno-assemblages (Krapovickas and Vizcaíno 2016), we consider below several of those that have been conducted.

9.6. Reconstruction of a Miocene landscape (Toro Negro Formation) from Vinchina, La Rioja Province, Argentina (from Krapovickas 2010). Paleoenvironmental interpretation was carried out by Ciccioli (2008); fossil traces of invertebrates and vertebrates were studied by Krapovickas (2010), who made inferences about the producers. Among mammals and birds, the latter author found tracks assignable to proterotheriid litopterns (*right foreground*), medium-sized rodents similar to cavids (*left*), and cursorial birds and medium-sized terrestrial sloths (*right background*). Courtesy of Verónica Krapovickas. Artist: Boris Budiša.

Krapovickas et al. (2009) used fossil tracks for inferring the paleoenvironmental features of the Toro Negro Formation (Miocene–early Pliocene; La Rioja Province, Argentina). The tetrapod fauna of the lower interval of this formation, revealed by its tracks, suggests an open environment, possibly related to a seasonal arid or semiarid climate, associated with closed forested zones (fig. 9.6).

Manera de Bianco and Aramayo (2004) studied the taphonomy of the ichnological site from Pehuen Có (late Pleistocene–early Holocene, Buenos Aires Province, Argentina) by considering the association of tracks, preservation quality, producers, and substrate. They found increased preservational quality in mudstones that lack cracks and ripple marks (fig.9.7), which reflects the potential for differential preservation related to the proximity to the water body of the flooded plain.

Most ichnologically based paleoautoecological studies are qualitative or biomechanical analyses of locomotion or other substrate uses of producers. Casamiquela (1974) analyzed the Pleistocene ichnofossil *Megatherichnum oportoi*, attributable to a megatheriid, and concluded that the absence of manus tracks reflected bipedal locomotion, in

accordance with many skeletal characteristics of megatheriids. Casinos (1996) studied the locomotory biomechanics of the Pleistocene megatheriid *Megatherium americanum* by means of body-mass estimates and the use of measurements of several tracks originally assigned to megatheriids (Frenguelli 1950; Bonaparte 1965; Casamiquela 1974, 1983; Aramayo and Manera de Bianco 1987). This author calculated several mechanical parameters, such as velocity, Froude number, athletic capacity markers, and flexion and resistance movements of the spine in bipedal and quadrupedal postures. Although Casinos's (1996) analysis could not discriminate whether *Megatherium* was better suited for one or the other posture during locomotion, it indicated that this megatheriid walked at low speed.

Blanco and Czerwonogora (2003) reconsidered the locomotory capacities of *Megatherium* based on tracks (fig. 9.8) from the Pehuen Có site that were assigned to *Neomegatherichnum pehuencoensis* by Aramayo et al. (2015). Blanco and Czerwonogora (2003) developed a geometrical model for calculating the percentage of body weight supported by each pair of limbs to estimate the pressure the animal exerted on the ground and calculated the locomotion

9.7. Paleoichnological site from Pehuen Có, located along the southeastern coast of Buenos Aires Province, Argentina. *A*, partial view of the site on the coastal abrasion platform, with Professor McNeill Alexander and M. Susana Bargo taking measurements of a trackway of camelid footprints. Detail of *B*, camelid footprints; *C*, bird footprints; and *D*, two glyptodont footprints. Photos by the authors (SFV and MSB).

velocity following Alexander (1976). Their conclusion was that *Megatherium* walked slowly as a strategy for minimizing transport costs and for better supporting the great lateral flexion moments that occur during bipedal walking.

Paleoburrows, with or without filling and of varied dimensions, represent another kind of frequently encountered trace fossil that has been ascribed to vertebrate activity. These biogenic structures can provide information about particular behaviors of the producer organism.

Vizcaíno et al. (2001) reported paleoburrows in Pleistocene deposits in the coastal zone of Buenos Aires Province (Argentina) and attributed them to the activity of extinct mammals based on physical aspects, transgressive boundaries with respect to sedimentary units, the presence of claw marks on walls and roof of the burrow, and congruence with morphological and biomechanical attributes of the fossil mammals. Burrow diameters ranged between 0.80 and 1.80 m, with width generally exceeding height.

9.8. Paleoichnological site at Pehuen Có, southeastern coast of Buenos Aires Province, Argentina. *A* and *B*, fossil trackways of a megatheriid, assigned to *Neomegatherichnum pehuencoensis* (Aramayo and Manera de Bianco 1987). *C*, detail of a footprint, left foot. Photos by the authors (SFV and MSB).

Some of these paleoburrows were previously attributed to large Pleistocene armadillos, such as *Propraopus*, *Eutatus*, and *Pampatherium*. Vizcaíno et al. (2001) proposed that some terrestrial sloths, such as *Scelidotherium* and *Glossotherium* (Mylodontidae), possibly constructed the largest structures. Similarities in the diameters of the paleoburrows and the sloths support this hypothesis (fig. 9.9), as do anatomical, allometric, and biomechanical studies on the limbs of these sloths, which indicated that they were well suited for excavating; in addition, the shape of some claw marks preserved on cave walls coincides with the morphology of the skeleton of the manus (Bargo et al. 2000).

Isotopic Paleontology

Isotopic paleontology uses the isotopic composition of the remains of fossil organisms for inferring aspects of the physical environment that sustained their development and for obtaining information on several biological

variables. Isotopes (Greek root *isos*, meaning equal) are variants, owing to differences in nuclear composition, of a chemical element. Isotopes of an element have the same number of protons but differ in their atomic mass due to the different number of neutrons in the nucleus. They also have the same chemical properties, except in very few cases, such as protium and deuterium (isotopes of hydrogen), in which the effect of the relation between neutrons and protons is sufficient to alter chemical properties (Urey 1947). Isotopes differ in their times of radioactive disintegration or decay; the overall decay rate can be expressed as a decay constant or as a half-life. When radioisotopes of an element disintegrate (or decay), they emit radioactivity and decay—that is, transform—to a daughter isotope. This process is successively repeated in a decay chain until a radioactive isotope transforms into a nonradioactive or "stable" form, though not necessarily of the same element (e.g., a radioactive form of carbon, ^{14}C, decays ultimately into a stable form of nitrogen, ^{14}N). The

9.9. Paleoburrows: biogenic structures attributed to vertebrate activity. *A*, silhouette of a human (for scale) and life reconstruction of the ground sloth *Scelidotherium*, one of the potential producers of the paleoburrows preserved in Pleistocene outcrops from Buenos Aires Province, Argentina (Vizcaíno et al. 2001). *B* and *C*, paleoburrows of different diameters in the coastal cliffs in the area of Mar del Plata city, Buenos Aires Province, Argentina. Photos by the authors (SFV and MSB).

elements are found in nature in varied isotopic forms, but the proportions of stable isotopes (i.e., nondecaying) do not change over time. For instance, carbon exists in nature as the radioactive isotopes ^{10}C, ^{11}C, and ^{14}C, and the stable isotopes ^{12}C and ^{13}C; of the latter type, ^{12}C constitutes 98.99% of stable carbon, with ^{13}C comprising the remaining 1.11%. Oxygen occurs as the stable isotopes ^{16}O and ^{18}O and the unstable isotope ^{17}O.

Stable isotope analyses of fossil remains are becoming increasingly important for compiling information on the diet and environment of extinct species in terrestrial and aquatic ecosystems. The advantages of these analyses result from the geochemical signal left in the animal's tissues, especially on bones and teeth, by the environment in which an animal lives. Studies carried out in extant mammals reveal that the stable isotope composition of

light elements (hydrogen, nitrogen, carbon, oxygen, and sulfur) and even some heavy elements (calcium and strontium) serve as markers of ecological and physiological information; similarly, many of these can be applied to the study of extinct mammals (Clementz 2012). For instance, the isotopic composition of carbon in an animal's tissues provides insight on the nature of its diet, whereas the oxygen isotopic composition of carbonates and phosphates in bones and teeth mainly reflects the availability of water (through drinking or from food). These stable isotopic markers for diet and habitat are independent of the inferences on diet and habitat based on morphological features; as such, they provide a means for complementing the ecological interpretations based on the form of fossil remains, or vice versa. Therefore, when well-preserved specimens are available, any feeding study on the fossil species should strongly consider including this approach (Clementz 2012).

The isotopes can be used in paleobiology and paleoecology in two general ways. The distribution of isotopes of light elements (e.g., H, C, N, O, S) in nature is a function of chemical, physical, and biological processes. The variations in the isotopic relations can be used for controlling the magnitude and rate of these processes. Differences in those associated with carbon fixation, for instance, produce a marked difference in the carbon isotope composition among plants that use the C_3 versus the C_4 photosynthetic pathways. Natural isotopic differences can be used among substances for tracing their flow through biological systems. A straightforward example of this approach is the use of carbon isotopes in vertebrate tissues for tracing the dietary proportion of C_3 and C_4 plants. Those plants that rely solely on the C_3 pathway are indicative of environments characterized by bushes and trees and cold-climate grasslands (i.e., by moderate to high sunlight intensities, moderate temperatures, and plentiful groundwater). C_4 plants, by contrast, are indicative of mainly tropical and warm-temperate regions with predominantly open grasslands (Sage 2016).

Many analyses, following the principle of form-function correlation, infer biological variables of extinct species by means of the study of an average morphology, and in most such studies the paleobiology of organisms of the same species is considered as equivalent. However, different individuals of a population or species have different behaviors, habitats, and histories that powerfully influence ecological and evolutionary processes, as illustrated, for example, by the seasonal biological changes recorded in the proboscideans *Mammut americanum* (a mastodontid) and *Mammuthus jeffersonii* (an elephantid, and now often considered a synonym of *Mammuthus columbi*; see Grayson 2016) by Koch et al. (1989). The isotopic composition of oxygen present in precipitation water varies seasonally with temperature, and mastodonts reflect this seasonal change in the growth layers of their tusks. In addition, there is evidence that humans slaughtered some individuals. By tracing isotopic oscillations in growth layers, Koch (1998) demonstrated that the slaughtered individuals tended to die massacred at the end of autumn, whereas those not slaughtered by humans died at the end of winter or beginning of spring, suggesting that different processes (hunting vs. natural death) were responsible for mortality. Another application of isotopes on proboscideans is the analysis by Wooller et al. (2021) on the sequential analyses of strontium isotope ratios along an entire tusk to reconstruct the movements of an individual of *Mammuthus primigenius* (Arctic woolly mammoth) that lived approximately 17,100 years ago in northwestern North America. This study is in part based on the fact that strontium isotope ratios ($^{87}Sr / ^{86}Sr$) in soils and plants exhibit strong and predictable geographic variations that change minimally at millennial time scales and primarily reflect the underlying bedrock geology. Such information can be used to generate isotopic maps. The ingested local strontium $^{87}Sr / ^{86}Sr$ patterns of the environment are reflected in an animal's tissues. This information may be used in conjunction with the isotope maps to trace an animal's movement. Wooller et al. (2021) made use of such methodology, augmented by stable oxygen isotope analyses, to determine the provenance of the mammoth, which was at least 28 years old at the time of its death, over four life stages.

Isotopes of nitrogen provide another example. Differences in proportions of these isotopes in bones and teeth reflect variation in the amount of animal protein in the diet (meat), which increases with higher trophic levels. Further, in the case of mammals (including humans), lactating individuals exhibit nitrogen isotopes levels equivalent to those of carnivores due to the presence of the mother's proteins in the milk; thus, if they are not identified as lactating juveniles, they may appear to represent a higher trophic level than their parents.

Isotope analysis has also been used for characterizing aspects of biology and ecology at the species level. For instance, by using oxygen isotopes, Barrick et al. (1996) evaluated whether various species of dinosaurs were homeotherms or heterotherms. Recently, the composition of oxygen isotopes in teeth of Mesozoic marine reptiles was used to assess strategies of thermoregulation in these vertebrates (Bernard et al. 2010).

The geochemistry of vertebrate isotopes also allows inferring the structure of terrestrial ecosystems and their processes, and how these characteristics have changed through Earth history. In many paleontological sites from Texas and Florida (US), organisms coincident with diets excluding one another have been found, either 100% C_4 grasses or 100% C_3 trees and bushes. If these deposits can be considered an ecologic snapshot (i.e., an instant in time), isotopic data may indicate that there might have been a mosaic of vegetation over the range of the preserved organisms. Alternatively, if the deposits represent the passage of several thousand years, they might indicate ecotonal change in the region (Koch 1998; Koch et al. 1998).

APPLICATIONS IN PALEOBIOLOGY

This section presents several examples that illustrate the use of isotopes to investigate the ecology of extinct animals, three as applied to Northern Hemisphere fossil proboscideans by Clementz (2012) and two to South American megafauna by Czerwonogora et al. (2011) and Dantas et al. (2020).

On the North American proboscideans, the first example is a study of serial samples of tusk enamel corresponding to the estimated annual growth in 17 individuals of the gomphotheriid *Gomphotherium productum* from the Great Plains of North America (fig. 9.10*A*). The samples were taken for determining the trophic ecology of this species and establishing the environmental conditions during the middle–late Miocene (around 15 to 8 Ma). Profiles of $\delta^{13}C$ suggest a diet consisting of C_3 vegetation, indicating that it was either a browser or a mixed feeder (browser and grazer). The values correspond to those at the upper extreme for a C_3 consumer, thus suggesting that the *Gomphotherium* individuals fed on partially open, and possibly arid, environments with scrub rather than closed forests. The variation in $\delta^{13}C$ values among tusk layers was low, suggesting little seasonal variation in diet. The values of the oxygen isotopes also varied slightly (around 1.5%), but they indeed exhibited cyclic changes of $\delta^{18}O$ from high to low and then again to high values, probably reflecting seasonal changes in rain and temperature (high $\delta^{18}O$ values indicate rain during the warm season, and low $\delta^{18}O$ values indicate rain during the cold season). This suggests that the seasonal changes in rain did not correspond to seasonal changes in the availability of C_3 vegetation. The lack of significant differences or of an evident trend in $\delta^{13}C$ and $\delta^{18}O$ values among individuals of *G. productum* from different times and localities implies that the environmental conditions in the Great Plains did not change or were not markedly degraded from 15 to 8 Ma and that the scrub habitats preferred by *G. productum* were present during this time interval.

In the second example, serial sampling of the tooth enamel of the mastodontid *Mammut americanum* allowed investigation of the effects of changing environmental conditions at the end of the Pleistocene on its distribution range and migration patterns (Clementz 2012). In the case of strontium isotopes, the proportions in consumers and producers track those of the soils and bedrocks of an organism's habitat (because the mass difference between isotopes is too small compared with the element's atomic mass to permit measurable biological fractionation) without variation based on trophic level. Thus, an animal's movement between regions with distinct $^{87}Sr / ^{86}Sr$ values is reflected by oscillations in the values recorded in tissues such as tooth enamel. These data may then be used to track the seasonal movements of an animal between these regions. To this end for their analysis of *Mammut*, Hoppe et al. (1999) constructed an isotopic map of the geographic

variation in the strontium isotope proportion ($^{87}Sr / ^{86}Sr$) in the composition of local parent rocks, soils, and plants from the southeastern US. The gray-shaded zone in fig. 9.10*B* marks the range expected in $^{87}Sr / ^{86}Sr$ values for consumers in Florida. The enamel of *M. americanum* showed a significant variation in $^{87}Sr / ^{86}Sr$ values in molar enamel, with values that exceed the range for plants from Florida. This suggests that these animals moved long distances throughout the year (around 120 to 300 km in one direction), as far as the highlands of Georgia, implying a home environment far larger than that of extant elephants.

The third case is a study on the lactation and time of weaning of a young Pleistocene woolly mammoth (*Mammuthus primigenius*) from Wrangel Island, Siberia (fig. 9.10*C*). Milk consumption by young mammals leaves a different isotopic marker in $\delta^{13}C$, $\delta^{15}N$, and $\delta^{18}O$ values in the tissues, since they feed at a higher trophic level than their mothers (owing to the mother's contribution by way of her milk). The analysis of $\delta^{15}N$ profiles of a young woolly mammoth tusk shows a cyclic pattern consistent with the seasonal changes in its diet and its mother's (as reflected in the milk). This pattern is also reflected in $\delta^{13}C$ values. Values of $\delta^{15}N$ also show a consistent decline in the maximum values during each cycle, which has been interpreted as reflecting a progressive decrease in the contribution of maternal milk to the juvenile's diet. This suggests a long weaning period for the mammoth calves that, based on the count of the number of recorded cycles, may have occurred over a period exceeding five years. The weaning period for extant elephants is about 3.5 years, but in stressful conditions, the maternal investment in its calf can be extended to 5.5 years. The interpretation of a long weaning period is consistent with the environmental reconstructions for its time period and location (i.e., the late Pleistocene of Wrangel Island). Adverse climatic conditions toward the end of the last ice age may have forced mammoth females to spend more energy and time on each calf, reducing the number of descendants that they could produce.

As for South America, Czerwonogora et al. (2011) analyzed the stable isotopes ^{13}C and ^{15}N to infer the diet of two late Pleistocene species of mylodontid ground sloths, *Lestodon armatus* and *Glossotherium robustum*, from moderate latitudes in Uruguay and the Pampean region of Argentina. The average results for *L. armatus* were -18.8‰ for $\delta^{13}C$ and +9.5‰ for $\delta^{15}N$, and for *G. robustum* -20.5‰ for $\delta^{13}C$ and +10.2‰ for $\delta^{15}N$. The results of $\delta^{13}C$ for *L. armatus* and *G. robustum* indicate a preference for C_3 vegetation in open environments similar to those in northern Patagonia today. According to the authors, the results of the $\delta^{15}N$ for *L. armatus* and *G. robustum* could be related

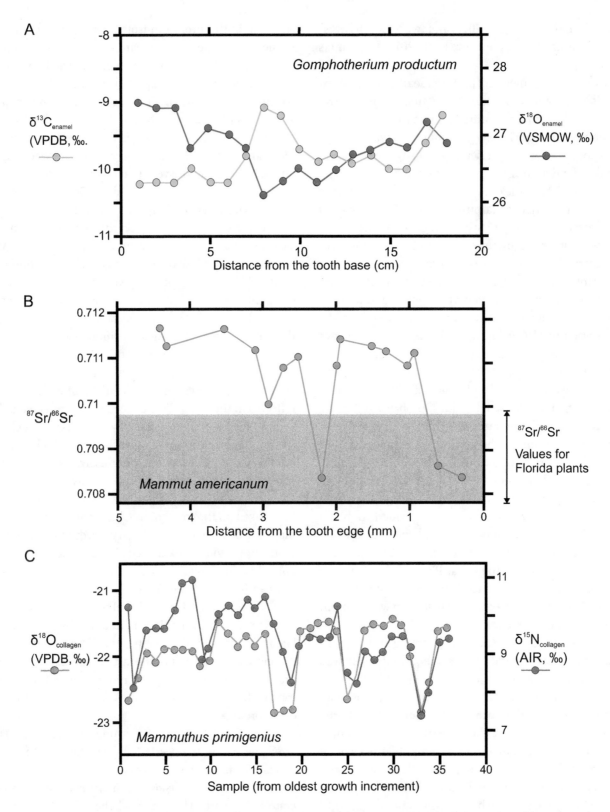

9.10. Three studies of fossil proboscideans that illustrate the application of isotope analysis to the understanding of the ecology of extinct animals (modified from Clementz 2012); see text for explanation. Carbon, nitrogen, and oxygen isotope values are reported as the international standards Vienna PeeDee Belemnite (VPDB), Atmospheric Air (AIR), and Vienna Standard Mean Ocean Water (VSMOW), respectively.

to a nonruminant herbivorous physiology or to the climate inferred for the Pampean region in the late Pleistocene, colder and drier than at present. Dantas et al. (2020, 138) studied isotopes from the late Pleistocene fauna from the Intertropical region of Brazil and evaluated the relative dietary contributions of leaves, fruits, and C_4 grasses for meso- and megaherbivores through employment of "a mathematical mixing model with the stable isotopic composition of one (carbon) and two elements (carbon and oxygen)." Among the taxa studied were the megatheriid *Eremotherium laurillardi*, mylodontid *Catonyx cuvieri*, glyptodontids *Glyptotherium* and *Panochthus*, pampatheriid *Holmesina*, toxodontid *Toxodon*, camelid *Palaeolama*, and equid *Equus*. Their analyses also permitted the authors to consider which of these herbivores might have been potential prey for *Smilodon populator* and *Caiman latirostris*. Further, Dantas et al. (2020) reconstructed the paleoenvironment of the vertebrate community, concluding that it was a more closed and drier landscape than the current African savanna. The results of the studies by Czerwonogora et al. (2011) and Dantas et al. (2020) also converge on the possibility that less marked differences between lower and higher latitudes of South America existed during the late Pleistocene as compared with this continent's current conditions.

Ecometrics

Eronen et al. (2010) applied the term *ecometrics* to a set of taxon-free analyses of the distribution within a community of ecomorphological traits such as body mass, limb proportions, and dental structure (as in our protocol defined at the end of chap. 1) or any aspect of anatomy or physiology that can be measured across some portion of the organisms in a community. According to these authors, ecometrics measures interactions, both biotic and abiotic, of organisms with their environment by focusing on traits that are easily measurable and structurally closely related to their function, and the function of which interacts directly with the local environment. Vermillion et al. (2018) reviewed the concepts and history of ecometric analysis and described practical methods for implementing ecometric studies. Ecometrics is the link between form and function in the example of paleoecological integration between form, function, and environment described in the following section.

Biota from the Santa Cruz Formation (Early–Middle Miocene, Patagonia) as a Paleoecological Case Study

Simpson (1980) emphasized the importance of the Santa Cruz Formation and its contained fossil remains for understanding the episode of South American mammalian history during which the faunas consisted of a complex mix of descendants of the continent's ancient lineages and new taxa (primates and rodents) from other land masses. The abundance and preservational quality of its specimens allow a profound and comprehensive understanding of the paleobiology and paleoecology of the Santacrucian fauna. Its thorough study, including efforts by Vizcaíno et al. (2010) and Kay et al. (2012b), among others (see also Kay et al. 2021), involved the selection of a series of localities representing a sufficiently narrow time interval and geographic area to permit the development of paleosynecological analyses of the Santacrucian. This (ongoing) body of research (considered in more detail below), however, has also led to the emergence of a singular investigative tool for other Miocene South American mammal communities: the Santacrucian as a model that may be applied in particular to the study of post–primate-and-rodent arrival and pre-GABI (Great American Biotic Interchange) South American faunas. In this regard, particularly meaningful aspects of the Santacrucian as a model are that its taxa are closer phylogenetically, morphologically, and temporally to those of other such faunas. The application of this approach is, as alluded to earlier in this chapter, an example of fossilism; to emphasize its narrower utilization as described here, it has been referred to as *Santacrucism* (Kay et al. 2021).

Much of the original research involved in allowing an understanding of the paleosynecology of the Santacrucian was presented together in the book, published in 2012, *Early Miocene Paleobiology in Patagonia: High-Latitude Paleocommunities of the Santa Cruz Formation* (Vizcaíno et al. 2012d). This volume was devoted to the comprehensive biology of the faunal elements of a past community of South America and aimed to interpret their ecological interactions (i.e., the paleocommunity's structure) and understand the environment that they inhabited. To these ends, the contributors of the volume gathered data from several methodological perspectives, including the novel incorporation of a form-function correlation approach.

The analyzed paleocommunity existed during the warmest period over the past 34 million years of Earth history, when South America was physically separated from other continental masses. The fossils are preserved in sedimentary rocks of the Santa Cruz Formation (Santacrucian Age; ~18 to 16 Ma), abundantly exposed throughout Santa Cruz Province, Argentine Patagonia. Outcrops of this formation along the Atlantic coast between the Río Coyle and Río Gallegos yield the most abundant and well-preserved vertebrate remains for paleobiological studies (fig. 9.11).

The outcrops and their contained fossil remains have been the focus of several past efforts, including those of perhaps the most renowned example, conducted by Princeton University under the leadership of the American paleontologist W. B. Scott (1858–1947) and published over several years as the *Reports of the Princeton University Expeditions to Patagonia, 1896–1899* (1903–1932). These efforts, however, now more than a century in the past, lacked the stratigraphic precision and controlled geological sampling required for a comprehensive understanding of past (as well as present) communities, particularly because many of the necessary conceptual frameworks had not yet been articulated.

To achieve their objectives, noted above, the editors of *Early Miocene Paleobiology in Patagonia* organized and conducted intensive and continuous recollection of fossils and other geological samples during the austral summers of 2002 to 2012. The study of the formation and its fossils was a collaborative effort carried out by a broad group of scientists from several institutions with specializations in various branches of geology and paleontology, including sedimentology, geochronology, ichnology, paleobotany, and invertebrate and vertebrate paleontology. Vertebrate paleontologists constituted the most diverse group, including experts on amphibians, reptiles, birds, metatherians, xenarthrans, ungulates, rodents, and primates. The contributions of the volume represent largely original research on the rocks and biota of the Santa Cruz Formation, including aspects that had received only cursory treatment and others that had been entirely ignored. They also represent an exhaustive example of Currie's investigative scaffolding and methodological omnivory cited in chapter 1.

The methodological thoroughness of the *Early Miocene Paleobiology in Patagonia* is evident in its organization and the nature of its contributions. Despite its main focus being on the vertebrates of the Santa Cruz Formation, nearly the first third of the volume deals with aspects that concisely place the fossil remains into the broader geological and temporal contexts required to carry out the detailed biological accounts that follow.

For example, following the introductory chapter, in which the editors provide the contextual framework for the book, two chapters summarize the geochronology of the Santa Cruz Formation based on new radiometric dating and tephrostratigraphic correlations among the fossiliferous localities throughout Santa Cruz Province. The following three chapters are devoted to establishing the paleoenvironmental scheme on which the paleoecological analyses of the Santacrucian faunas in the study area were carried out, including the stratigraphic framework for the abundant fossil vertebrate remains and descriptions of the sedimentary environments of the Santa Cruz Formation along the Atlantic coast between the Río Coyle and Río Gallegos. The subsequent three chapters, respectively, describe the marine invertebrate fauna adapted to the intertidal paleoenvironment at the base of the formation, the paleoenvironmental evidence based on fossil traces, and the previously poorly known paleoflora.

Only then, in the following nine chapters, are the vertebrates treated, including their diversity as recorded in the lower member of Santa Cruz Formation and paleobiological reconstructions of each taxon (paleoautoecology). One chapter provides a general overview of the scarce information available on heterothermic vertebrates (frogs and lizards) and their paleoenvironmental significance, with the next eight chapters summarizing and providing new data on the proposed habits of birds and mammals based on their morphology and the protocol outlined in this book: body-size estimation, inferences about dietary and other feeding habits through the analysis of cephalic structures, and inferences on substrate use based on the appendicular skeleton. A diversity of vertebrates is covered, including birds, noncarnivorous small (paucituberculatans and microbiotheres) and carnivorous (sparassodont) metatherians, armored (armadillos and glyptodonts) and pilosan (anteaters and sloths) xenarthrans, South American native ungulates (astrapotheres, notoungulates, and litopterns), caviomorph rodents, and primates. In the final chapter, the editors summarize their findings within the chronologic framework established and with the environmental information provided by geology, trace fossils, and paleobotany. Then they present a synthesis of the implications

---

Facing, **9.11.** Geographic location of the fossiliferous coastal localities of the Santa Cruz Formation (Santa Cruz Province, Argentina). Investigated localities are indicated with an *asterisk* (see Vizcaíno et al. 2012c). *Inset* images of in situ specimens reveal the exceptional preservation of the fossils. A, articulated skull and mandible of a sloth. B, partial skeleton of another sloth that includes the mandible, part of the articulated vertebral column, humerus, ribs, femur, and other bone fragments. C, skull of the notoungulate *Adinotherium* (Toxodontidae). D, skull of the notoungulate *Nesodon* (Toxodontidae).

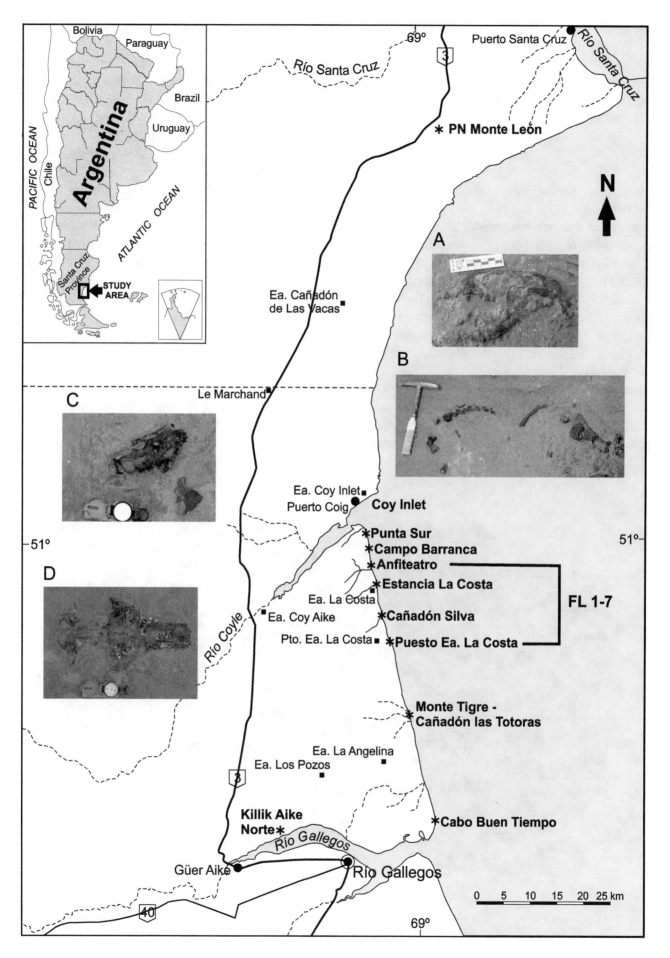

Bolivia
Paraguay
Brazil
Uruguay

Argentina

PACIFIC OCEAN

Chile

ATLANTIC OCEAN

Santa Cruz Province

STUDY AREA

N

Río Santa Cruz
69°
Puerto Santa Cruz
Río Santa Cruz

* PN Monte León

A

B

Ea. Cañadón de Las Vacas

C

Le Marchand

Ea. Coy Inlet
Puerto Coig
Coy Inlet

* Punta Sur
* Campo Barranca
* Anfiteatro
* Estancia La Costa
Ea. La Costa
Río Coyle
Ea. Coy Aike
* Cañadón Silva
Pto. Ea. La Costa
* Puesto Ea. La Costa

FL 1-7

D

51°
51°

Monte Tigre -
* Cañadón las Totoras

Ea. La Angelina
Ea. Los Pozos

3

Killik Aike Norte *

* Cabo Buen Tiempo

Río Gallegos

Güer Aike
Río Gallegos

0 5 10 15 20 25 km

40

69°

Paleoecology 205

of the vertebrate assemblage for the paleoecology (paleo-synecology) of the Santa Cruz Formation, comparing their results with those of equivalent analyses from biotas of other latitudes and ages (e.g., the middle Miocene fauna from La Venta badlands in Colombia; Kay and Madden 1997a, 1997b).

Contextual Information

Outcrops of the Santa Cruz Formation (early–middle Miocene) between 50.3° S and 51.6° S on the Atlantic coast of Santa Cruz Province have yielded an enormous quantity and variety of partial to entirely articulated (fig. 9.12) skulls and skeletons, probably more abundant than from any other region of South America. Santacrucian fossils represent the best record for interpreting the biological diversity and ecological dimensions of mammal communities in the southern part of South America (Patagonia) before the GABI. They also constitute the Earth's southernmost continental biota during the time interval known as the Miocene Climatic Optimum (MCO, also referred to as the Middle Miocene Climatic Optimum by some authors; Steinthorsdottir et al. 2021). During the early–middle Miocene, between 17 and 15 Ma, warm superficial seawaters were transported southward by the

Brazil Current and allowed the extension of subtropical conditions from Amazonia to the south. Moreover, the greater part of Andean uplift had not occurred yet, and the rain-shadow effect that eventually led to the present-day aridity in most of Patagonia had not yet been established. This expansion of warm and wet climates toward the pole coincides with the appearance of mammalian taxa adapted to these climates, such as primates, erethizontid rodents (tree porcupines) with low-crowned teeth, and anteaters, among other faunal elements. Later, during the middle Miocene, the propagation of colder and more arid conditions commenced as a consequence of global cooling and the beginning of glaciations, as well as Andean uplift, causing the regional extirpation of these faunal elements beginning about 15.5 Ma. Because the Santa Cruz Formation was deposited during the MCO, the climatic conditions during its deposition must have been extraordinary and without modern equivalents for its latitude at nearly 50°S, with the seasonality in the energy availability having profoundly affected biotic productivity.

The most conspicuous vertebrates of this fauna were birds and mammals. Among birds, the most significant are the phororracoids, terrestrial animals that must have played an important role as scavengers, carnivores, or both. The mammalian community consisted of a complex

3 cm

9.12. Nearly complete specimen of the armadillo *Prozaedyus proximus* recovered from outcrops of the Santa Cruz Formation (early Miocene) along the Atlantic coast. Photo by the authors (SVF and MSB).

mixture of descendants of ancient lineages of the South American continent (Metatheria, Xenarthra, Litopterna, Notoungulata, and Astrapotheria) and the new forms from other continents (Rodentia and Primates). Metatherians (sparassodonts, paucituberculatans, and microbiotheres) and xenarthrans (sloths, anteaters, armadillos, and glyptodonts) have very different morphologies from their extant relatives, and "archaic" South American native ungulates (notoungulates, astrapotheres, and litopterns) have left no descendants; it is thus difficult to reconstruct their life habits (Vizcaíno and Bargo 2014; Vizcaíno et al. 2017). In the case of caviomorph rodents and platyrrhine primates, it is much easier to find extant models, but it is still difficult to interpret some aspects of their paleobiology. For instance, it is unclear which adaptations allowed primates to survive in an apparently highly seasonal environment at some 20° south of the present distribution of members of this group.

Paleoecology of Lineages and Guilds

Before continuing the paleosynecological summary, the following sections present several detailed examples of lineage and guild paleoecology treated in Vizcaíno et al. (2012d) to illustrate the application of the protocol for

paleobiological studies adopted here. In particular, we will focus on endemic South America lineages: sloths (Bargo et al. 2012) and native ungulates (Cassini et al. 2012a) among primary consumers, and sparassodont metatherians (Prevosti et al. 2012) among secondary consumers.

SLOTHS

Most of the sloths from the Santa Cruz Formation belong to Megatherioidea, with fewer representatives of Mylodontoidea. With body masses between 40 and 150 kg, these Miocene sloths were much larger than the extant *Bradypus* and *Choloepus*, both with body mass less than 10 kg.

According to Gaudin's (2004; fig. 9.13) phylogeny, *Bradypus*, lacking fossil representation, is sister taxon to all remaining sloths, with *Choloepus* well nested within Megalonychidae. More recent molecular-based phylogenies (Delsuc et al. 2019; Presslee et al. 2019) suggest that *Choloepus* is a mylodontid and *Bradypus*, rather than sister taxon to all other sloths, is closely related to Nothrotheriidae and Megatheriidae (i.e., Megatherioidea). Either way, the two extant sloth genera are thus only distantly related, so that their convergence in morphology, function, and ecology is astounding.

Santacrucian sloths are characterized by elongated, rather tubular skulls generally small in proportion to body size when compared with other mammals. As in other xenarthrans, the teeth are hypsodont and lack enamel, and the dental formula is 5/4. In megalonychids and basal megatherioids, the first upper and lower teeth are caniniform and of variable size (fig. 9.14). The appendicular skeleton differs considerably from that of extant sloths, resembling more that of anteaters, such as *Tamandua* and the giant anteater *Myrmecophaga*, and in certain features to some extant armadillos, such as the giant armadillo *Priodontes* (fig. 9.15).

In the Miocene, megalonychids were represented by *Eucholoeops*, of which numerous remains have been recovered, and *Megalonychotherium*, which is less well known. Mylodontidae, which includes well known Pleistocene taxa such as *Scelidotherium*, *Glossotherium*, and *Mylodon*, is represented by *Analcitherium* and *Nematherium* during the Miocene. Megatheriidae, better known from the Pleistocene giant ground sloths *Megatherium* and *Eremotherium*, is represented by *Prepotherium* and *Planops* in the Miocene. A diversity of basal Megatherioidea is also known, including *Hapalops*, *Hyperleptus*, *Analcimorphus*, *Schismotherium*, and *Pelecyodon*. Among them, *Hapalops* is the best represented, with numerous specimens collected and more than 20 named species (see Bargo et al. 2019).

Bargo et al. (2012) offered original contributions on body-size estimates, appendicular skeleton function, and

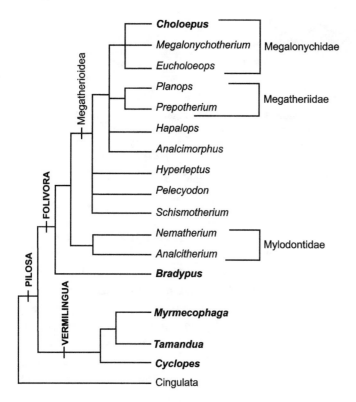

9.13. A phylogeny of Pilosa, a clade that comprises sloths (Folivora) and anteaters (Vermilingua); only Santacrucian sloths and the two extant genera, *Bradypus* and *Choloepus*, are included. Based on Gaudin (2004).

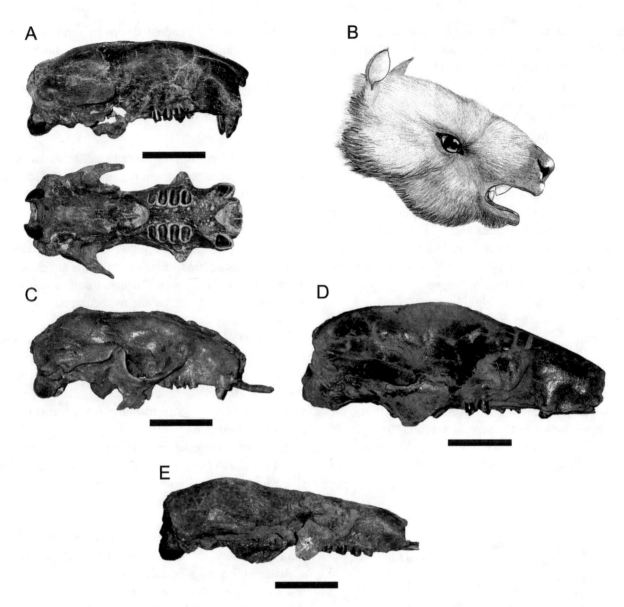

9.14. Santacrucian sloths. *A*, Megalonychidae *Eucholoeops ingens*, skull of the specimen MPM-PV 3451 (Museo Regional Provincial Padre M. J. Molina, Río Gallegos, Santa Cruz Province, Argentina) in right lateral and palatal views. *B*, life reconstruction of *E. ingens*. *C*, basal Megatherioidea *Hapalops longiceps*, skull of specimen YPM-VPPU 15523 in left lateral view (reversed from original right; Yale Peabody Museum, New Haven, CT). *D*, Megatheriidae *Planops magnus*, skull of specimen YPM-VPPU 15346 in right lateral view (reversed from original left). *E*, Mylodontidae *Nematherium* sp., skull of specimen YPM-VPPU 18009 in right lateral view. Scale bars = 5 cm. Photographs by the authors; life reconstruction by N. Toledo.

substrate use by Santacrucian sloths that were supported by subsequent publications (Toledo et al. 2013, 2014, 2015; Toledo 2016). This research included descriptions of mandibular and appendicular skeletal elements and emphasized features of functional relevance, such as articular facets, muscular attachment sites, and dental wear facets, among others. For dietary aspects, Bargo et al (2012) relied on previous research (Bargo et al. 2009), considered here in chapter 8, in which the dental morphology and mechanical components of masticatory movements of *Eucholoeops* were analyzed and compared with those of other Santacrucian sloths.

The analysis of the appendicular skeleton generated results for two components, estimation of body size and substrate use, of the basic protocol followed here. As explained in chapter 4, estimation of body size was based on comparison of the linear measurements of functionally relevant features and proportions and absolute lengths of the limb bones of Santacrucian sloths with those of various extant mammals exhibiting diversity in body size, substrate preferences, and locomotory mode (taxa from 0.5 kg to 350 kg; arboreal, semiarboreal, and ground dwelling; diggers and nondiggers).The body-mass estimates were obtained by means of allometric equations (see chap. 4).

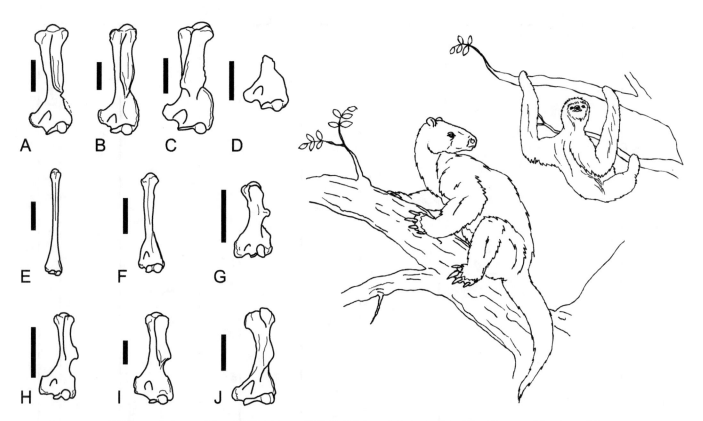

9.15. Comparison of left humeri in anterior view of Santacrucian (*A–D*) and extant sloths (*E–F*) with other extant Xenarthra, anteaters (*G–I*) and an armadillo (*J*). A, *Hapalops.* B, *Eucholoeops* (reversed from original right). C, *Nematherium.* D, *Schismotherium.* E, *Bradypus.* F, *Choloepus.* G, *Cyclopes.* H, *Tamandua.* I, *Myrmecophaga.* J, *Priodontes.* Scale bars = 5 cm, except in *G*, scale bar = 1 cm. Modified from Toledo et al. (2013). On the *right*, reconstruction of *Hapalops* next to *Bradypus*, at about the same scale. Illustrations by N. Toledo.

Further, reconstruction of the musculature of the appendicular system was carried out, based on that of extant xenarthrans and other mammals.

The incorporation of descriptions and reconstructions of soft tissues allowed qualitative assessment of the morphological features of functional interest in comparison with the homologous features of extant xenarthrans. Following the methodological concepts detailed in chapter 2, each feature was functionally characterized using the most closely related extant taxa as functional models, and the results were then used to construct a mechanical profile of each limb. When required, comparisons were made with distantly related taxa (e.g., wombat, a diprotodont marsupial) and even mechanical analogs. In a second stage, causal associations between these functional patterns and biology were sought (e.g., substrate preference and use), according to the definitions of faculty and biological role. Ecomorphological analyses (PCA and DA; Toledo 2012, 2016; Toledo et al. 2012) were carried out as well to explore the morphometric similarity with extant mammals of the sample, producing a posteriori functional hypotheses that could explain instances of similitude. Likewise, varied biomechanical and functional indices were applied (fig. 9.16).

A summary of the results includes:

1. Santacrucian sloths were large mammals in comparison with extant sloths. Among basal megatherioids, *Hapalops* reached between 30 and 80 kg, *Analcimorphus* ~67 kg, and *Schismotherium* ~44 kg; the megalonychid *Eucholoeops* around 70 kg; the largest genera include the megatheriid *Prepotherium* (~123 kg) and the mylodontids *Analcitherium* (~88 kg) and *Nematherium* (~89 kg).

2. The proportions of bony elements of the limbs and the development of the features for the origin and insertion of muscles and ligaments (entheses) are more similar to those of anteaters and other extant digging mammals than to those of extant sloths.

3. The musculature of the appendicular system was quite developed, allowing the exertion of considerable force during limb adduction, antebrachial and crural extension and flexion, manual flexion and prehension, and pedal extension and prehension.

4. The appendicular skeleton permitted bent elbow and knee stances and a wide range of hip and shoulder abduction. The capacity for supination was limited, and the ankle was restricted to flexion-extension. The foot would have had some prehensile capacity.

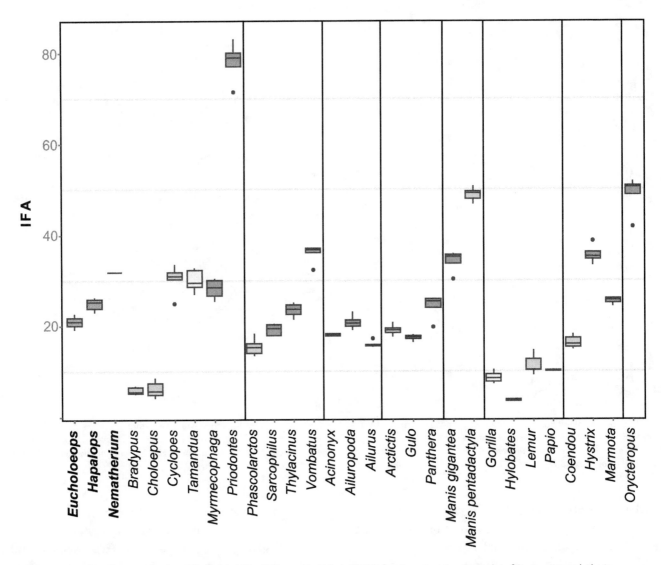

9.16. Boxplot of the functional index of fossorial ability (IFA) used by Toledo (2016) for characterizing the limbs of Santacrucian sloths in comparison with a sample of extant mammals. IFA describes the relative mechanical advantage of forearm extension. *Gray*, Santacrucian sloths; *green*, arboreal taxa; *yellow* and *orange*, semiarboreal and extant terrestrial mammals, respectively. Modified from Toledo (2016).

The analysis of these functional features and the development of hypotheses on faculties indicate that Santacrucian sloths might have been arboreal with regard to substrate preference. However, their climbing abilities would be somewhat limited due to their relatively large body size, as mentioned in chapter 6. In all cases, they would have climbed using slow and careful movements. These sloths might have moved with flexed limbs, maintaining their position on a support by means of gripping with their curved claws and powerful muscular strength. The smaller genera, such as *Analcimorphus*, *Schismotherium*, *Pelecyodon*, *Eucholoeops*, and smaller species of *Hapalops*, possibly exhibited more frequent climbing behavior than *Prepotherium* and the mylodontids *Analcitherium* and *Nematherium*. These two mylodontids would have used an arboreal substrate only for particular activities, such as resting or escaping from predation, whereas on land, they would have moved quadrupedally and with a plantigrade posture, thus fitting the description of *ambulatory* as defined in chapter 6 (Polly 2007).

From a paleosynecological perspective, Bargo et al. (2012) analyzed the position of the Santacrucian sloths in the inferred ecological guilds and their relationship with other guilds within the Santacrucian fauna (see also Toledo 2016). With regard to substrate preference, Santacrucian sloths were classified into two groups:

1. Arboreal and semiarboreal forms between 30 and 80 kg (*Hapalops*, *Eucholoeops*, *Pelecyodon*, *Analcimorphus*, and *Schismotherium*).
2. Terrestrial, facultative semiarboreal forms between 70 and 150 kg (*Prepotherium*, *Analcitherium*, and *Nematherium*).

This categorization of substrate preference frames *Hapalops, Eucholoeops, Pelecyodon, Analcimorphus,* and *Schismotherium* within the diversity of arboreal Santacrucian mammals that includes the anteater *Protamandua* (Myrmecophagidae), the primate *Homunculus* (Ceboidea), the porcupine *Steiromys* (Rodentia, Erethizontidae), and several taxa of carnivorous sparassodont metatherians. Sloths would have been the largest arboreal mammals, completely dominating the 30 to 80 kg range, with only the sparassodont *Prothylacynus* (37 kg) approaching the body mass of the smallest sloths.

With reference to the third biological attribute, diet or feeding, the megatherioids *Eucholoeops, Hapalops, Pelecyodon,* and *Prepotherium* were probably mainly folivorous, with predominantly orthal (vertical) masticatory movements and a dental morphology specialized for cutting and shearing. The mylodontids *Nematherium* and *Analcitherium* resemble Pleistocene mylodontids. Based on this resemblance and an analysis of the masticatory apparatus in the Pleistocene mylodontids (Bargo and Vizcaíno 2008), Bargo et al. (2012), employing the fossilism approach noted earlier in this chapter, concluded that the Santacrucian mylodontids had, in contrast to the megatherioids mentioned above, a mainly lateral component to their masticatory movements and an occlusal morphology more suitable for crushing or grinding. This would indicate a more varied diet including, in addition to leaves, fibrous and turgid items such as fruits or tubers.

The inferred feeding habits, together with mass estimates and inferences of substrate use and preference, suggest that at least *Eucholoeops* and *Hapalops* spent most of their time in trees, possibly feeding on leaves. Mylodontids, larger and spending far more time as ground dwellers, were only facultatively arboreal, and possibly used their greater digging capacities for obtaining food from the soil. Finally, *Prepotherium*, also folivorous and the largest of Santacrucian sloths, was mainly ground dwelling and possibly only an occasional digger. This ecological characterization suggests that all Santacrucian sloths inhabited forested areas, or at least areas with plants that could provide support of necessary diameter for use as substrates.

In short, the inferences of body size, dietary habits, and substrate use and preference allow dividing Santacrucian sloths into three ecological groups:

1. Arboreal and semiarboreal, strictly folivorous forms between 30 and 80 kg (*Eucholoeops, Hapalops, Pelecyodon, Schismotherium,* and *Analcimorphus*).

2. Terrestrial and possibly digging forms around 80 kg, consumers of leaves, tubers, and fruits (mylodontids *Analcitherium* and *Nematherium*).

3. Folivorous terrestrial forms between 120 and 200 kg (*Prepotherium* and other megatheriids).

With regard to ecological characterization (fig. 9.17), at least three genera (*Eucholoeops, Hapalops,* and *Analcimorphus*) in the first group are recorded in the same locality and fossiliferous levels, indicating that they were probably coeval, which raises interesting questions about possible competition for trophic and spatial resources. Among these genera, some functional differences possibly indicate slight differences in locomotory styles and access to different levels of vegetation; *Hapalops* was a robust form with a more angled limb posture that might have moved

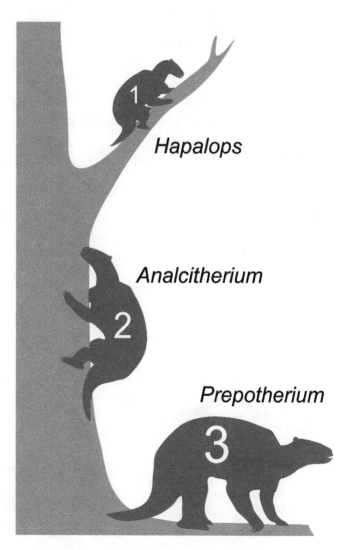

9.17. Ecological groups of Miocene sloths, each represented by a characteristic taxon. (1) Arboreal and semiarboreal between 40 and 80 kg, strictly folivorous; (2) semiarboreal around 80 kg, consumers of leaves, tubers, and fruits; (3) ground dwellers between 120 and 200 kg, folivorous. Modified from Toledo (2016).

along vertical or steeply inclined supports, whereas *Analcimorphus* and *Eucholoeops* were more robust forms with more extended limbs that might have ventured over less steeply oriented branches and were also more agile on land.

Likewise, as arboreal and semiarboreal herbivorous animals, sloths would have competed for trophic resources with the porcupine *Steiromys* and the primate *Homunculus*. *Steiromys* was a semiarboreal rodent of 10 to 15 kg (Candela and Picasso 2008; Candela et al. 2012)—about half the weight of the smallest sloths. Its diet would have consisted of conifer leaves, seeds, and bark (fig. 9.18). This animal might have fed on items obtained from thinner branches and thus probably did not compete with sloths of the first group, which, as just noted, were larger climbing folivores restricted to thicker branches. *Homunculus* was an arboreal frugivore and folivore (Kay et al. 2012a). However, its much smaller size (about 2.5 kg) and apparently low population densities (based on its rarity in the fossil record) would have precluded it as a serious competitor of sloths.

The genera of ecological groups 2 and 3 (fig. 9.17) would likely have shared dietary preferences with a greater diversity of ground-dwelling herbivorous mammals, from armadillos of a few kilograms (*Peltephilus*, 11 kg and *Proeutatus*, 15 kg), proterotheriid litopterns (30 to 80 kg), to glyptodonts of about 80 to 100 kg (Vizcaíno et al. 2006a; Vizcaíno et al. 2011a, 2011b; Vizcaíno et al. 2012b), and macraucheniid litopterns, which exceeded 130 kg (litopterns are considered again in the following section). On the use of trophic resources, it is likely that mylodontids competed to some extent with larger litopterns (e.g., *Theosodon*) of closed habitat and with the glyptodonts *Propalaehoplophorus* and *Cochlops*, selective-feeding herbivores in moderately open environments (Vizcaíno et al. 2011b; Vizcaíno et al. 2012b; fig. 9.18). However, the digging abilities of mylodontids might have allowed them access to feeding resources probably not exploited by other herbivores of similar size, such as glyptodonts. With regard to the third ecological group, the large ground dweller and folivorous *Prepotherium* might have used its forelimbs for accessing leaves of low branches, without having to share trophic resources with glyptodonts, although it may have faced strong competition from larger litopterns. In venturing into open areas, *Prepotherium* must have shared trophic and spatial resources with toxodontids between 100 kg and 600 kg and with a homalodotheriid of more than 400 kg (see following section).

In reference to the Santacrucian sloths as a feeding resource for the guild of coeval carnivorous animals, it is likely that their main predators were the larger sparassodonts, such as *Prothylacynus*, a semiarboreal taxon

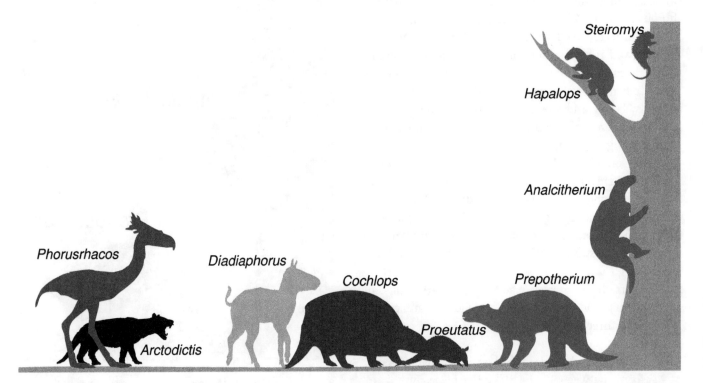

9.18. Santacrucian sloths in relation to the remaining Santacrucian vertebrate paleocommunity (each represented by a characteristic taxon) of the Santa Cruz Formation (Toledo 2016). As potential competitors for the trophic and spatial resources, the tree porcupine *Steiromys*, the armadillo *Proeutatus*, the glyptodont *Cochlops*, and the litoptern *Diadiaphorus* are represented. Among potential sloth predators, the carnivorous marsupial *Arctodictis* and the phororracoid bird *Phorusrhacos* are illustrated. The silhouettes are drawn at an approximate scale. Modified from Toledo (2016).

weighing about 35 kg, and *Borhyaena* and *Arctodictis*, ground-dwelling predators weighing 36 and 50 kg, respectively (Prevosti et al. 2012; see carnivorous metatherians below). However, considering the enormous muscular strength of these sloths and their greatly developed curved claws, it is plausible that they were difficult to bring down and that they possibly developed defensive behaviors, such as those observed in extant anteaters, which stand on their hind limbs and use their manual claws as weapons (Taylor 1978; Nowak 1999). It is more likely that sparassodonts of larger size sporadically hunted smaller sloth adults of *Pelecyodon* and *Schismotherium*, and perhaps also juvenile, old, or weakened individuals from the larger sloth genera, whereas arboreal sparassodonts probably hunted and consumed juveniles of the smaller sloth genera. Lastly, among Santacrucian carnivores, terror birds (Phorusrhacidae) are conspicuous candidates as sloth predators, with one genus, *Phorusrhacos*, weighing more than 150 kg and standing about 2.5 m in height (Degrange et al. 2012). This animal was a strong and agile predator that might have hunted and consumed Santacrucian sloths on land, taking advantage of its height and speed to avoid defensive behaviors (fig. 9.18).

Nowadays, there are no mammalian communities that include an arboreal guild with such a high diversity of large-sized taxa. Although there are arboreal and semiarboreal primates of comparable size in Africa, their ranges do not generally overlap (Nowak 1999). Other regions include ursids and felids of large size and excellent climbing abilities, but without a comparable diversity. In addition, the mammalian communities of the Santa Cruz Formation include a diversity of taxa (xenarthrans and metatherians) with low basal metabolism, which complicates comparisons with most of the extant mammalian communities based on a strict actualistic approach.

The only widespread extant community largely (and historically) comprising mammals with low basal metabolism is the Australian marsupial fauna, and it is the only current example that can be considered as an analog for comparative purposes for the Santacrucian community (Vizcaíno et al. 2010). The arboreal marsupials from Australia include a great diversity of taxa in a wide range of body size, from tiny animals such as possums (frugivores and nectarivores) to forms of about 10 kg (Nowak 1999), such as the arboreal kangaroo *Dendrolagus* (frugivore and folivore; Smith and Ganzhorn 1996) and the koala *Phascolarctos cinereus* (folivore; Smith and Ganzhorn 1996). In particular, the koala is a robust herbivore that climbs trees with relatively slow movements, opposing its hands and feet on either side of a support and using its claws for increasing grip (Smith and Ganzhorn 1996; Nowak 1999;

Toon and Toon 2004), strategies similar to those inferred for Santacrucian sloths, which also share similarities with it in features of the appendicular skeleton. In addition to the Australian community, there are examples of paleocommunities that apparently included a guild of climbing, or more strictly arboreal, herbivorous mammals of large body size. These are the communities from the Pleistocene of Madagascar, with huge arboreal lemurs (e.g., *Megaladapis* of 60 kg and *Archaeoindris* of 150 kg; Jungers et al. 2002), and the Miocene of Australia, with arboreal diprotodont marsupials of great size (*Nimbadon* of 70 kg; Black et al. 2012).

SOUTH AMERICAN NATIVE UNGULATES

Cassini et al. (2012a) considered the paleobiology of the assemblage of South American native ungulates (SANU) that evolved within the geographic context of South American isolation during much of the Cenozoic (Bond 1986). In the Santa Cruz Formation, SANU are represented by the following groups: Astrapotheria, with Astrapotheriidae; Litopterna, with Proterotheriidae and Macraucheniidae; and Notoungulata, with Typotheria (including Hegetotheriidae and Interatheriidae) and Toxodontia (including Toxodontidae and Homalodotheriidae).

Astrapotheriids, highly peculiar and among the largest animals of the Cenozoic of South America (fig. 9.19), lacked upper incisors but possessed hypertrophied canines as defensive weapons, premolars that were reduced in size and number, and molars with lophodont crowns, similar to those of rhinos (Ameghino 1894; Kramarz 2009; Kramarz and Bond 2009). The Santa Cruz Formation has yielded *Astrapotherium*, the wide and retracted nasal cavity of which has been interpreted as indicating the presence of a proboscis. Based on its dentition, limb morphology, and geological context, it has been considered an herbivore inhabiting low and wet environments and frequenting lakes and wetlands (Scott 1928; Riggs 1935). More recently, Townsend and Croft (2005) pointed out that it exhibits dental microwear features expected from a typical browser.

Santacrucian litopterns are more numerous and diverse (fig. 9.20). Macraucheniids are represented only by *Theosodon*, characterized by its nasal retraction, similar to that exhibited by some artiodactyls (e.g., saiga antelope *Saiga tatarica* or the dik-dik *Madoqua kirkii*). It was described in the older literature as an herbivore of medium to large size and ecologically similar to the guanaco. By contrast, proterotheriids have been described as light and slender ungulates, digitigrade, and generally with three digits in each foot, the middle one much larger than those on either side, imitating the condition in miniature horses

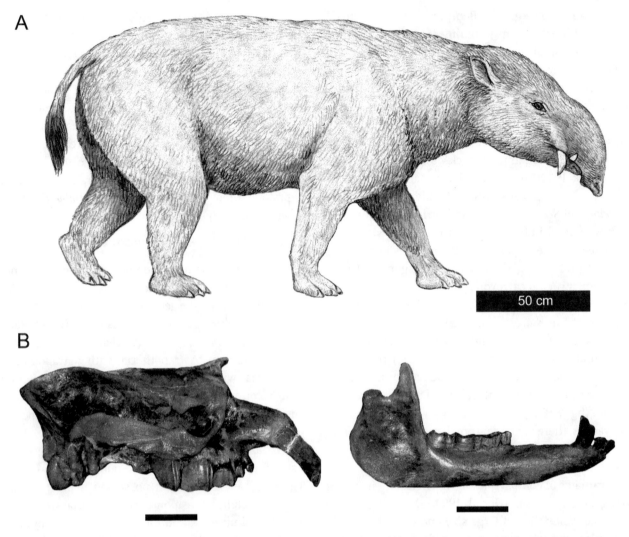

9.19. Astrapotheria, *Astrapotherium magnum*. A, life reconstruction. B, skull and mandible in right lateral view (skull reversed from original left; AMNH 9278, American Museum of Natural History, New York). Scale bar = 10 cm. Adapted from Cassini et al. (2012a). Photographs by the authors. Life reconstruction by and used courtesy of Marcelo Canevari.

(Ameghino 1898). Four genera of Santacrucian proterotheriids are known: *Anisolophus*, *Tetramerorhinus*, *Diadiaphorus*, and *Thoatherium*, traditionally considered as herbivores of open environments due to their monodactyly (either anatomical or functional). More recently, proterotheriids were considered as browsing animals mainly due to their brachydont dentition and because they are often recorded in faunas associated with forested environments (Bond 1986; Cifelli and Guerrero 1997; Bond et al. 2001; Kramarz and Bond 2005; Villafañe et al. 2006).

Santacrucian notoungulates are even more abundant and diverse. Toxodontia is represented by the nesodontine toxodontids *Nesodon* (fig. 9.21) and *Adinotherium*, and the homalodotheriid *Homalodotherium*, all with a complete dentition lacking a diastema. Nesodontines have hypertrophied lateral incisors and high-crowned cheek teeth but with limited growth (protohypsodonty sensu Mones 1982), whereas *Homalodotherium* has low-crowned dentition of

limited growth (brachydonty sensu Mones 1982) without hypertrophied incisors. Nesodontines have been considered open-plain inhabitants feeding mainly on Gramineae, whereas browsing habits have been suggested for *Homalodotherium*.

Santacrucian typotheres were small animals represented by the interatheriids *Interatherium* and *Protypotherium* and the hegetotheriids *Hegetotherium* and *Pachyrukhos* (fig. 9.22). Interatheriids have a complete dentition (without diastema) and subequal incisors. In hegetotheriids, by contrast, the upper and lower central incisors are hypertrophied and the lateral incisors are atrophied, a condition resembling that of extant hares (Leporidae). Earlier interpretations of typotheres, based mainly on their hypsodont cheek teeth, suggested grazing habits in open plains. However, Townsend and Croft (2008), based on analyses of microwear grooves and facets, suggested that *Protypotherium* browsed on soft fruits.

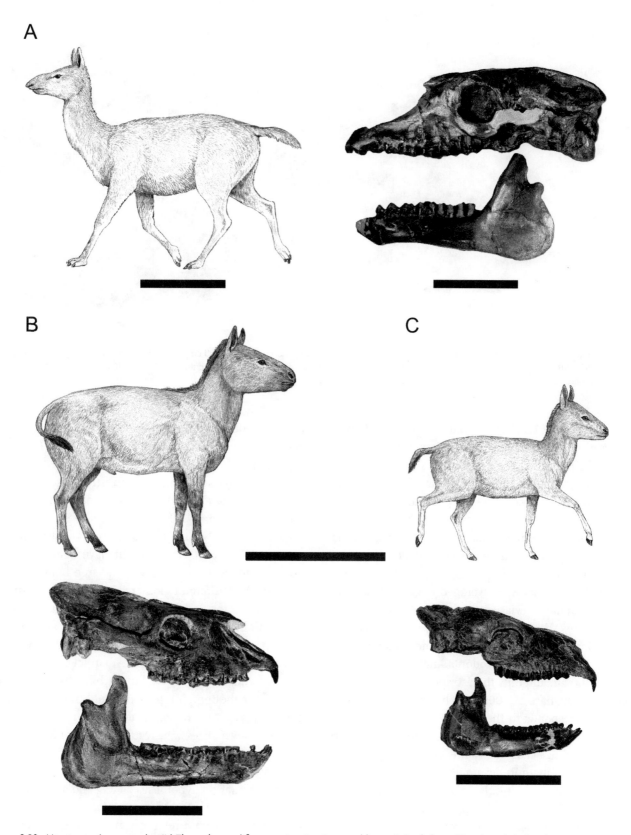

9.20. Litopterna. *A*, macraucheniid *Theosodon sp.*, life reconstruction in an ambling gait (scale bar = 50 cm), and skull and mandible of *T. lydekkeri*, left lateral view (MACN-A 9269-88, Museo Argentino de Ciencias Naturales B. Rivadavia, Buenos Aires, Argentina; scale bar = 10 cm). *B*, proterotheriid *Diadiaphorus majusculus*, life reconstruction (scale bar = 50 cm) and skull and mandible, right lateral view, reversed from original left (MPM-PV 3397, Museo Regional Provincial Padre M. J. Molina, Río Gallegos, Santa Cruz Province, Argentina; scale bar = 10 cm). *C*, proterotheriid *Thoatherium minusculum*, life reconstruction (scale bar = 50 cm) and skull, right lateral view (MPM-PV 3529), and mandible, right lateral view, reversed from original left (YPM-VPPU 15719, Yale Peabody Museum, New Haven, CT; scale bar = 10 cm). Adapted from Cassini et al. (2012a). Photographs by the authors. Life reconstructions by and used courtesy of Marcelo Canevari.

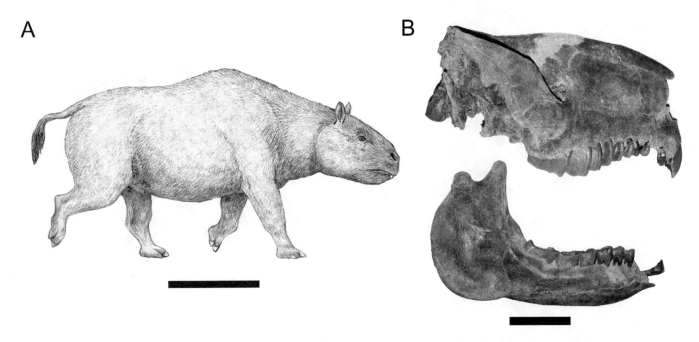

9.21. Notoungulata, Toxodontia. Toxodontid *Nesodon imbricatus*. *A*, life reconstruction (scale bar = 50 cm). *B*, skull and mandible, right lateral view (MPM-PV 3659, Museo Regional Provincial Padre M. J. Molina, Rio Gallegos, Santa Cruz Province, Argentina; scale bar = 10 cm). Adapted from Cassini et al. (2012a). Photographs by the authors. Life reconstruction by and used courtesy of Marcelo Canevari.

Cassini et al. (2012a, 2012b) conducted several analyses for obtaining estimations of body mass, a characterization of substrate use and habitat use, as well as the type of diet and hypotheses on digestive physiology. For body-mass estimations, they employed linear craniomandibular measurements in bi- and multivariate equations, as well as centroid size-based regressions (chap. 4), of extant artiodactyl and perissodactyl ungulates, the analogs proposed for these native lineages. For each specimen, they obtained several estimates that were averaged to generate a mean mass and standard deviation for each species. Santacrucian ungulates displayed a wide range of body size. Typotheres varied between 2 and 10 kg and proterotheriids between 20 and 100 kg, toxodonts and macraucheniids exceeded 100 kg, and only the astrapotheriid would have reached a body mass of 1,000 kg. The greatest taxonomic richness corresponds to the size range between 20 and 100 kg. Based on these values, the authors placed native ungulates into size categories adjusted to the logarithmic scale in order of magnitude: small ungulates (between 1 and 10 kg), which include only typotheres, medium ungulates (between 10 and 100 kg), represented by proterotheriids and some specimens of the toxodontid *Adinotherium*, and large ungulates (between 100 and 1,000 kg), which include the macraucheniid *Theosodon*, the toxodontids *Nesodon* and the larger specimens of *Adinotherium*, the homalodotheriid *Homalodotherium*, and the astrapotheriid *Astrapotherium*. Some specimens of the latter might

have exceeded 1 tonne, making it the only megaherbivore (sensu Owen-Smith 1988) of the Santa Cruz Formation (fig. 9.23).

For the characterization of substrate use, Cassini et al. (2012a) conducted a study based on the calculation of indices (see chap. 6). All ungulates of the Santa Cruz Formation were likely quadrupedal, from walkers to runners, though *Homalodotherium* might have adopted bipedal postures for feeding on leaves and tree buds (Elissamburu 2010). Although typotheres included the most gracile and agile forms among the Santacrucian ungulates, some might have been occasional diggers, perhaps more so than proterotheriids, owing to their smaller body size (see chap. 6, fig. 6.24). Two quite distinct forms, *Astrapotherium* and *Interatherium*, might have had swimming habits. For the remaining typotheres, and the toxodonts and litopterns, a more precise determination of gait is not possible. Nevertheless, the results of these authors suggest greater variation in locomotory styles than had previously been supposed.

In formulating hypotheses on diet and digestive physiology, Cassini et al. (2012a) executed an ecomorphological approach based on a reference frame comprising a vast sample of extant ungulates of known habitat and feeding behavior. First, they conducted a hierarchical analysis to identify the correlation between craniomandibular morphology and habitats and then analyzed feeding behavior within each habitat. The extant species were classified

A   B

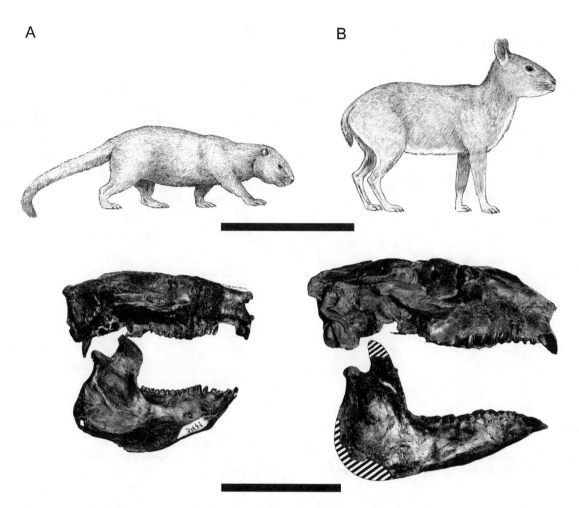

9.22. Notoungulata, Typotheria. *A*, interatheriid *Interatherium robustum*, life reconstruction (scale bar = 25 cm) and skull and mandible in right lateral view (skull, reversed from original left; MPM-PV 3471; Museo Regional Provincial Padre M. J. Molina, Río Gallegos, Santa Cruz Province, Argentina; scale bar = 5 cm). *B*, hegetotheriid *Hegetotherium mirabile*, life reconstruction (scale bar = 25 cm) and skull and mandible in right lateral view (mandible reversed from original left; MPM-PV 3526; MPM-PV 4316; scale bar = 5 cm). Adapted from Cassini et al. (2012a). Photographs by the authors. Life reconstructions by and used courtesy of Marcelo Canevari.

by preference for three environment categories—open, mixed, and closed—and according to the type and proportion of plant cover (Mendoza et al. 2005). Within each one of these categories, the species were subdivided into five different feeding behaviors according to the plant composition of the diet: grazers, intermediates, browsers, frugivores, and omnivores. The authors then calculated the occlusal surface area (OSA) of the upper cheek teeth, following the methodology proposed by Vizcaíno et al. (2006a; see chap. 8), and analyzed its relationship to body mass by means of allometric equations of the standardized major axis.

Feeding behavior was evaluated in an ecomorphological context through DA and classification trees based on 14 craniodental measurements that were size-adjusted by dividing each of them by the length of the lower molar series (LMRL) and the hypsodonty index (HI). Based on HI and

relative mandible length (JAW; length of the horizontal ramus divided by LMRL), notoungulates were classified as mammals of open environments, and litopterns and astrapotheres of closed environments (fig. 9.24). In extant ungulates, the relationship between these variables allows a reliable discrimination among the three types of habitat (i.e., open, mixed, and closed). When analyzing within each habitat type, notoungulates were characterized as grazers, and litopterns and astrapotheres as browsers (fig. 9.25). Toxodontids and proterotheriids more closely fit a model of herbivorous ungulates using artiodactyls and perissodactyls as analogs. Likewise, by combining the results with those of substrate use, the hypothesis that *Homalodotherium* was a tree browser seems plausible. Among typotheres, it is noteworthy that such small ungulates were characterized as grazers (small extant ungulates are folivorous); and in some cases, they even exaggerate

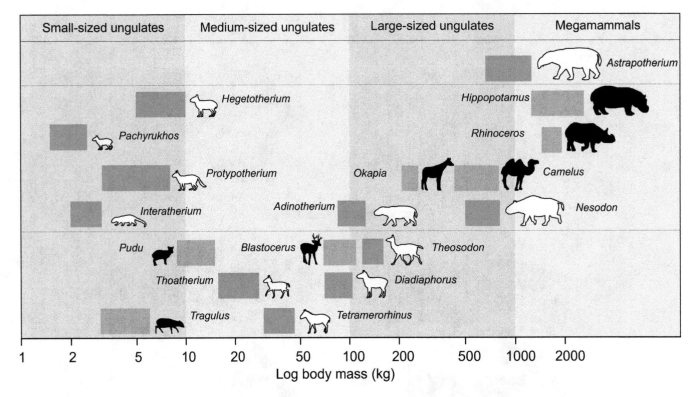

9.23. Representation of logarithmically scaled body masses of Santacrucian ungulates, silhouetted in *white*: *Interatherium*, *Protypotherium*, *Hegetotherium*, and *Pachyrukhos* (Typotheria), and *Adinotherium* and *Nesodon* (Toxodontia) among Notoungulata; *Diadiaphorus*, *Tetramerorhinus*, and *Thoatherium* (Proterotheriidae) and *Theosodon* (Macraucheniidae) among Litopterna; *Astrapotherium* (Astrapotheria); and Artiodactyla and Perissodactyla, silhouetted in *black*, for comparison and reference. The horizontal length of *rectangles* represents the body-mass range. Silhouettes are not to scale. Modified from Cassini et al. (2012a).

the morphological features of extant ungulates associated with this behavior. As has been observed in studies of substrate use, these animals would have had the capacity of digging, a behavior that can affect certain variables. For instance, hypsodonty might have increased largely as a response to abrasive elements of the substrate rather than to the silica of grasses. Moreover, Strömberg et al. (2013) recently demonstrated that in South America this morphological feature appeared before the generalization of grasslands and is more likely correlated with the presence of abrasive elements in the environments resulting from volcanism. Indeed, typotheres seem rather removed from the artiodactyl-perissodactyl morphotype and more strongly resemble leporids, caviomorph rodents, and hyraxes (Hyracoidea). It is for this reason that the comparative reference frame of extant analogs should be widened, in contrast to the more common practice of restricting analogs to those assumed to represent a particular type of behavior based on cursory analyses (see also Cassini et al. 2011).

The case of *Astrapotherium*, with its peculiar morphology, is also worth noting. It has been traditionally classified as an omnivore, as it shares with extant pigs (Suidae) similarities in the relative length of the premolars and of the masseteric fossa. However, as other features of its dentition resemble those of herbivores, Cassini et al. (2012a) considered its potential as a nonomnivore. Thus, the astrapotheres occupy a morphospace intermediate between those characterizing browsing artiodactyls and perissodactyls of closed environments.

The results of their analyses of dental OSA (see chap. 8) allowed Cassini et al. (2012a) to explore two possible interpretations. On the one hand, they evaluated whether the slope between a variable that increases as a cubic function (mass) versus one that increases as a squared function (area) follows that expected for isometry (i.e., a ratio of 0.6666, see chap. 4). The regression lines of the entire sample (i.e., all herbivorous eutherian mammals), of the SANU sample, and of the notoungulate sample only exhibit positive allometry; that is, the largest animals have a larger OSA per unit of mass than that expected for an isometric ratio (table 9.1). These values are reasonably similar to (though differ significantly from) those predicted by the 0.75 power of Kleiber's law (see chap. 4); in other words, by the correlation of these variables with metabolism (see also Gould 1975). As mentioned in chapter 8, there is a relationship between OSA and metabolic rate, and Cassini et al. (2012a) therefore interpreted that Santacrucian

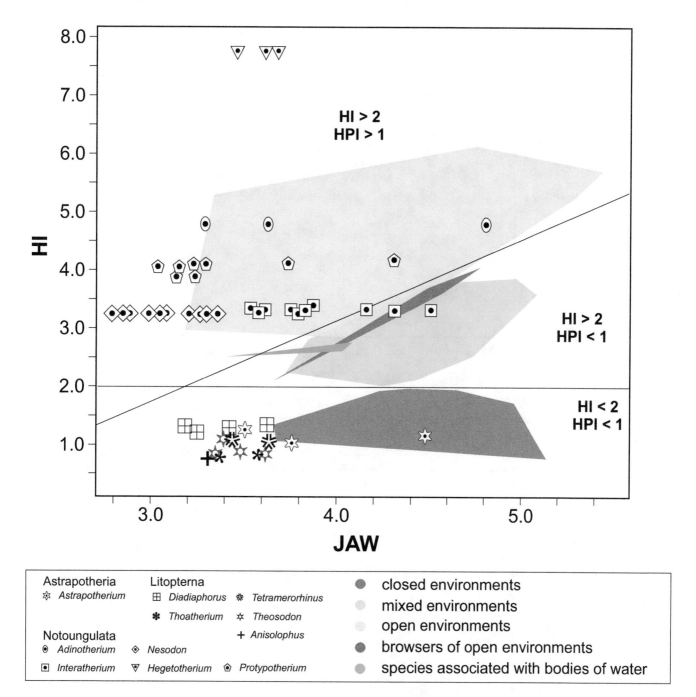

9.24. Distribution of extant ungulates (enclosed by *color polygons*) and Santacrucian ungulates (*symbols*) in the morphospace defined by the hypsodonty index (HI) and relative length of the mandible (JAW: length of the horizontal ramus divided by length of the lower molar series, LMRL), which, together with habitat preference index (HPI), allows the characterization of species according to the indicated habitats. Modified from Cassini et al. (2011).

ungulates (unlike xenarthrans; see Vizcaíno et al. 2006a) would have had basal metabolic rates similar to those of other eutherian mammals.

On the other hand, by analyzing the regression residuals (part of the data not explained by the dependent variable), Cassini et al. (2012a) found that, among extant ungulates, perissodactyls exhibit positive values, whereas those corresponding to ruminant artiodactyls are negative (fig. 9.26). This was interpreted as evidence of a greater

processing capacity in the oral cavity for perissodactyls. In these animals, which are hindgut fermenters, food is processed in the oral cavity only once, whereas ruminants, which are foregut fermenters, have a more efficient digestive system where the food is regurgitated for further oral processing ("chewing cud"). The authors postulated that the similarities of litopterns, particularly proterotheriids, to perissodactyls in regression results suggests that litopterns may have required longer processing time in the

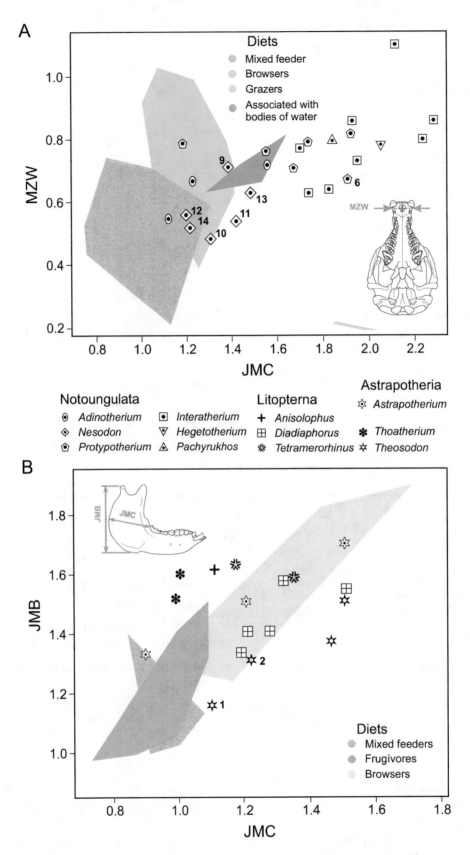

9.25. *A*, distribution of extant ungulates (enclosed by *color polygons*) that are inhabitants of open and mixed environments and Santacrucian notoungulates (*symbols*) in the morphospace defined by the relative muzzle width (MZW) and the relative maximum width of the mandibular angle (JMC), which allows characterization of species according to the indicated feeding habits and habitats. *B*, distribution of extant ungulates of closed environments (enclosed by *color polygons*) and Santacrucian astrapotheres and litopterns in the morphospace defined by the relative depth of the mandibular angle (JMB) and the maximum width of the mandibular angle (JMC), which allows reliable characterization of browsers with respect to frugivores and intermediate (mixed) diet. Modified from Cassini et al. (2011, 2012a).

Table 9.1. Results of the regression for log-transformed cheek tooth occlusal surface area versus body mass

Taxon	No. of spp.	Mass range	R^2	y-Int.[a]	Slope	Trend	Residuals
Mammalia	51	0.084–4637	**0.958**	**1.448**	**0.716**	+	X < G, Af, P
Artiodactyla + Perissodactyla	30	12.00–3729	**0.845**	**1.518**	**0.691**	iso	A < P
† SANU	17	1.67–922	**0.953**	**1.453**	**0.795**	+	N < L
† Litopterna	7	14.74–158	**0.694**	**2.074**	**0.526**	iso	Pr = Ma
† Notoungulata	9	1.67–738.6	**0.991**	**1.407**	**0.788**	+	Tx = Ty

Note: Most common parameters of standardized major axis (SMA) regressions for each group: R^2, estimators of the allometric coefficient (slope) and the constant of normalization (the y-intercept, y-Int.[a]). The residual analysis expresses significant differences ($p < 0.001$) among the groups. Due to the use of logarithms, the straight-line intercept corresponds to a mass of 1 kg (0 is the decimal logarithm of 1). Trend: allometric tendency, with *iso* indicating isometry (slope does not differ from the expected value of 2/3), + indicating positive allometry (slope significantly higher than the value of 2/3). A = Artiodactyla, Af = Afrotheria, G = Glires, L = Litopterna, Pr = Proterotheriidae, Ma = Macraucheniidae, N = Notoungulata, P = Perissodactyla, Tx = Toxodontia, Ty = Typotheria, X = Xenarthra. † designates extinct taxa.

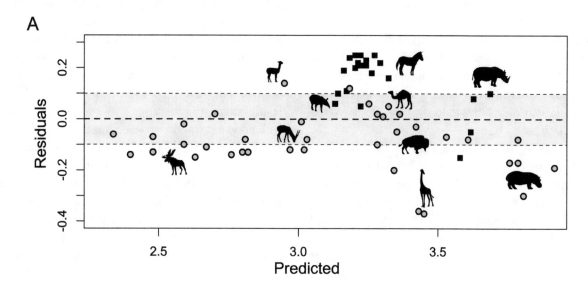

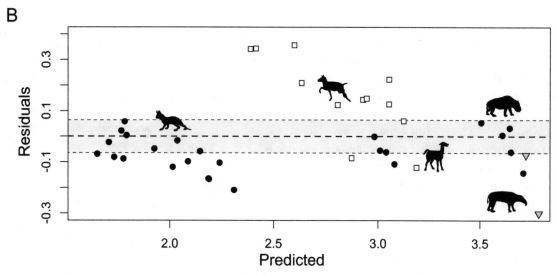

9.26. Graphic analyses of regression residuals. *A*, extant Artiodactyla (*green circles*) and Perissodactyla (*black squares*). *B*, Santacrucian ungulates, Notoungulata (*black circles*), Litopterna (*white squares*), and Astrapotheria (*green inverted triangles*). Modified from Cassini (2011) and Cassini et al. (2012a).

oral cavity than notoungulates and astrapotheres, which resemble more the artiodactyls in OSA regression values, suggesting relatively less oral processing.

In summary, partitioning of the herbivore niche in the native ungulate assemblage is based on the differential use of open, mixed, and closed environments and in the differentiation of diet as reflected mainly by three biological attributes: (1) body size, (2) form and function of craniodental features, and (3) energetic requirements or digestive physiology. The hypotheses and paleobiological reconstructions by Cassini et al. (2012a) have been evaluated in several more recent articles by Cassini and contributors (Cassini et al. 2012b; Cassini and Vizcaíno 2012; Cassini 2013; Cassini et al. 2017a, 2017b), in which methods based on landmarks were incorporated for estimating body mass, biomechanical analyses of the masticatory system, ecomorphological and tridimensional morphogeometric analyses of the skull and mandible, morphological integration of the mandible, and the relationships between several dental variables (hypsodonty, OSA, and enamel crest crenulations).

As noted earlier, stable isotopic analyses can provide an independent line of evidence (i.e., independent of form-function approaches) for evaluating hypotheses of diet and environment. Recently, enamel isotopic analyses by Trayler et al. (2020) revealed similar $\delta^{13}C$ values for *Homalodotherium*, *Nesodon*, and *Adinotherium*, compatible with a browsing diet for all these taxa. Their data also support the interpretation of brachydont litopterns as browsers. However, as the carbon isotope composition for C_3 grasses falls in the middle of the overall range in C_3 plants, they could not exclude the possibility of C_3 grazing. They also found that $\delta^{13}C$ values for *Astrapotherium* do not differ significantly from those of the other taxa, but its mean $\delta^{18}O$ value is lower than the mean for all other taxa. As increased water availability can allow rapid turnover of body water, which drives $\delta^{18}O$ values (lower) toward local water $\delta^{18}O$ (Kohn 1996; MacFadden 1998; Clementz et al. 2008), Trayler et al. (2020) interpreted the low $\delta^{18}O$ values for *Astrapotherium* as supporting the lines of evidence mentioned above for a semiaquatic lifestyle.

CARNIVOROUS METATHERIANS

During the Santacrucian, the guild of mammalian predators was occupied mainly by sparassodont metatherians, which reached their greatest diversity during this time period (Prevosti and Forasiepi 2018). Their paleoecology was reviewed by Prevosti et al. (2012). In general, sparassodonts were mainly hypercarnivores, with different substrate preferences and locomotory abilities (from climbers

to cursorial terrestrial) and a wide range of body sizes (from 1 kg to almost 100 kg).

In the Santa Cruz Formation, 11 species belonging to 10 genera are recorded (fig. 9.27): six Hathliacynidae (*Acyon tricuspidatus*, *Cladosictis patagonica*, *Sipalocyon gracilis*, *Sipalocyon obusta*, *Pseudonotictis pusillus*, and *Perathereutes pungens*) and five Borhyaenoidea (*Prothylacynus patagonicus*, *Lycopsis torresi*, and three Borhyaenidae, *Borhyaena tuberata*, *Acrocyon sectorius*, and *Arctodictis munizi*).

For species with reasonably well-preserved remains, Prevosti et al. (2012) derived body-mass estimates by means of allometric equations using the centroid size from morphogeometric analysis of postcranial elements (Ercoli 2010; Ercoli and Prevosti 2011); for those represented by fragmentary or only cranial remains, the authors based the estimates on dentition following Vizcaíno et al. (2010). Further, they employed regression analysis of phylogenetic eigenvectors to correct for bias imposed by phylogenetic patterns (PVR; Diniz-Filho et al. 1998). Their results suggest a size range between 1 kg and about 40 kg, with good separation between species (table 9.2).

The largest species are more clearly separated by size than the smaller species (*Sipalocyon* species are similar in size, as are *Perathereutes* and *Pseudonotictis*). The authors emphasized that during the Santacrucian, there might not have been mammalian predators with body masses larger than 100 kg; this pattern contrasts with that of most extant and extinct communities of eutherian mammals (e.g., van

Table 9.2. Body mass (BM, in kg) and relative grinding area (RGA) of the last lower molar (m4) of Santacrucian sparassodonts

Species		BM	RGA
Borhyaenidae	*Arctodictis munizi*	> 37.00[a]	0*
	Borhyaena tuberata	36.40[b]	0*
	Prothylacynus patagonicus	31.79[b]	0.17
	Lycopsis torresi	20.07[a]	0.30
	Acrocyon sectorius	11.49[a]	0*
Hathliacynidae	*Cladosictis patagonica*	6.60[b]	0.17
	Acyon tricuspidatus	4.30[a]	0.30
	Sipalocyon gracilis	2.11[b]	0.33
	Sipalocyon obusta	2.06[a]	0.27
	Pseudonotictis pusillus	1.17[b]	0.30
	Perathereutes pungens	1.00[a]	0.34

Note: Arctodictis munizi is the largest species, followed by *Borhyaena tuberata* and then by species of *Prothylacynus*, *Lycopsis*, *Acrocyon*, *Cladosictis*, *Acyon*, *Sipalocyon*, *Pseudonotictis*, and *Perathereutes*.
[a] Estimates based on dental measurements.
[b] Estimates based on centroid size.
* These taxa have virtually no talonid.
** The specimen is a juvenile and the m4 had not yet erupted.

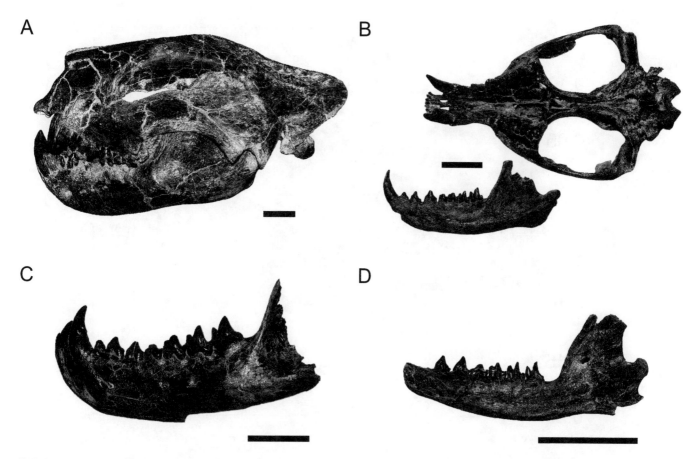

9.27. Santacrucian sparassodont metatherians. *A*, skull and mandible of *Artodictis munizi* (CORD-PZ 1210; Museo de Paleontología, Facultad de Ciencias Exactas, Físicas y Naturales, Universidad Nacional de Córdoba, Argentina) in left lateral view. *B*, skull in ventral view and mandible in left lateral view of *Cladosictis patagonica* (MPM-PV 4326; Museo Regional Provincial Padre M. J. Molina, Rio Gallegos, Santa Cruz Province, Argentina). *C*, mandible of *Prothylacynus patagonicus* (MPM-PV 4318) in left lateral view. *D*, mandible of *Perathereutes pungens* (MPM-PV 4322) in left lateral view. Scale bars = 3 cm. Photos courtesy of Analía M. Forasiepi.

Valkenburgh 1985, 1999; Fariña 1996; Prevosti and Vizcaíno 2006) but does resemble some extant ecosystems of South America, such as Patagonia.

For inferring substrate preference and use, Prevosti et al. (2012) employed geometric morphometry and a large database of extant mammals, including more than 200 specimens distributed among more than 100 species of marsupials (Didelphimorphia and Dasyuroidea) and eutherian carnivorans (Canidae, Ursidae, Felidae, Mustelidae, Mephitidae, Viverridae, and Herpestidae). They classified the extant species as follows: arboreal (dwelling mainly on trees and rarely descending to the ground), scansorial (moving on the ground and trees; with good climbing abilities, but spending most their time on the ground), terrestrial (mainly moving on the ground and never, or only occasionally, climbing or digging), cursorial (moving almost exclusively on the ground, capable of fast, long-distance running and occasional digging), semiaquatic (frequently or always swimming), and semifossorial (frequent digging). They proposed that Santacrucian

sparassodonts might have been diverse in their locomotory capacities and habitat use (fig. 9.28). Species of *Prothylacynus*, *Cladosictis*, *Sipalocyon*, and *Pseudonotictis*, for instance, might have had considerable mobility at the joints, especially for forelimb supination, resembling extant climbing carnivorans such as procyonids (e.g., coati, *Nasua*) and viverrids (e.g., binturong, *Arctictis*). *Prothylacynus* might have had prehensile capacity in the forelimbs, which were even more developed than in *Cladosictis*, *Sipalocyon*, and *Pseudonotictis*. However, the tibial morphology of *Prothylacynus* suggests more restricted movements than in entirely arboreal extant taxa (e.g., binturong; de Muizon 1998). The limb morphology of *Borhyaena* and *Artodictis* is comparable to that of strictly ground-dwelling taxa, due to the small medial epicondyle and a larger humeral trochlea.

For inferring diet, Prevosti et al. (2012) carried out a geometric morphometry analysis of the mandible and applied a morphometric index to the last lower molar. To these ends, they used a wide sample of predominantly

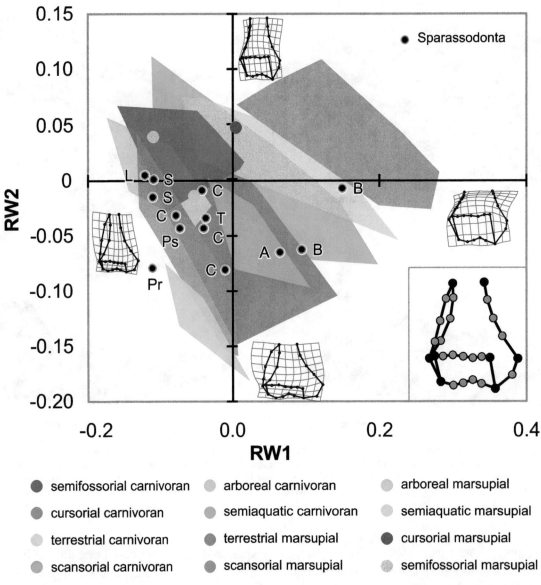

9.28. Morphospace of the analysis of anterior view of the distal humeral epiphysis of extant carnivorans and marsupials (enclosed by *color polygons*; *color key* in the figure) and sparassodont metatherians (*black* and *gray circles*) defined by the first two relative warps (RW). A, *Arctodictis sinclairi*. B, *Borhyaena tuberata*. C, *Cladosictis patagonica*. L, *"Lycopsis" longirostrus*; Pr, *Prothylacynus patagonicus*; Ps, *Pseudonotictis pusillus*; S, *Sipalocyon gracilis*; T, *Thylacosmilus* cf. *T. atrox*. Modified from Ercoli et al. (2012).

carnivorous extant mammalian lineages (i.e., Carnivora among eutherians and Didelphimorphia and Dasyuroidea among marsupials) that were classified as hypercarnivores (feeding mainly on meat from other vertebrates), mesocarnivores (feeding mainly on other vertebrates, generally smaller than themselves, but also on plants and invertebrates), omnivores (feeding mainly on plants and invertebrates), herbivores (feeding on plants), and insectivores (feeding on insects). These categories were used in the discriminant and canonic analyses (same probability for each diet group), using values of the relative warps (a type of principal components of the distributions of the landmark configurations) for inferring the diet of

Santacrucian sparassodonts. Two sets of analyses were conducted, one that included only marsupials (because both sparassodonts and marsupials are metatherians), and another that also included the eutherian carnivorans. In analyses using only extant marsupials, the sparassodonts fell clearly among two groups (fig. 9.29): one, including species of *Borhyaena*, *Artodictis*, *Cladosictis*, and *Prothylacynus*, is characterized by short and deep dentaries and an anteriorly displaced masseteric fossa, and overlaps the grouping of extant marsupial hypercarnivores; the other, including the species of *Perathereutes*, *Sipalocyon*, and probably *Acyon* and *Pseudonotictis*, is characterized by shallower dentaries and overlaps with extant marsupial

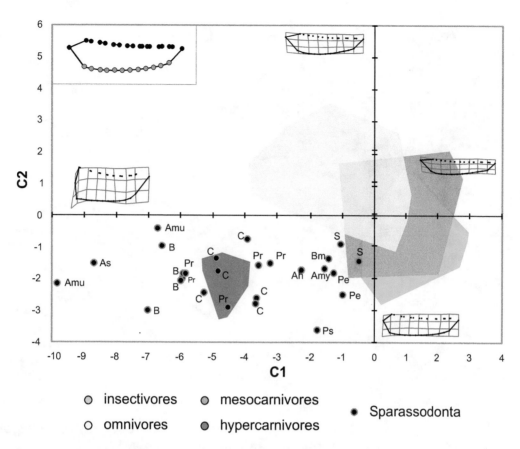

| ○ insectivores | ○ mesocarnivores | |
| ○ omnivores | ● hypercarnivores | ● Sparassodonta |

9.29. Morphospace defined by the first two canonical axes (C) of the discriminant analysis of landmark configuration of the mandibular alveolar ramus in sparassodont metatherians (*black* and *gray circles*) compared to extant carnivorans, which allows reliable characterization of hypercarnivores (*red*), insectivores (*light blue*), mesocarnivores (*orange*), and omnivores (*yellow*). Ah, *Acyon herrerae*; Amu, *Arctodictis munizi*; Amy, *Acyon myctoderos*; As, *Arctodictis sinclairi*; B, *Borhyaena tuberata*; Bm, *Borhyaenidium musteloides*; C, *Cladosictis patagonica*; Pr, *Prothylacynus patagonicus*; Pe, *Perathereutes pungens*; Ps, *Pseudonotictis pusillus*; S, *Sipalocyon gracilis*. Modified from Prevosti et al. (2012).

omnivores, mesocarnivores, and insectivores. This pattern is clear using the marsupial sample, but in the analysis that also included eutherian carnivorans (Carnivora), *Cladosictis* and *Sipalocyon* fell in an area dominated by nonhypercarnivores, whereas the others are placed with hypercarnivores. Discriminant functions, with correct reclassification between 69.59% and 79.39%, classify the members of the first group as hypercarnivores and those of the second group partially as omnivores or insectivores. Dentary form is the most important discriminant parameter in these functions, and therefore it explains why sparassodonts with a hypercarnivorous dentition but with more generalized dentaries are not classified as hypercarnivores. According to Prevosti et al. (2012), their results coincide with previous hypotheses that *Artodictis munizi* and probably *Borhyaena tuberata* (Marshall 1977, 1978; Argot 2004; Forasiepi et al. 2004) might have been scavengers, at least opportunistically.

The morphometric index that they used, relative grinding area (RGA, adapted from van Valkenburgh 1989) measures the relative development of the talonid in the distal lower molar (m4):

RGA = *square root of the talonid area / length of trigonid.*

In the case of extinct species, a taxon is considered hypercarnivorous when the RGA index is lower than 0.48, mesocarnivorous when the index falls between 0.48 and 0.54, and omnivorous when it is higher than 0.54 (Prevosti et al. 2013). The results indicate that all Santacrucian sparassodonts are comparable to extant hypercarnivorous marsupials and eutherians (table 9.2).

Prevosti et al. (2012) concluded that the guild of Santacrucian mammalian predators was formed by hypercarnivorous taxa with morphological specializations that suggest partitioning of the trophic resource use. Most hathliacynids (e.g., *Sipalocyon*, *Perathereutes*, and *Pseudonotictis*) have long and shallow mandibles and well-developed talonids, which indicates a less hypercarnivorous diet than most of the borhyaenids. On the other hand, large borhyaenids (such as *Borhyaena* and *Artodictis*) have more robust mandibles and lower molars practically lacking talonids, which correlates with extreme hypercarnivory and the capacity, at least opportunistically, for scavenging. These authors did not consider analogies of sparassodonts with

1. *Arctodictis munizi*
2. *Borhyaena tuberata*
3. *Cladosictis patagonica*
4. *Lycopsis torresi*
5. *Prothylacynus patagonicus*
6. *Pseudonotictis pusillus*
7. *Sipalocyon gracilis*

9.30. Reconstruction of the guild of Santacrucian sparassodonts. Modified from Prevosti et al. (2012).

eutherian carnivorans (e.g., canids, ursids, and mephitids) particularly convincing, although they concluded that the diversity of Santacrucian sparassodonts is similar to that observed in communities of extant and extinct eutherian hypercarnivores. Incidentally, these conclusions provide an example of the limitations of using modern analogs to reconstruct the past (see chap. 1).

Last, Prevosti et al. (2012) moved toward paleosynecological integration. By considering the body sizes of the primary consumers studied in other chapters of the Vizcaíno et al. (2012d) volume, they proposed hypotheses on the trophic relationships of Santacrucian sparassodonts (fig. 9.30). They interpreted that *Arctodictis munizi* and *Borhyaena tuberata* might have hunted mammals of a wide range of sizes, from rodents (e.g., *Neoreomys* of about 4 kg), small xenarthrans (e.g., armadillos of 8 to 11 kg), small notoungulates (e.g., interatheriids of 3 kg), to juveniles of larger Santacrucian mammals (e.g., *Astrapotherium*, with adults of almost 1 tonne). According to the authors, the

most accessible prey might have occupied a body size range between 20 and 150 kg, including sloths (e.g., *Hapalops* and *Eucholoeops*, 35–85 kg), toxodontids (e.g., *Adinotherium*, 100 kg) and litopterns (e.g., *Thoatherium* and *Diadiaphorus*, 24–82 kg). *Prothylacynus* and *Lycopsis* might have had diets similar to *Artodictis* and *Borhyaena*, but adult toxodontids and astrapotheres were probably beyond their capture capacities, and the most probable common prey might have been sloths, armadillos (e.g., *Proeutatus*, 15 kg), litopterns, and hegetotherids (e.g., *Hegetotherium*, 8 kg). The potential prey of *Acrocyon* might have included small litopterns (e.g., *Licaphrium* and *Thoatherium*, 18–24 kg), hegetotheriids, interatheriids, rodents, small armadillos (e.g., *Stenotatus* and *Prozaedyus*, 1–4 kg), and primates (*Homunculus*, 2.7 kg). Small sparassodonts probably concentrated their diet on small rodents (e.g., *Eocardia* and *Stichomys*, 0.16–0.79 kg), interatheriids, and small marsupials (e.g., *Palaeothentes*, 0.08–0.36 kg). In particular, the largest hathliacynids, *Cladosyctis* and *Acyon*, might have preyed more frequently on small armadillos, larger rodents (e.g., *Neoreomys*), and primates (*Homunculus*).

An additional line of evidence on the diet of sparassodonts is provided by the recent analysis by Tomassini et al. (2019) on trace fossils consisting of numerous coprolites (fossilized feces) recovered from localities and levels of the Santa Cruz Formation. Several of the coprolites include bones and teeth belonging to small mammals (e.g., octodontoid or chinchilloid rodents). The size, shape, chemical composition, and taphonomy of the bone remains suggest that the coprolites were produced by sparassodonts, probably hathliacynids and/or small borhyaenoids.

Paleosynecology

In the last chapter of the volume edited by Vizcaíno et al. (2012d) on paleoecological synthesis, as also noted earlier in this chapter, Kay et al. (2012b) pointed out the distinction, often undervalued, between the two perspectives from which paleoecological reconstruction is approached: (1) that based on biotic and abiotic evidence (e.g., fossils, sedimentological characteristics) and used to reconstruct the environmental parameters of (in this case) the Santa Cruz Formation; and (2) that based on form-function features (e.g., body size, diet, and substrate preference and use; i.e., paleoautoecology) of the different species, and the reconstruction from the general characteristics of these niches of the Santa Cruz Formation as a whole (i.e., paleosynecology).

To avoid the effects of averaging, Kay et al. (2012b) focused their study area on a series of geographically and temporarily restricted localities on outcrops extending along approximately 15 km of the intertidal zone of the Atlantic coast, south of the Río Coyle (Santa Cruz Province, Argentina). The tephrochronological information indicated that the localities in this area, referred to as FL 1–7 (fig. 9.11), represent a narrow temporal period (~ 17.4–17.5 Ma). These authors used two approaches, specific and synthetic, for the paleoenvironmental and paleoecological reconstruction of the FL 1–7 biota.

SPECIFIC APPROACHES

The specific approaches consisted of examining abiotic and biotic markers, such as ichnofossils, taxa, or taxon groups with similar habits, and considering their distributional abundance in comparison with that of phylogenetically close extant representatives. This approach seeks answers to questions concerning, for example, the possible "ecological" causes for the presence of myrmecophagous mammals (anteaters and armadillos) or the absence of

crocodiles in the Santacrucian. As this approach is beyond the scope of this book, it is dealt with only briefly.

Several recorded taxa such as palm trees, the frog *Calyptocephalella*, the lizard *Tupinambis* (= *Salvator*), the anteater *Protamandua*, and the primate *Homunculus* clearly indicate that the climate at FL 1–7 was much warmer and wetter than today. The presence of trees, birds, and mammals that nowadays inhabit tropical and subtropical forests in South America (laughing falcons, porcupines, spiny rats, sloths, scansorial marsupials, and monkeys) supports this conclusion. The record of calcareous root molds in paleosols indicates high rainfall seasonality, with cold, wet winters and warm, dry summers. Also recorded are Gramineae and a series of vertebrate taxa (giant terrestrial birds, many notoungulates, glyptodonts, and armadillos) that were adapted to open environments. Consideration of taphonomical (see Montalvo et al. 2019, above), sedimentological, ichnological, floral, and faunal elements taken together suggests a landscape for FL 1–7 consistent with a mosaic of open temperate humid and semiarid open forests with ponds in some areas and seasonal flooding in others, promoting the formation of marshlands with a mix of grasses and forbs.

SYNTHETIC APPROACHES

As a first step, Kay et al. (2012b) compared species richness of FL 1–7 with that of 25 extant faunas of mammals from different regions in South America, from the equator to Patagonia, in environments that vary in rainfall, temperature, and plant composition. Of these, only five lie south of the Tropic of Capricorn, outside the boundaries of tropical zones. These tropical and extratropical localities and their faunas are markedly distinct; in the tropics, there is a strong correlation between rainfall and total richness of species, whereas no correlation was found in the faunas farther south. In FL 1–7, 49 species of mammals were identified, a much greater richness than in any of the extant Patagonian faunas considered, regardless of differences in the rainfall of these extant faunas. Comparison of the Santacrucian fauna with tropical ones suggest an annual rainfall of around 1,500 mm (fig. 9.31A).

These authors then focused on reconstructing the niche structure by applying an ecometric approach (see above), identifying the number of species and grouping them in categories according to body size, substrate preference and substrate use (expressed as locomotion), and feeding (see the text box "Categories and Indices"); in other words, this synthetic approach relies on establishing parameters for the fossil taxa based on form-function studies. They used these categories for calculating four

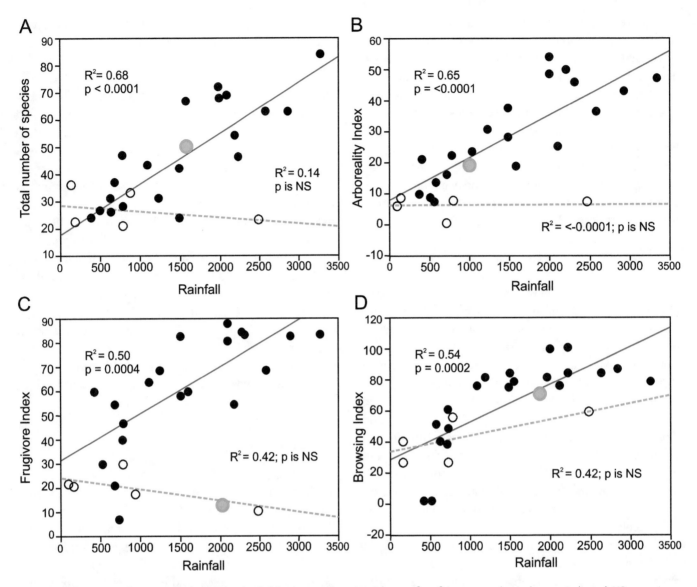

9.31. Rainfall, species richness, and niche indices. *A*, rainfall versus total species richness of nonflying mammals; *p* values are indicated; NS = nonsignificant. *B*, rainfall versus arboreality index. *C*, rainfall versus frugivore index. *D*, rainfall versus browsing index. *Black circles* represent faunas within the tropical zone (*N* = 20) and *white circles* represent extratropical faunas (*N* = 5); the lines represent the adjustment by least squares for the tropical species (*solid red line*) and extratropical species (*dashed green line*). The FL 1–7 fauna is indicated by the *blue circle*.

parameters of the number of species within a guild (i.e., with a particular niche specialization) based on the total number of species and on proportions between primary and secondary consumers: arboreality index, frugivore index, browser index, and predator/prey ratio.

Tropical faunas with higher rainfall have more arboreal and scansorial species, as revealed by the arboreality index, but this relationship does not hold for extratropical localities (fig. 9.31*B*). The FL 1–7 fauna follows the pattern for tropical faunas with around 1,500 mm of rainfall per year.

As regards diet, the number and species proportions in different feeding niches among the extant communities vary considerably with respect to rainfall. The proportion of frugivorous species compared to other primary

consumers is higher in wet climates than in dry climates (fig. 9.31*C*). Likewise, the number of browsers (folivores) in relation to the number of herbivores is higher in wet localities than in dry ones (fig. 9.31*D*).

The browser index in FL 1–7 reflects rainfall in the range of 1,500 mm per year, but the low frugivore index (fig. 9.31*C*) suggests that if the FL 1–7 fauna were tropical, rainfall would have been lower than 500 mm per year. This index is very similar to that of any fauna of high latitude, independently of rainfall. Kay et al. (2012b) speculated, by analogy with low-latitude faunas, that the low number of frugivores may have been a consequence of extreme seasonality in fruit production in these latitudes, rather than being a marker of rain scarcity, or, alternatively, that fruits may have been present, but the frugivores were birds (rather than mammals), many of which were migrating species. This would explain, to a large extent, the scarcity of their fossil record, apart from the much lower preservation potential of birds as compared to mammals.

With regard to proportions between predators and prey, predator richness is proportionally higher when the annual rainfall decreases in extant South American faunas; the driest localities have a higher number of predator species compared to prey species (fig. 9.32). The predator/prey ratio for FL 1–7, although statistically less strongly supported than the others, nonetheless suggests rainfall values higher than 1,000 mm per year. This result is reasonable, given that species richness and other niche parameters suggest similar rainfall levels.

Indices and variables could be related with each other and with rainfall and, therefore, would not be statistically independent estimators of the latter. Kay et al. (2012b) investigated this possibility in the 25 extant faunas by means of a PCA that considered the following eight factors associated with rainfall: abundance of primary consumers, the proportion of predator to prey, three niche indices, and three cenogram results (not described here). The PCA analysis revealed that two factors explained nearly 82% of the total variance, with the first explaining 71.2% of

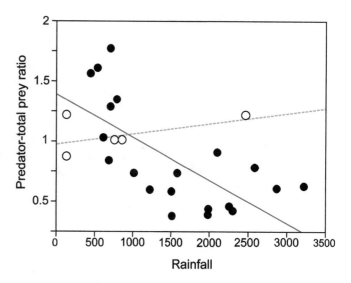

9.32. Rainfall versus predator/prey ratio. *Black circles* represent faunas within the tropical zone (*N* = 20) and *white circles* represent extratropical faunas (*N* = 5); the lines represent the adjustment by least squares for the tropical species (*solid red line*) and extratropical species (*dashed green line*).

CATEGORIES AND INDICES USED FOR RECONSTRUCTION OF NICHE STRUCTURE OF THE SANTA CRUZ FORMATION

Body-mass categories: (I) 10 to 100 g, (II) 100 g to 1 kg, (III) 1 to 10 kg, (IV) 10 to 100 kg, (V) 100 to 500 kg, and (VI) > 500 kg.

Substrate preference and use categories **(the secondary habit is set in parentheses, whereas the primary habit is not):** A, arboreal; A(T), arboreal and terrestrial (scansorial); SAq, semiaquatic; T, terrestrial; T(A), terrestrial and ambulatory; T(C), terrestrial and cursorial; T(F), terrestrial and fossorial.

Feeding categories: F(I), fruits and invertebrates, F(L), fruits and leaves; G, grass stems and leaves (grazers); I(F) insects and fruits or nectar; L, leaves (= dicot leaves, buds, shoots, sprouts not including grass, which have a high silica content); MYR, termites and ants; S, small seeds of grasses and other plants, S(I), scavenging and insects, S(L), scavenging and leaves, S(Tu), scavenging and tubers; V, vertebrate prey.

Arboreality index expresses the proportion of arboreal species to the total number of nonflying species:

$$[(A + 0{:}5\ A(T))/(A + A(T) + SAq + T)] \times 100.$$

Frugivore index expresses the proportion of frugivorous species to the total number of herbivorous species in a fauna:

$$[(F(I) + S + F(L))/F(I) + S + F(L) + L + G)] \times 100.$$

Browser index expresses the proportion of browsing or leaf-eating species to the total number of herbivorous species in the fauna:

$$[(L)/(L + G)] \times 100.$$

Predator/prey ratio expresses the proportion of secondary consumers:

[Number of insectivores (including anteaters) + carnivores + scavengers]/number of herbivores.

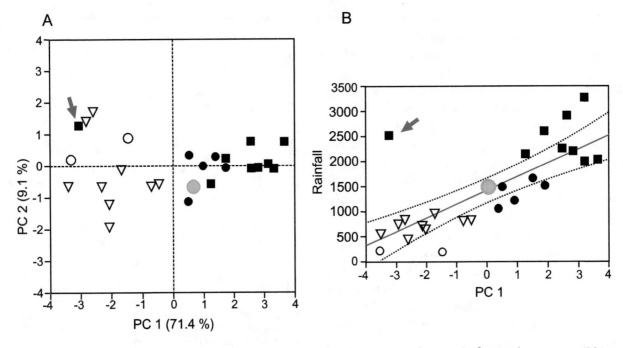

9.33. Principal component analysis of eight variables and 25 communities of extant mammals. *A*, graph of principal components (PCs) 1 and 2. Least squares regression for predicting rainfall in mm from PC1 = 322.7 (PC1) + 1,368. Predicted rainfall at FL 1–7 = 1,512 mm/year. *B*, graph of rainfall versus PC1. The *blue circle* indicates the adjusted value for FL 1–7. *Black squares* indicate rainfall higher than 2,000 mm/year. *Black circles* indicate rainfall between 1,000 and 2,000 mm/year. *Inverted white triangles* indicate rainfall between 1,000–500 mm/year. *White circles* indicate rainfall less than 500 mm/year. The *red arrow* indicates the atypical fauna of the humid forest in Nahuel Huapi (Río Negro Province, Argentina).

Table 9.3. Factor loadings (eigenvectors) for principal components analysis on scaled correlations among eight variables

Variable	Principal component 1	Principal component 2
Number of primary consumers	.386	.005
Arboreality index	.388	.031
Browsing index	.363	.186
Frugivore index	.362	.122
Predator/prey ratio	-.355	-.355
Absolute value of ±500-g interval	-.313	.589
Absolute value of slope for > 500 g	-.334	.489
Absolute value of slope for < 500 g	-.320	-.320

Table 9.4. Values of the first and second principal components in a principal components analysis of 25 extant faunas

Locality	PC1	PC2
Acurizal, Brazil	.50	-1.15
Belém, Brazil	1.77	.20
Caatingas, Brazil	-2.32	-.64
Chaco, Argentina	-2.11	-1.23
Chubut, Argentina	-1.45	.85
Cocha Cashu, Peru	3.35	-.11
Ecuador Tropical, Ecuador	3.20	.02
Esmeralda, Venezuela	3.70	.76
Federal district, Brazil	1.43	.26
Guatopo, Venezuela	1.77	-.04
Manaus, Brazil	2.84	-.04
Masaguaral, Venezuela	0.99	-.05
Misiones, Argentina	1.26	-.60
Nahuel Huapi humid forest, Argentina	-3.10	1.30
Nahuel Huapi steppe, Argentina	-2.80	1.73
Pampas, Argentina	-1.67	-.12
Península Valdés, Argentina	-3.37	.20
Puerto Ayacucho, Venezuela	2.58	.77
Puerto Páez, Venezuela	.55	.36
Río Cenepa, Perú	2.60	-.08
Río Teuco, Argentina	-2.10	-1.96
Salta low montane, Argentina	-.48	-.60
Salta transitional forest, Argentina	-.73	-.64
Tucumán subtropical forest, Argentina	-2.93	1.41
Tucumán thorn forest, Argentina	-3.45	-.63
Santa Cruz FL 1–7	.56	-.78

Note: Values of PC1 and PC2 for FL 1–7 extrapolated from factor loadings of extant faunas.

the variance (fig. 9.33; tables 9.3, 9.4). The positive factor loads in PC1 included all the indices and total richness of primary consumers. The negative factor loads were for the absolute values of the cenograms and for the predator/prey ratio. In other words, localities with more rainfall tend to have more frugivores, fewer herbivores, and more arboreal species, but correspondingly lower numbers of prey species in comparison with the number of predators and lower absolute values of the three cenogram results. The value of the factor scoring in PC1 showed a strong correlation with rainfall (R^2 = 0.56) and predicted an annual rainfall of 1,579 mm (fig. 9.33*B*). Several indices and the richness measurements correlated significantly with rainfall, although no index seems to be correlated to mean annual temperature or how open the environment is. In the PCA multivariate space, FL 1–7 was much closer to four of the extant faunas from tropical sites, with a plant mosaic of savannas and gallery forests, often with seasonal flooding, but also with six months of dry season.

By means of DA, the success of the classification of the same eight variables was analyzed for assigning existing localities to one of the four following rainfall categories: 0 = less than 500 mm per year, 1 = between 500 and 1,000 mm per year, 2 = 1,000 to 2,000 mm per year, and 3 = above 2,000 mm per year. According to DA, the combined variables correctly assigned 92% of the 25 extant faunas to rainfall groups. The remaining cases were also correctly assigned, but with a confidence level below 95%. Surprisingly, the FL 1–7 fauna was assigned to category 3 (> 2,000 mm of rainfall per year) with a confidence level of 95%, a rainfall level that seems high based on all other results.

OTHER ASPECTS OF PALEOSYNECOLOGICAL RELATIONSHIPS OF FL 1–7

Further resolution of aspects of the paleosynecological relationships of the FL 1–7 fauna has been provided by recent investigations based on approaches other than from a form-function perspective. For example, Trayler et al. (2020) provided insight into the paleoenvironmental aspects. These authors' isotopic analyses on tooth enamel of the SANUs *Nesodon*, *Adinotherium*, *Homalodotherium*, and *Astrapotherium* from localities with a similar faunal composition to that of FL 1–7 resulted in $\delta^{13}C$ values consistent with a C_3-dominated ecosystem with moderate precipitation and a mix of wooded and more open areas, supporting the conclusions of Kay et al. (2012b) in these parameters.

Another aspect was considered by Rodríguez-Gómez et al. (2020). These authors analyzed the food web of the large mammals assemblage from FL 1–7 using a model that estimates the biomass of the primary consumers of the paleoecosystem and the biomass required by the secondary consumers to calculate the degree to which the dietary requirements of the latter are satisfied. The model distributes the biomass available to the secondary consumers among the members of the carnivore paleoguild according to their prey preferences described above (Prevosti et al. 2012; Ercoli et al. 2014), which in turn allows estimation of their sustainable densities (Ds). This methodology has

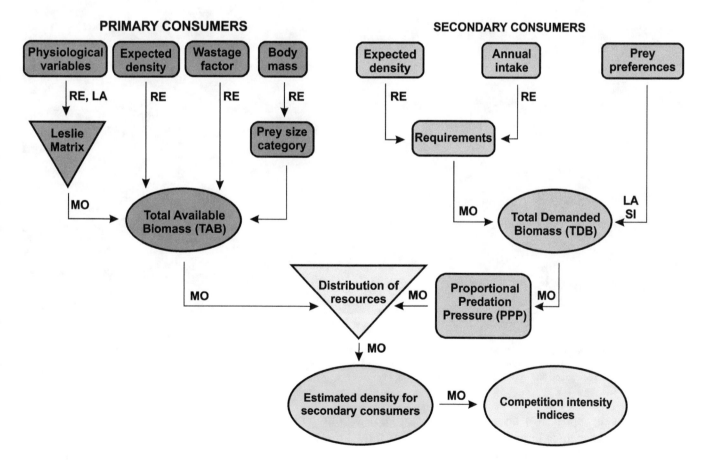

9.34. Flowchart diagram showing the components of the model used to evaluate trophic resource availability of carnivores and competition intensity in the carnivore guild of FL 1–7 from the Santa Cruz Formation. RE, regression-derived estimates; MO, model output data; LA, inferences made on living analogs; SI, inferences made on stable isotopes. Modified from Rodríguez-Gómez et al. (2020).

9.35. Artistic reconstruction of the fauna and flora of FL 1–7, Santa Cruz Formation, early Miocene, Patagonia, Argentina. In the *foreground*, the ground sloth *Nematherium* (Mylodontidae). On the *right*, the tree porcupine *Steiromys duplicatus* (Erethizontidae). On the *left*, the laughing falcon *Thegornis musculosus* (Falconidae). *On the tree branch* the primate *Homunculus patagonicus* (Platyrrhini). *Lower left*, near the river, a group of *Nesodon imbricatus* (Notoungulata, Toxodontidae). Illustration by and used courtesy of Manuel Sosa.

been previously applied to a number of European Pleistocene sites that include human remains for measuring the level of competition intensity of their carnivore paleoguilds (Rodríguez-Gómez et al. 2013, 2014, 2017). In the case of FL 1–7, Rodríguez-Gómez et al. (2020) tested different scenarios to assess whether competition within the carnivore paleoguild was intense or the resources available were in excess of those expected for satisfying the nutritional requirements of all species of secondary consumers (predators and/or scavengers). The model considered: (1) the biomass of prey resources available in the paleocommunity, which depends on the demographic structure of the prey species and their population densities; (2) the ability of each secondary consumer to obtain and process these resources; and (3) the level of competition intensity among the members of the carnivore paleoguild. The model computes the biomass of herbivores in the ecosystem as the total available biomass (TAB) and the meat necessary to sustain the carnivorous species as the total demanded biomass (TDB; fig. 9.34). Then, the distribution of meat resources among the predator/scavenger species is modeled, which provides an estimate of their sustainable densities. The results provided estimates of the nutritional requirements from the predator/scavenger guild under maximum and minimum quantities of meat offered by the prey community, indicating a well-balanced paleocommunity. The competition indices point to a relatively high level of competition for small to medium-sized prey, although competition for large mammal prey was rather low, suggesting insufficient carnivore species to fully consume the megaherbivore biomass.

FINAL CONSIDERATIONS

Kay et al. (2012b) emphasized the distinction, noted at the beginning of this chapter, between two general sorts of paleoecological approaches. One was as a tool to reconstruct environmental parameters of the Santa Cruz Formation, such as mean annual rainfall and temperature. The other was to understand aspects of the paleoecology of individual species based on their adaptive characteristics, such as body size, diet, and substrate preference and use (paleoautecology), and to reconstruct from this information the overall niche characteristics of the Santacrucian fauna as a whole (paleosynecology).

The several approaches outlined in the preceding sections of this chapter reveal that the FL 1–7 fauna would have had a species richness and niche structure analogous to those of a modern subtropical fauna of limited geographical distribution, with a plant mosaic of savannas and gallery forests, with abundant rain, seasonal flooding, and long dry intervals. Most niche markers resulting from extant neotropical communities (e.g., high species richness of primary consumers, frequent occurrence of arboreal and scansorial species and low percentage of grazing species among primary consumers) suggest a wet environment. An exception is the surprisingly low percentage of frugivores, which may be explained by the southern location of FL 1–7 and the resulting seasonality of this locality,

at least with regard to daylight cycle, possibly meaning that they were not permanent residents and thus reducing the likelihood of their fossilization. Likewise, the frugivorous guild in extant faunas in mild southern climates is dominated by birds, especially migratory species, which have a poor Miocene fossil record.

In short, the evidence indicates that the FL 1–7 fauna developed in an area with forests similar to those present today in the Andes and Andean foothills within Patagonian latitudes, coexisting with grasslands and permanent and temporary body waters (fig. 9.35). The zone was wetter and warmer than today, with more than 1,000 mm of annual rainfall, wet winters and dry summers, with mean annual temperatures above 14° C and marked seasonality in daylight.

Further, the estimations of niche parameters for individual taxa and the reconstruction of the community structure provide insight on the ecological relationships of each taxon with other components of the FL 1–7 fauna, identifying potential competition for resources, as noted above in the section on the paleoecology of lineages and guilds (e.g., among the numerous arboreal and semiarboreal plant feeders). With respect to the interguild relationships, the food-web analysis based on the biomass of the primary consumers and the biomass required by the secondary consumers provides support for a relatively high level of competition among secondary consumers for small to medium-sized prey, whereas competition for large mammal prey would have been rather low.

Finally, in the introductory chapter of *Early Miocene Paleobiology in Patagonia*, Vizcaíno et al. (2012c) noted their intention to expand their form-function based paleoecological approach to encompass a wider geographic and chronologic range of the Santa Cruz Formation. To these ends, they have since conducted further field work to recover assemblages from several localities and levels and evaluate the faunal and ecological changes that occurred over the depositional timeframe of the formation in the different areas, with the FL1–7 assemblage serving as a contextual model to assess similarities and differences. Examples of such efforts include investigations in the cliffs along the Río Santa Cruz (Kay et al. 2021) and the Andean foothills (Cuitiño et al. 2019; Vizcaíno et al. 2022). Studies of additional series of localities are in progress and will enhance our understanding of the evolution of the paleoecological scenarios that played out during that part of the MCO in Patagonia encompassed by the Santacrucian biota.

Epilogue

Some four score years ago, Efremov (1940) considered that paleontology was already past the stage of primary data collection and that the older and stereotypical geology-based articles were being replaced by a growing number of works that incorporated paleontological data within the framework of evolutionary theory, a gauntlet that Osvaldo Reig took up and then threw down to Rosendo Pascual, as we described in the Prologue. Issues such as the phylogeny of groups of organisms, biological interpretation of extinct forms, influence of the organisms on the environment over time, and their past geographical distribution were becoming more commonly addressed, at least among the more influential paleontological publications. Efremov held that, for the fullest development of this "theoretical" aspect, paleontology should be subdivided into subdisciplines: biostratigraphy, paleoecology (paleobiology), and paleopathology, among others. Each subdiscipline has its own distinct conceptual and empirical issues, requiring specialized approaches.

The homepage of the journal *Paleobiology* (http://www.cambridge.org/core/journals/paleobiology/), arguably the most emblematic scientific publication of this subdiscipline of paleontology, states that it publishes articles dealing with any aspect of biological paleontology, with emphasis on biological and paleobiological processes and models, including natural selection and patterns of variation, abundance and distribution in time and space, diversification, speciation, phylogeny, biogeography, functional morphology, taphonomy, paleoecology, macroevolution, and extinction, among others. Such is the breadth of paleobiological interests that they overlap to a great extent (or totally?) those of paleontology in general. Indeed, from our perspective it seems that only such topics as alpha taxonomy and biostratigraphy would be excluded from paleobiology, although a modern view should consider seriously intraspecific variation, which is densely packed with biological implications. However, following Efremov's suggestions, whether consciously or unconsciously, paleontologists who have focused their work along the first-noted spheres of research have identified themselves as paleobiologists so that the discipline would not remain

essentially a tool subsidiary to geology for determining depositional environments and rock ages.

In the first chapter of *The Paleobiological Revolution: Essays on the Growth of Modern Paleontology* (Sepkoski and Ruse 2009), Sepkoski provided a historical overview of paleobiology, suggesting that while its roots were firmly established by the work of the previous generation of paleontologists, paleobiology experienced an accelerated period of activity during the 1970s and 1980s. MacLeod (2014) noted that the earnest incorporation of a modern paleobiological mindset in paleontology research programs began during the early 1970s, especially with the publication of "Evolutionary paleontology and the science of form" (Gould 1970) and *Models in Paleobiology* (Schopf 1972), a volume then considered both seminal and promising but also questioned over interpretation of the potential value of the application of its models (see Demar 1973; Rowell 1973; Deevey 1974). This period did indeed witness a profound transformation in paleontology, one that influenced "the way broad questions about the history of life were incorporated into the developing field of evolutionary biology" (Sepkoski 2012, 1).

Such works, as well as many others, clearly established the unveiling of major features of life and its environments through time as the paramount goal of paleobiology, making it thus central to the study of biological evolution. In this view, the reconstruction of the lives of extinct organisms from the available fossils, including interactions with other organisms and the environment, may seem only a narrow component of paleobiology. However, although we agree wholeheartedly with Dobzhansky's (1973, 125) assertion that "Nothing in biology makes sense except in the light of evolution," we consider organismic biology, with its intricate combinations of structure, function, physiology, behavior, and ecology, as a major realm of knowledge and research in itself, and the endeavor to reveal past organismic biology as a significant challenge that requires considerable intellectual efforts. As alluded to throughout the current text, the 1980s witnessed the profound influence of Radinsky's work and its contribution to the application and establishment of the form-function approach to paleobiological investigations—one, of course, that we follow in this book.

Modern paleobiology is thus already some five decades old, and since the 1980s has undergone considerable methodological progress thanks to developments in computer-based processing and the increased ability to handle large databases and advanced multivariate statistical software. However, drawing together the often disparate approaches employed and aspects investigated by paleobiologists into integrated research outcomes has not usually fully materialized. This is possibly because of the combination of the complexity of its subject matter and the lack of awareness, by the paleontologists who practice paleobiology, of its unique constellation of causal associations and unifying potential as a scientific discipline with regard to a comprehensive understanding of the lives, relationships, functioning, and ecological contexts of the extinct forms with which it deals.

Unlike Efremov, we do not believe that the time for collecting data has ended. Moreover, it may never end, as the development of new techniques and methodologies, the reformulation of paradigms or, simply, the formulation of new hypotheses quite commonly require the acquisition of new data. In this book, we have attempted to encourage new paleontologists to envision paleobiology in the sense of the reconstruction of the lives of extinct organisms from the available fossils as an intellectually challenging subject of research, and as a prior step in the contribution to the development of more comprehensive topics such as paleoecology and macroevolution.

Appendix

Notions of Anatomy

Although molecular-based analyses have recently made important contributions to the study of extinct organisms, paleontology largely remains a morphological science, and morphology is a descriptive discipline. For anatomical descriptions to meet the requirements of being unambiguously communicable and repeatable, technical terminology for designating direction and position of and within the body are employed, based on a reference frame defined by polarity axes and sectional planes. This appendix develops these concepts and complements the characterization of the most relevant tissues and organs for paleobiological studies in fossil vertebrates dealt with in the main text: skeleton, muscles, and teeth.

Axes of Polarity and Planes of Section

Vertebrates are nearly bilaterally symmetrical (with some exceptions, such as the flatfish or the skull of many cetaceans). With the body oriented with its long axis horizontal and parallel with the substrate, as in nearly all vertebrates, three axes of polarity are defined: anteroposterior (or craniocaudal), dorsoventral, and transverse. A vertebrate body may, in theory and in practice, be sectioned along any of a number of planes, but three main or standard planes of section, projections of the three axes of polarity, are recognized. Two of them, the sagittal and transverse, are oriented vertically, whereas the third, the frontal, is oriented horizontally. A sagittal plane extends along the anteroposterior axis (i.e., from head to tail) and divides the body into left and right parts. The plane passing through the midline is the midsagittal plane (or *the* sagittal plane) and subdivides the body equally into left and right halves, which, as noted, are nearly symmetric; the midsagittal plane is thus considered a plane of symmetry. Sagittal planes to one side of the midsagittal plane are termed *parasagittal*. The transverse plane divides the body into anterior and posterior (i.e., front and back) parts, and the frontal plane into dorsal and ventral (i.e., top and bottom) parts (fig. A.1); these planes are not planes of symmetry. These axes and planes form directions of polarity that must be considered in carrying out morphological descriptions.

Based on these axes and planes and with the body in its typical (i.e., standard) anatomical position, a terminology of adjectives and adverbs has been established for describing position and direction. These terms come in pairs or couplets that denote opposing anatomical directions. Among the common terms are *anterior* and *posterior* (or anteriorly and posteriorly), meaning toward the head and tail, respectively, along a sagittal or frontal plane; *medial* and *lateral* (or medially and laterally), meaning toward or away, respectively, from the midsagittal plane along a frontal or transverse plane; and *dorsal* and *ventral* (or dorsally and ventrally), meaning toward the back and belly, respectively, along a sagittal or transverse plane. Naturally, structures and directions of interest may not lie neatly along these planes, and terms may be combined (e.g., anteroventrally, mediodorsally) to accommodate such directions. Another pair of terms is *proximal* and *distal*, toward and

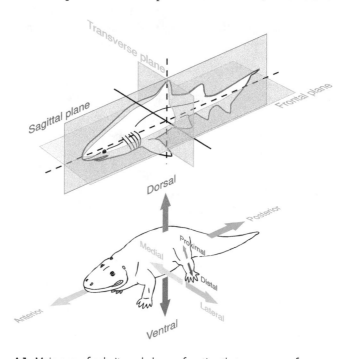

A.1. Main axes of polarity and planes of section that serve as a reference for describing anatomical direction and position in a vertebrate body. Axes of polarity: anteroposterior, dorsoventral, and transverse. Planes of section: sagittal, frontal, and transverse. Directional terms, as depicted in an early tetrapod: anterior, posterior; medial, lateral; dorsal, ventral; proximal, distal.

away, respectively, from a particular frame of reference; these are commonly used to describe position along the long axis of an appendage, with the trunk of the body as the reference frame.

The terms are used for all vertebrates, but some of the terms differ in directional designation when applied to humans. This is due to the anatomical position (i.e., posture) of humans, in which the long axis of the body is vertical with respect to the substrate. Thus, for example, anterior is synonymous with ventral, and denotes the direction toward the belly. Further, superior and inferior are used for humans, the former meaning toward the head, the latter toward the feet. A further difference is in the orientation of the frontal and transverse planes, which in humans are respectively vertical and horizontal to the substrate. Although the terms, for both humans and other vertebrates, may coincide with axes of space, it is important to remember that the directional terms are used with respect to the body. Thus, superior always means toward the head of a human and not necessarily up (unless of course the body is in anatomical position, in which case the two coincide).

Anatomical Terminology

For the purpose of unifying terminology, the guidelines for veterinary and human anatomical description are often followed. As anatomy, however, began as a subdiscipline of medicine and art, the human body was the initial reference frame. This often resulted in the adoption of human-based terms for the anatomy of other vertebrates, even when dealing with clearly different anatomical forms or functions, and has given rise to considerable confusion. For instance, cranial nerve VII is called the facial nerve because in humans (and other mammals) it innervates, among other structures, muscles of the face, whereas in other vertebrates (which lack facial musculature) it innervates muscles of the pharyngeal region. Guidelines devoted to other vertebrates are scattered in scientific articles, textbooks, and designated volumes or *Nomina*, but most deal only with mammals and birds (e.g., *Nomina Anatomica Veterinaria, Nomina Anatomica Avium*). Unfortunately, there are no formal guidelines for the conventional use of terms for the anatomy of anamniotes and nonavian reptiles, but several comprehensive, updated, and well-illustrated dissection manuals are available for vertebrates (e.g., Diogo and Abdala 2010; Diogo et al. 2018; De Iuliis and Pulerà 2019).

In the scientific literature, formal anatomical terms are usually derived from Latin or Greek roots and frequently written in Latin and italicized. For instance, the biceps muscle (in the vernacular) is designated formally as *musculus biceps brachii* and often as the *biceps brachii* muscle. In many cases, a term's etymology conveys useful descriptive information. For example, for many muscles the names are based on their shape (biceps—two bellies; deltoid—triangular in shape, named after the uppercase Greek letter delta, Δ), function (*supinator longus*—long supinator), or position or movement (*caudofemoralis*—from tail to femur). However, this is not always the case for many other anatomical terms, some of which are named after the anatomist who first noted or described them (e.g., foramen of Monro, Leydig's gland, Glaserian fissure).

Further, different authors have used different terms for referring to the same organ or structure, which has given rise to numerous synonyms and considerable confusion. For instance, the vagus nerve or cranial nerve X, which carries autonomic fibers to thoracic and abdominal structures, has at times been referred to as the pneumogastric nerve. Whenever possible, it is better to use concise and more universally accepted names (e.g., cranial nerve VII instead of facial nerve, interventricular foramen instead of foramen of Monro, and petrotympanic fissure instead of Glaserian fissure).

Tissues

Tissues can be classified into four main categories: epithelial, connective, muscular, and nervous. These tissue types react quite differently to taphonomic processes (see chap. 9) and, in most cases, only the hardest and most mineralized connective tissues are preserved. As connective and muscular tissues provide a great part of the information for paleobiological reconstructions based on the form-function correlation paradigm, particular attention will be paid to them.

Among vertebrates, particular connective tissues—bone and cartilage—form an organized system of hard structures that fossilize better than any of the other tissue types. Such structures include teeth and bony scales, the latter being an example of an integumentary bony covering termed the *exoskeleton*. In several early vertebrates, such as ostracoderms and placoderms, the exoskeleton was extensive and formed from both large bony plates and smaller discrete elements. Small dermal bony elements (osteoderms or osteoscutes) are present among some extant vertebrates as well, such as the osteoderms of crocodiles and armadillos. An internal skeletal frame for the body, the *endoskeleton*, is formed from numerous skeletal elements, many of which have been noted in the main text of this book. Endoskeletal elements are usually formed from bone but may be cartilaginous in some vertebrates (e.g., extant chondrichthyans). Because such elements, particularly bones and teeth, are the most commonly fossilized vertebrate structures, their analysis is essential for reconstructing the history of vertebrates. A composite skeleton may be assembled, for instance, by combining the preserved parts of different specimens of the same taxon. Missing portions of a skeletal element may be reconstructed based on elements with which it articulates or by comparison with phylogenetically close extant species, and even fossils can act as models for other fossils. The more complete the skeletal reconstruction, the more that can be inferred from it about the soft parts and the organism's habits, as the morphology of skeletal elements often yields clues on such structures as nerves, glands, blood vessels, and muscles. As noted in the text (see chap. 2), the form of the articular surfaces of limb bones, combined with the analysis of proportions of their segments, can provide relevant information about the posture of an animal and its modes of substrate use; similarly, mandibular proportions and dental features provide information about the possible food habits. Muscle attachments, as well as ligaments, leave marks on the bone surface. With this information, in combination with knowledge of the musculature of closely related extant species, the musculature of an extinct taxon can be reconstructed to provide information on how the muscular system moved the skeletal elements.

Connective Tissue

A connective tissue is characterized by a considerable amount of matrix, its extracellular, nonliving component, and includes bone, cartilage, fibrous, and adipose tissue, as well as blood. The composition of the matrix—that is, the kinds and proportions of substances that compose it—determines the physical and biological properties of a particular connective tissue and, consequently, its functional capacities (see chap. 3). For instance, the matrix is hard in bone, gelatinous in loose connective tissues, and fluid in the blood. Three components (ground substance, fibers, and cells) are almost invariably present in a connective tissue, with the cells separated from each other by the matrix; in other words, the cells "reside" in the matrix. The fibers are proteins and include collagenous, elastic, and reticular fibers, with collagen being the most abundant of the proteins—indeed, it is the most abundant protein in the body. Collagenous fibers are flexible but resist stretching. Reticular fibers, also formed from collagen, are linked together into a network. Elastic fibers, made of the protein elastin, are capable of stretching and, as their name implies, returning to their original shape. Ground substance, composed of large molecules (e.g., glycosaminoglycans and proteoglycans) that attract and retain water, is a gel-like substance that varies from gelatinous to rubbery in texture. It can thus absorb compressive forces and is particularly suited to a protective role, particularly loose fibrous connective tissues such as areolar tissue (see below). In histological sections of many connective tissues, the ground substance is barely noticeable, because other substances crowd the matrix.

Connective tissues may be classified in several ways; indeed, some authorities even differ on which tissues are considered connective tissue. For example, because blood does not contain fibers except during clotting, it is sometimes excluded as a connective tissue. Here, we recognize two broad categories, proper or general connective tissue and special connective tissue, subdivided as explained below (also see table A1), and focus on those most relevant to the subjects covered in this book.

Table A1. Types and subtypes of connective tissue and its basic functions

General connective tissue	Loose connective tissue		Papillary dermis and mucous membranes
	Dense connective tissue	Regular dense connective tissue	Tendons and ligaments
		Irregular dense connective tissue	Reticular dermis; muscular fasciae
	Elastic connective tissue		Tunica media of large arteries
	Reticular		Stroma of organs such as the spleen
Special connective tissue	Adipose tissue		Mechanical and thermal insulation; energy conservation
	Cartilage	Hyaline	Endoskeleton of most vertebrates, at least in some stages of ontogeny; joint capsules; nasal cartilage
		Elastic	Auricular cartilages of many tetrapods
		Fibrous	Intervertebral discs
	Bone	Compact	Cortex of most bones
		Spongy or cancellous	Inner part of most bones
	Hematopoietic tissue		Formation of blood cells
	Blood		Transport

Note: Modified from Scanlon and Sanders (2007).

General connective tissue includes loose and dense fibrous connective tissues, the former further subdivided into areolar, reticular, and adipose (classified as a special connective tissue by some authorities), and the latter into regular and irregular. Special connective tissue includes cartilage (hyaline, elastic, and fibrous), bone (compact and spongy or cancellous), hematopoietic tissue, and blood (table A.1).

GENERAL CONNECTIVE TISSUE

These connective tissues are widely distributed over the body (fig. A.2). The main cellular type is the fibroblast, which gives rise to most components of the extracellular matrix. Several other cell types (e.g., macrophages, plasma cells, and mast cells, among others) are also generally present. The ground substance is relatively prominent, although in histological sections the space it occupies in life, amid the fibers and cells, can appear empty. Loose and dense connective tissues are aptly named, as they differ primarily in the density of fibers. The fibers of areolar tissue, mainly collagenous, vary in thickness and are loosely organized; this tissue is widespread in the body, surrounding blood vessels and nerves and underlying many epithelia. The fibers of reticular connective tissue are linked to form the netlike (hence its name) supportive stroma (or framework) of several organs, such as bone marrow, lymph nodes, and the spleen and thymus.

Dense connective tissues, as might be expected, are densely packed with collagen fibers; indeed, they occupy more space than the cells and ground substance

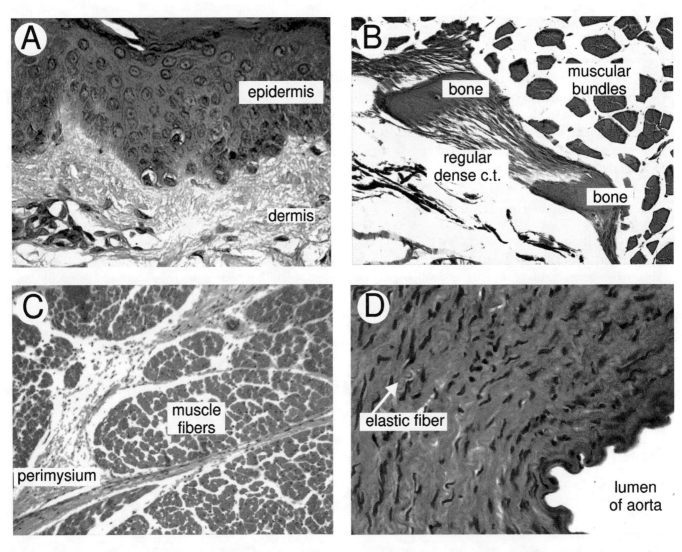

A.2. General connective tissue. *A*, mammalian epidermis, formed by epithelium, and dermis, formed largely by connective tissue (40×, haematoxylin-eosin stain). *B*, dense fibrous connective tissue joining skeletal elements (decalcified bone) of a fish larva (or alevin; 40×, decalcification and haematoxylin-eosin stain). *C*, irregular dense connective tissue surrounding the muscular fascicles in mammalian striated skeletal muscle (10×, haematoxylin-eosin stain); note the connective envelope of the muscle bundles (perimysium). *D*, elastic fibers in the middle layer (tunica media) of a mammalian aorta (40×, haematoxylin-eosin stain). In all images, *white* areas are empty spaces formed due to shrinkage of cells during histological preparation. Photos courtesy of the Cátedra de Histología y Embriología Animal, Facultad de Ciencias Naturales y Museo, Universidad Nacional de La Plata, Argentina.

and appear to crowd them out in histological sections. A main difference between regular and irregular dense connective tissues is the arrangement of the fibers, and the functional consequences of these different designs wonderfully illustrate the relationship, here at a microscopic rather than macroscopic level, between form and function. Regular dense connective tissue, in which fibers are arranged parallel to each other, compose structures such as tendons and ligaments. Tendons (generally though not exclusively) join muscles to skeletal structures, such as a bone, whereas ligaments (again, generally though not exclusively) link together skeletal elements; these structures are formed largely of collagenous fibers, but some ligaments may also contain elastic fibers. Fasciae and aponeuroses are special types of tendons that connect muscles to each other. These structures undergo repeated tensile forces in a predictable direction, explaining the orientational arrangement of their design; that is, the long axis of the fibers is arranged along the direction of force. By contrast, the fibers of irregular dense connective tissue are arranged in different directions, allowing the tissue to resist forces from multiple directions. It forms most of the dermis and the protective capsule around organs such as the kidney and spleen. A modified form of regular dense fibrous connective tissue is elastic connective tissue, which includes a great amount of elastin. It can be found, for example, in walls of arteries and surrounding the alveoli of lungs.

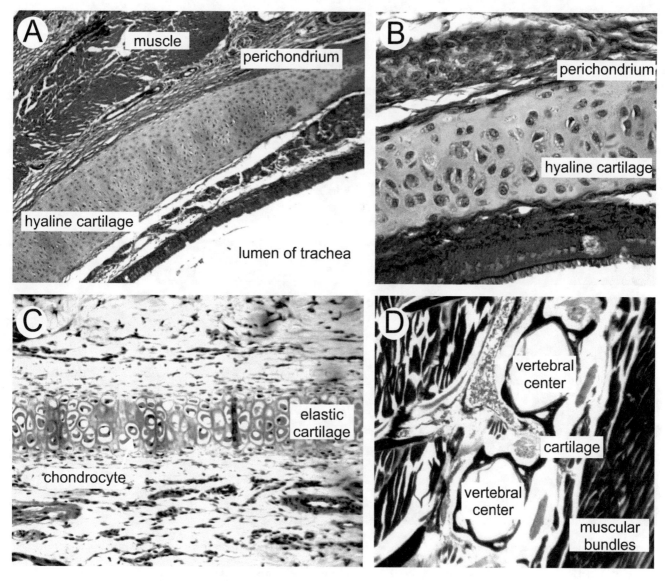

A.3. Cartilaginous tissue. *A*, hyaline cartilage of the tracheal ring of a mammal (10×, trichrome stain). *B*, detail (i.e., higher magnification image) of a tracheal ring of the same individual, with individual chondrocytes readily distinguishable within the matrix; note the perichondrium, which surrounds and nourishes the cartilage (40×). *C*, elastic cartilage of a mammalian ear; note chondrocytes surrounded by elastic fibers (10×, haematoxylin-eosin stain). *D*, fibrous cartilage in the intervertebral discs of a fish larva (or alevin; 4×, ferric haematoxylin). Photos courtesy of the Cátedra de Histología y Embriología Animal, Facultad de Ciencias Naturales y Museo, Universidad Nacional de La Plata, Argentina.

SPECIAL CONNECTIVE TISSUE

Among the special connective tissues, cartilage and bone differ in their cellular types (chondrocytes and osteocytes, respectively), their matrix composition, and vascularization (avascular and vascular, respectively, meaning that cartilage is less metabolically active and, incidentally, explains why it heals slowly). The surface of bones and most cartilages is covered by irregular dense fibrous connective tissue, respectively the periosteum and perichondrium.

Cartilage is a firm but flexible tissue (fig. A.3). Its physical properties are mainly determined by the proportion of ground substance and type of prevailing fibers. In cartilage, crystals of calcium minerals can be deposited in its extracellular matrix, increasing its hardness (e.g., in cartilaginous fishes or in late ontogenetic stages in tetrapods). As the chondrocytes deposit the matrix, they become entrapped in spherical spaces termed *lacunae* (sing. *lacuna*). There are three types of cartilage:

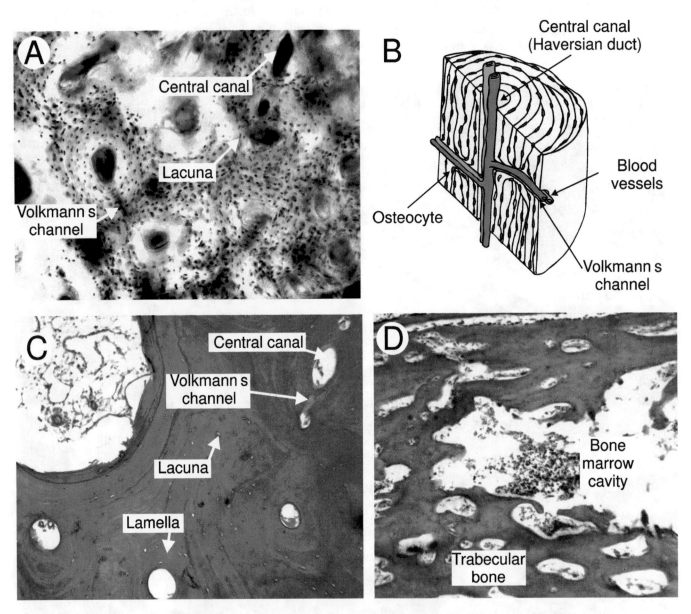

A.4. Osseous tissue. *A*, transverse section of mammalian compact bone, showing several osteons. Each osteon is composed of matrix layers (technically termed *lamellae*) arranged concentrically around a central canal (or Haversian duct). The matrix was deposited by osteoblasts that have become entrapped within the matrix and transformed into osteocytes; here, however, only the lacunae (spaces occupied by osteocytes) are observed. A Volkmann's channel may be detected due to translucency of the matrix (10×, wear). *B*, diagrammatic section of an osteon. *C*, decalcified section of a mammal bone. Note concentric arrangement of lamellae around the central canals; owing to decalcification, only the ground substance and collagenous fibers remain of the originally calcified matrix (10×, decalcification). *D*, section through a mammalian long bone showing trabecular bone (4×, haematoxylin-eosin stain). Photos courtesy of the Cátedra de Histología y Embriología Animal, Facultad de Ciencias Naturales y Museo, Universidad Nacional de La Plata, Argentina.

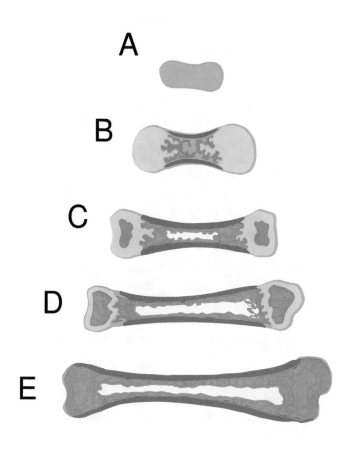

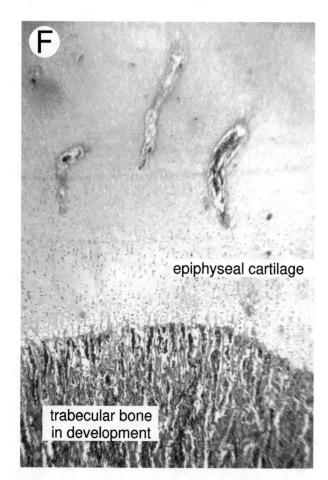

epiphyseal cartilage

trabecular bone in development

A.5. Bone growth. *A–E*, schematic sequence of endochondral ossification in a mammalian long bone, from embryo (*top*), to adult (*bottom*). *A*, initially, the entire element is formed in cartilage (*blue*). *B*, an osseous bony collar forms, and cartilage is first broken down and replaced by bone in the region, termed the *primary center of ossification*, that will form the diaphysis (or shaft) of the element; this is mainly compact (or cortical) bone (*red*). At either end, the matrix is still cartilaginous. *C*, the diaphysis becomes progressively larger. Ossification begins at either end; these are referred to as secondary centers of ossification, forming the epiphyses. *D*, the diaphysis and epiphyses are now largely composed of bone, but a thin cartilaginous zone, the epiphyseal (or growth) plate, remains interposed between the diaphysis and either epiphysis. It is at these plates that most of the element's increase in length is achieved; hyaline cartilage is continuously formed in the middle of the plates, which effectively pushes either epiphysis farther away from the diaphysis. While new cartilage is being formed, some of the cartilage adjacent to the bony margin of the diaphysis and epiphysis is broken down and replaced by bone; this transitional zone, where cartilage is transformed to bone, is termed the *metaphysis*. This process continues such that the zone of cartilage is maintained as a thin plate. *E*, as adult size is reached, cartilage production at the epiphyseal plates ceases; the remaining cartilage is broken down and replaced by bone, thus fusing the epiphyses to the diaphysis, marking the termination of growth, leaving only a cap of articular cartilage. Throughout this process, *B–E*, spongy or trabecular bone (*orange*) is also formed, particularly near the epiphyses. Hematopoietic tissue or red bone marrow (*pink*) is commonly found in these regions. The center of the diaphysis becomes hollow, forming the medullary cavity, which is typically filled with yellow bone marrow (*yellow*). Not all endochondral skeletal elements have defined epiphyses at either end (modified from Scanlon and Sanders 2007). *F*, epiphysis of a developing long bone in a mammalian embryo. Note the trabecular bone and epiphyseal cartilage; the interface between them is the metaphysis, the transitional zone where progressive calcification of the matrix occurs (4×, haematoxylin-eosin stain). Photo courtesy of the Cátedra de Histología y Embriología Animal, Facultad de Ciencias Naturales y Museo, Universidad Nacional de La Plata, Argentina.

Hyaline has a prevalence of amorphous ground substance, which allows accumulation of proportionately large amounts of water, providing it hydrostatic properties; its collagenous fibers are extremely fine, giving hyaline cartilage a clear, vitreous appearance. It is found, for instance, on the articular surfaces of long bones, tracheal rings, and parts of the skull.

Fibrous (or fibrocartilage): coarse collagenous fibers prevail in the extracellular matrix, giving fibrous cartilage considerable tensile strength and stress resistance; it is present, for instance, in intervertebral discs, the pubic symphysis, and menisci of the knee.

Elastic has conspicuous elastic fibers in its matrix that make it flexible and elastic; it helps support, for example, the ear and epiglottis.

The term *bone*, of course, denotes both a skeletal structure and the osseous tissue from which it is made. As a tissue, bone is characterized by the capacity of accumulating organic salts, mainly calcium phosphate or apatite (specifically, hydroxyapatite), in its matrix (see chap. 5). Three cell types are commonly involved in bone: osteoblasts, osteocytes, and osteoclasts. A general method of bone formation involves osteoblasts that produce a gelatinous ground substance containing collagenous fibers; apatite crystals are then deposited around the fibers to form an osseous structure. The osteoblasts become entrapped in the bony matrix and transform into osteocytes, which are mainly involved in maintenance. Osteoclasts are cells that break down osseous tissue.

Bone may be classified by several criteria, such as appearance (compact and spongy, also known as cancellous or trabecular), location (cortical and medullary), embryonic development (endochondral and intramembranous), and particular histological structure (cellular, acellular, lamellar, woven). Compact bone, which usually forms the periphery of an osseous skeletal structure and thus is cortical based on position, is organized into regular and ordered units termed *osteons* (or Haversian systems; fig. A.4). Each osteon comprises a series of concentric cylinders constituted by osteocytes and layers of a bony matrix that surround a central canal (or Haversian canal) through which pass blood vessels, lymph vessels, and nerves. The passages that extend perpendicularly to the central canal and connect osteons are called Volkmann's channels. As with chondrocytes, the spaces occupied by the osteocytes are termed *lacunae*, but they are stellate and communicate with each other through tiny canals termed *canaliculi* (sing. *canaliculus*). Whereas compact or cortical bone is ordered into osteons, the mineralized structure of trabecular or spongy bone is less regularly arranged, but it nevertheless shows a pattern related to stress dissipation (see Wolff's law of bone transformation in chap. 3).

A skeletal structure may be ossified by two main processes: endochondral and intramembranous ossification. During the development of endochondral bone, a cartilaginous model or precursor of the future bony skeletal structure forms first and is subsequently replaced by osseous tissue. In some bones (e.g., limb long bones), two regions are recognized: diaphysis (shaft) and epiphyses (ends; fig. A.5). A thin cartilaginous zone, the epiphyseal or growth plate, remains between the diaphysis and epiphysis until adulthood and is the site at which most of the increase in length of the skeletal structure occurs. Intramembranous bone forms directly from mesenchyme or from fibrous connective tissue, without the development of a cartilaginous precursor (e.g., many cranial bones, scapular girdle bones of fishes, amphibians, and reptiles, some sesamoid bones).

Muscular Tissue

Skeletal muscular tissue (fig. A.6), relatively abundant in a vertebrate body, is essential for body movements (see chap. 3, figs. 3.7, 3.8). It is formed by multinucleated cells specialized for contraction called muscle fibers or myocytes. The most important ultrastructural feature of these cells is the presence of actin and myosin filaments organized into structural proteins to form functional units termed *sarcomeres*. A muscle fiber is surrounded by a thin layer of connective tissue, the endomysium. Groupings of muscle fibers form a fascicle (wrapped by perimysium), and the groupings of fascicles form a muscle (wrapped by epimysium).

Generally among basal vertebrates, muscles are arranged metamerically, although during the evolution of more derived vertebrates, this segmental pattern has been significantly modified. The muscular tissue may be classified into two large groups according to the characteristics of its cells:

Striated muscle: sarcomeres exhibit a characteristic transverse striation, visible microscopically, with bands perpendicular to the longitudinal axis of the cell. These mainly make up muscles of fast but relatively short contractions. Striated muscles are subdivided into skeletal muscle, capable of voluntary contraction (mainly associated with skeletal elements), though it may also be contracted involuntarily (i.e., somatic reflexes, such as the patellar ligament reflex), and cardiac muscle, which forms most of the heart and contracts involuntarily.

Smooth muscle: its cells lack transverse striation and undergo slow, long-lasting, and involuntary contraction. They help form the walls of blood vessels and viscera.

Skeletal System

The skeleton of vertebrates is formed mainly from bone and cartilage. It gives shape to the body, supports its weight, and protects the soft parts (organs), among its varied functions. Many skeletal elements form systems of levers that together with skeletal musculature enable movement (and also nonmovement, as when maintaining a posture), participating in vital aspects such as locomotion (chaps. 5 and 6), feeding (chaps. 7 and 8), and breathing. The skeleton also houses myeloid or red bone marrow,

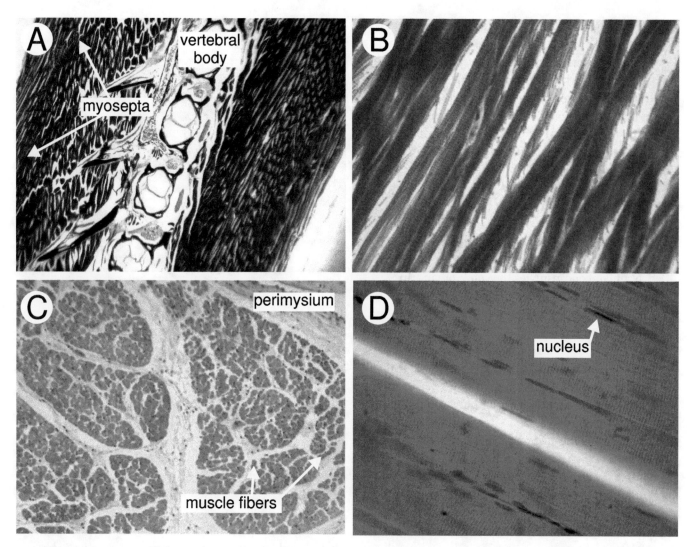

A.6. Striated skeletal muscle tissue. *A*, metameric (segmentally arranged) muscular bundles, termed *myomeres*, in a fish in longitudinal section; the myomeres are separated by the diagonally oriented connective-tissue partitions, termed *myosepta* (sing. *myoseptum*), that are apparent on the right (dorsal) and left (ventral) sides of the vertebral column, which is undergoing ossification (4×, ferric haematoxylin). *B*, in longitudinal section; note the striated appearance (alternating thin/light and thicker/dark bands perpendicular to the longitudinal axis of the fibers), a manifestation of the arrangement of myofibrils to form functional units or sarcomeres (40×, ferric haematoxylin). *C*, mammalian striated skeletal muscle in a cross section. Fibers are bundled into fascicles, which are wrapped by perimysium; endomysium is not visible (40×, haematoxylin-eosin stain). *D*, mammalian striated skeletal muscle in longitudinal section; note the striated (alternating light and dark banding) pattern of sarcomeres and the peripheral arrangement of the nuclei of these multinucleated cells. The *thick white band* represents empty space due to histological preparation (40×, haematoxylin-eosin stain). Photos courtesy of the Cátedra de Histología y Embriología Animal, Facultad de Ciencias Naturales y Museo, Universidad Nacional de La Plata, Argentina.

a hematopoietic tissue, and serves as a reservoir for phosphorus and calcium (in the form of calcium phosphate). As explained in chapter 5, phosphorus is not a particularly abundant element in nature but is highly important. Given vertebrates' highly active nature, it plays an essential role (among its several functions) in neutralizing the lactic acid produced by anaerobic activity.

The skeleton is an intricate and complex system and is thus often studied in parts or smaller structural units. For instance, it can be divided into a superficial skeleton

(dermal skeleton or exoskeleton) and a deep skeleton (endoskeleton). The dermal skeleton is formed from the integument; integumentary mesoderm and migrating neural crest cells can produce bone and other osteogenic materials (in general, forming bony sheets). The endoskeleton constitutes the largest skeletal component of vertebrates (at least, among most vertebrates; recall the extensive exoskeleton of ostracoderms and placoderms) and is derived largely from mesoderm, although neural crests cells, ectodermally derived, are involved in skull bone formation. The tissues that form the endoskeleton include fibrous connective tissue, bone, and cartilage.

Another practical and useful way of subdividing the skeleton for study is by regions, with the cranial skeleton (skull and mandible) and postcranial skeleton (the rest of the body) being one of the most common schemes; another recognizes an axial skeleton (skull, vertebral column, ribs, and sternum) and appendicular skeleton (girdles and limbs; fig. A.7).

The cranial skeleton includes the parts of the skull listed below and the mandible (fig. A.8):

Neurocranium (or braincase): its function is protection and support, as it covers most parts of the brain and encapsulates the sense organs. In all vertebrates, at some time during ontogeny, it involves a cartilaginous stage, termed the *chondrocranium* in adult chondrichthyans, which later ossifies (at least partially) in bony vertebrates.

Splanchnocranium: also called the visceral skeleton or viscerocranium; in basal vertebrates, it functions mainly to support soft structures related to gas exchange (gills), but among gnathostomes, it also constitutes the jaws and hyoid arch.

Dermatocranium: superficial dermal bones that almost completely roof over the neurocranium and the anterior part of the splanchnocranium (jaws, hyoid arch, and in some cases also the gills), including the musculature, the eyeballs, and other soft structures.

The postcranial skeleton (fig. A.7) is subdivided into:

Axial skeleton: includes the notochord, vertebral column, sternum and ribs, and the median fins (dorsal, anal, and caudal) when they are present.

Appendicular skeleton: includes the paired fins and limbs and the girdles supporting them.

Axial Skeleton

The longitudinal axis of the vertebrate body is defined by two structural elements: the notochord and vertebral column. The notochord is a longitudinal rod with a fibrous connective tissue sheath enveloping a cellular fluid. The vertebral column consists of a repeated series of discrete cartilaginous or osseous elements (vertebrae), which form around the notochord and, in most vertebrates, functionally replace it partially or completely. The original function of vertebrae was to protect the spinal cord and dorsal

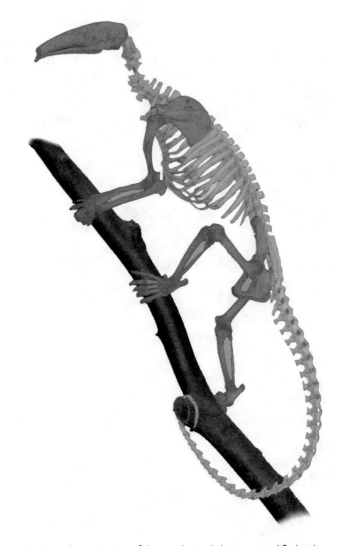

A.7. General organization of the vertebrate skeleton, exemplified with a skeleton of the lesser anteater (*Tamandua*). The axial skeleton includes the skull, mandible (in *purple*), and hyoid (not illustrated), and the vertebral column, ribs, and sternum (in *blue*). The appendicular skeleton (pectoral and pelvic girdles and limbs) is depicted in *red*.

aorta, but they evolved to become sites of attachment for musculature involved in locomotion. In tetrapods, they also became involved in supporting the body (chap. 5). The structure of vertebrae varies considerably among body regions and different groups of vertebrates. The common components include a vertebral body or centrum, a neural arch surrounding the spinal cord dorsally, intervertebral foramina for the passage of nerves, and a hemal arch, which wraps ventrally around the caudal artery and vein and can extend to form a hemal spine.

The neural and hemal arches serve also for the attachment of ligaments and muscles that participate in body support and locomotion. In many organisms, the area available on the arches for muscular attachment and the

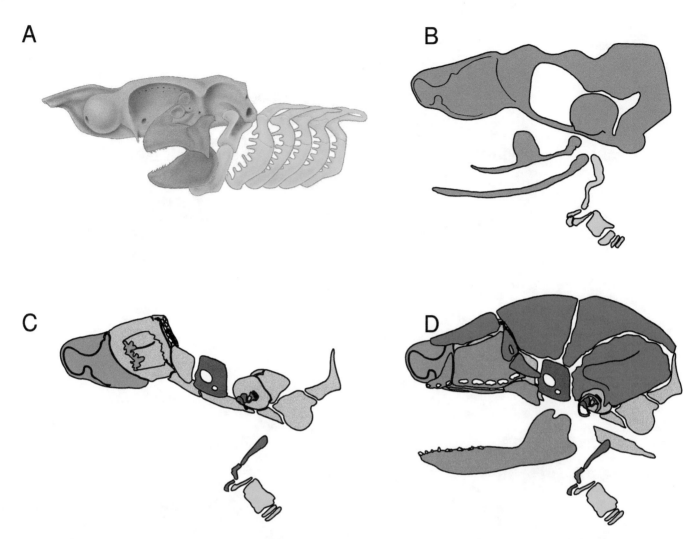

A.8. Cranial components in *A*, an adult shark, and *B*, *C*, and *D*, a mouse embryo. In the shark (*A*), which lacks bone in its skeleton, the components are cartilaginous and therefore lack the dermal osseous components present in bony vertebrates. The adult has only the neurocranium (or braincase, in *purple*) and palatal complex (upper jaw); the latter is formed from the mandibular arch (a visceral arch), which also forms the lower jaw (*light purple*). The hyoid arch (also a visceral arch, *light blue*) helps support the jaws, and the remaining parts of the splanchnocranium (branchial arches, *light green*) function in supporting branchial structures. In ossified vertebrates, the chondrocranium (*B*) is replaced by endochondral osseous elements during embryogenesis (*C*: gray, dark purple, and pink, ossifications from the neurocranium, mandibular, and hyoid arches, respectively) and bony elements of dermal origin (*D*: red and orange, cranial and visceral dermatocranium, respectively). Some may remain cartilaginous into adulthood (e.g., nasal cartilages, larynx). *A*, © Dino Pulerà and Gerardo De Iuliis (2019), reproduced by kind permission of Elsevier Inc. *B*, *C*, and *D* modified from Depew et al. (2002).

mechanical advantage of the arches as lever arms are increased by processes (neural and hemal spines) that extend dorsally and ventrally, respectively, from the arches. In tetrapods especially, vertebrae bear apophyses, processes that project from the centrum and arches. They include diapophyses and parapophyses, which articulate with ribs; basapophyses, paired ventral remnants of the base of the hemal arch; and pre- and postzygapophyses, processes that limit torsion and reinforce the joints between adjacent vertebrae (fig. A.9). In general, the musculature that supports and moves the vertebral column (and hence the trunk) is relatively simple in fish, whereas it becomes extremely complex in tetrapods with the appearance of the rib cage and lung ventilation for gas exchange.

Appendicular Skeleton

The appendicular skeleton includes paired appendages (fins, or pterygia, and limbs, or chiridia) and the pectoral and pelvic girdles supporting them (see chap. 5). In general, among bony pisciform vertebrates, the pectoral girdle is linked to the axial skeleton (connected to the skull), whereas in derived tetrapods it is not, permitting a mobile cervical region. On the other hand, the pelvic girdle is not linked to the axial skeleton in fishes but it is in tetrapods.

The fish fin or pterygium has a rather simple architecture, forming a single foil with restricted movement among its skeletal elements. A considerable part of a fin's surface is usually formed by some type of dermal rays (fig.

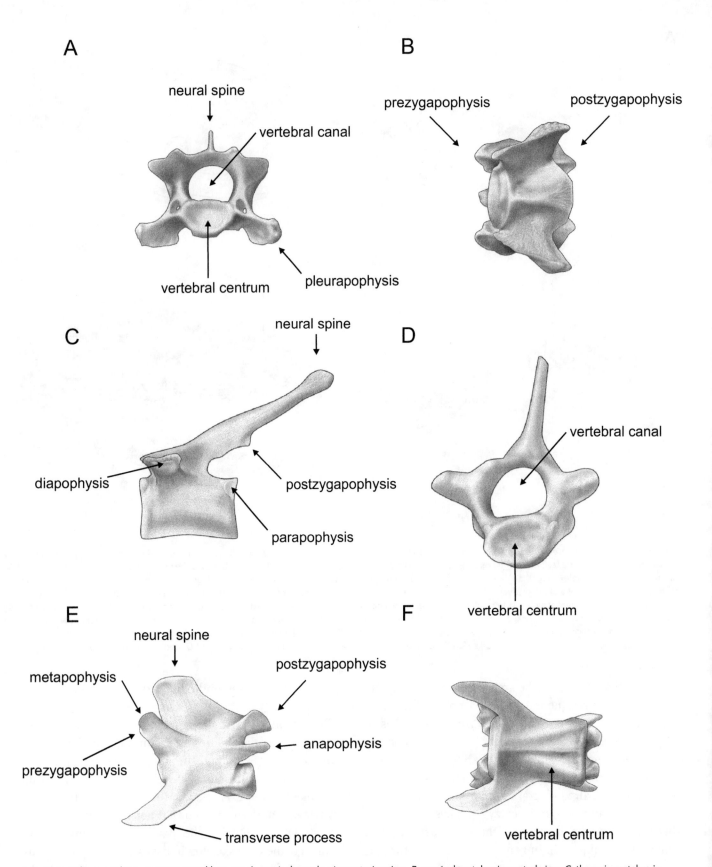

A.9. Mammalian vertebrae, as represented by a cat. *A*, cervical vertebra in posterior view. *B*, cervical vertebra in ventral view. *C*, thoracic vertebra in lateral view. *D*, thoracic vertebra in anterior view. *E*, lumbar vertebra in lateral view. *F*, lumbar vertebra in ventral view. © Dino Pulerà and Gerardo De Iuliis (2019), reproduced by kind permission of Elsevier Inc.

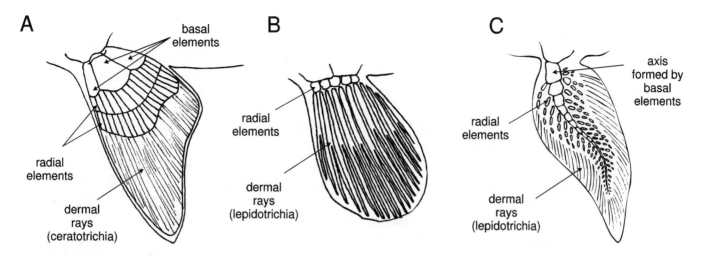

A.10. Fish fins. *A*, pectoral fin of a shark. *B*, pectoral fin of a modern teleost fish (which lacks basal elements). *C*, pectoral fin of a lungfish (where basal elements are arranged forming an axis).

A.10). The musculature that moves fins is also relatively simple and consists of two main masses: one (an abductor) to pull the fin dorsally, the other (an adductor) to pull the fin ventrally.

The chiridium (fig. A.11) of tetrapods is more complex. It is composed of discrete segments, with joints between adjacent segments and a more complex musculature permitting movement. In nearly all tetrapods, there are three segments: the stylopodium (proximal part), with the humerus in the forelimb and femur in the hind limb; zeugopodium (middle part), with the ulna and radius in the forelimb and the tibia and fibula in the hind limb; and the autopodium (distal part), a manus or hand in the forelimb and a pes or foot in the hind limb. The proximal part of the autopodium (basipodium) includes the carpal (wrist) and tarsal (ankle) elements of the manus and pes, respectively.

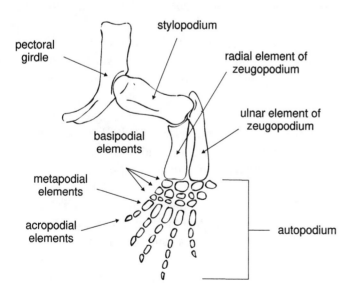

A.11. Idealized pectoral chiridium of a basal tetrapod. Modified from Liem et al. (2001).

The distal part of the autopodium is formed by several digits with metapodial (metacarpals in the manus and metatarsals in the pes) elements, followed by phalanges (the fingers and toes).

Relationship between Skeletal Elements

The elements of the vertebrate skeletal system may be linked in several ways (with some changing over the lifetime of an individual). The meeting of skeletal elements is termed an *articulation* or *joint*, and the following types are commonly recognized.

- *Synostosis:* the elements are fused together into a single unit, as in bones of the pelvic girdle in mammals and some bones of the skull. Humans, for instance, are born with left and right frontal bones, but these soon fuse at the midline to produce a single frontal bone.
- *Synarthrosis or fibrous:* in a synarthrosis, adjacent skeletal elements are bound tightly together by collagen fibers and movement is nearly absent (collagen fibers are extremely short) or limited (collagen fibers are slightly longer). Three types of fibrous joints—sutures, gomphoses, and syndesmoses—are commonly recognized, as indicated below.
- *Suture* occurs between some bones of the skull, where the skeletal elements are joined without fusing, although with age fusion may occur in some individuals; the joint surfaces between elements are generally quite complex, severely restricting movement, with several types being recognized (e.g., squamous, beveled).

Gomphosis describes the attachment between a tooth and its alveolus (although not a bone, the attachment of a tooth to its socket is classified as a joint); teeth are held in place by periodontal ligaments, which permit some movement in response to forces generated during mastication.

Syndesmosis: in this type, collagen fibers are longer and permit more movement, but the elements are still tightly bound together; examples include the diaphyses of the ulna and radius, held together by a fibrous interosseous membrane, which in humans allows pronation and supination, and the distal ends of the tibia and fibula, between which less movement can occur.

Amphiarthrosis: a joint with very limited mobility, in which adjacent elements are bound together by cartilage; in general, an amphiarthrosis occurs between elements that require some flexibility between them, and two types are generally recognized: synchondroses and symphyses.

Synchondrosis: an amphiarthrosis formed by hyaline cartilage, as between the transient epiphyseal cartilage between the epiphysis and diaphysis, and between the sternum and first rib.

Symphysis: an amphiarthrosis formed by a combination of hyaline and fibrous cartilage, as occurs at the pubic symphysis (between right and left coxal bones) and the intervertebral discs between adjacent vertebrae; although movement is limited between adjacent elements, in the case of the vertebral column, the slight movement between vertebrae allows considerable collective flexibility of the spine.

Diarthrosis: an articulation that permits considerable movement between elements. The joint surface of each element is covered by articular cartilage, and the elements are held in place by ligaments. The joint is wrapped in a connective tissue sheath, the articular capsule, that is filled with synovial fluid, a lubricant secreted by the synovial membrane lining the articular capsule; these joints are thus also referred to as synovial joints.

Types of Joints and Degrees of Freedom

Joints allow the skeletal elements to move relative to one another. In almost all cases, such movement is produced by muscle contraction. There are three basic types of joints (fig. A.12): hinge joint, universal joint, and ball-and-socket joint:

Hinge joint allows rotation around a single axis, with the relationship between skeletal elements described by one quantity, the angle formed between them; a hinge joint is thus considered to have one degree of freedom. An example is the elbow joint in humans.

Universal joint allows rotation around the two axes, and two angles serve to describe its position between elements; that is, it has two degrees of freedom of movement. Several such joints have complex "saddle-shaped" articular surfaces, such as between the cervical vertebrae of birds.

Ball-and-socket joint allows rotation around any axis that passes through the center of the ball, and is defined with three degrees of freedom of movement because any rotation that is produced can be described as a combination of rotations around the three axes to yield three angles. Examples include the shoulder and hip in humans.

These three basic joint types become more complex when, in addition to rotational movements, sliding components occur between joint surfaces, as for example in the knee or the craniomandibular joints. Some joints have internal connective tissue elements that promote joint stability by guiding rotational and/or sliding movements; a well-known example is that of the menisci of the knee.

Degrees of freedom of movement can be computed also for entire limbs. For example, the forelimb of a mammal generally has six degrees of freedom: three at the shoulder, one at the elbow, and two at the wrist. In humans, it has seven degrees of freedom because there is an additional degree of freedom between the ulna and radius of the antebrachium that allows turning one over the other (pronation-supination).

Joints retain their integrity due to the combined actions of connective and muscular elements, such as articular cartilages, synovial capsules, ligaments, and bursae (fig. A.13).

Articular cartilage: composed of (with few exceptions) hyaline cartilage, it covers joint surfaces, providing resistant and low-friction surfaces well suited to movement between skeletal elements. The cartilage itself contains considerable fluid (mainly water), which helps in the diffusion of the compressive loads generated by weight and/or movement. As it is avascular, its capacity to heal is rather poor (its wear causes a pathology termed *arthrosis*).

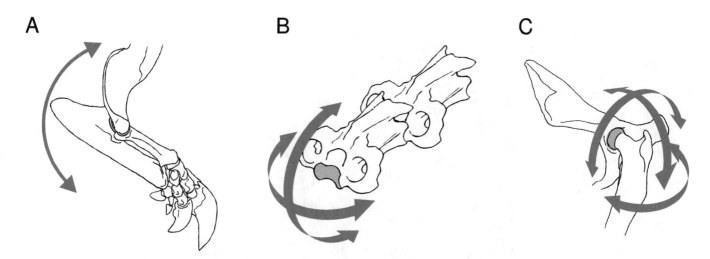

A.12. Types of joints. *A*, hinge joint, with one degree of freedom, in the elbow of an armadillo. *B*, saddle-shaped joint, with two degrees of freedom, in cervical vertebrae of a bird. *C*, ball-and-socket joint, with three degrees of freedom, in the hip of a horse.

Synovial capsule: a fibrous connective tissue sac that tightly surrounds the joint and contains synovial fluid, which functions as a lubricant for the joint.

Ligaments: quite resistant connective structures that join skeletal elements. They are commonly positioned adjacent to joints and limit their mobility within a range of safety tolerance. Ligaments may be an integral part of and help form the synovial capsule (e.g., the ankle ligaments) or easily distinguished discrete structures (e.g., the cruciate ligaments of the knee). Sports injuries frequently involve ligament damage caused when a joint is forced to move beyond the limits imposed by the ligaments.

Bursae: saclike connective structures containing synovial fluid and interposed between mobile soft structures, such as ligaments and tendons, that are subjected to great compressional stresses during movements of skeletal structures (e.g., the bursa under the patella).

Muscular System

Muscles are organs, composed primarily of muscle tissue, that provide the force that can result in movement. Together with skeletal elements, skeletal muscles form lever systems that allow a vertebrate to carry out activities such as locomotion and mastication (chap. 2). Equally important, however, is their role in restricting or modulating movement; that is, muscular effort can result in nonmovement, as when maintaining a posture (e.g., standing still).

A muscle does not attach, and thereby transmit its force, to a skeletal structure by way of its cells, but through tendons, which are composed, as noted earlier in this appendix, of fibrous connective tissue. As also noted, tendons are composed largely of collagenous fibers, and it is these fibers that penetrate into and become firmly anchored to the osseous tissue of a bone. The same is true for ligaments, which are histologically very similar to tendons. Recall that muscles also contain considerable connective tissue: the endo-, peri-, and epimysium that surround muscle fibers, fascicles, and the entire muscle. These connective tissue structures extend beyond the muscular tissue itself and are gathered together to form a tendon. Tendons may vary considerably in form; they can, for example, be short and inconspicuous or extensive and broad.

Not all muscles attach to a skeletal structure. For example, muscular strength may be transmitted by way of a broad, thin, sheetlike tendon, termed an *aponeurosis* (chap. 3, fig. 3.9). Such structures may attach a muscle to the surface of another muscle or to another aponeurosis, as in the abdominal muscles of mammals. The aponeurosis of the left and right side external obliques, for example, extends from the ventral margin of the muscular tissue. The two aponeuroses approach each other at the midventral line and fuse together to form a raphe. Other connective structures, also composed largely of collagen, surround a muscle. Such structures, termed *fasciae* (sing. *fascia*), separate muscles from other muscles or link them, as the case may be, to the underside of the skin.

The attachment of a tendon or ligament to a bony element is termed an *enthesis* (pl. *entheses*; fig. A.14), which denotes more than the mere attachment site. Indeed, the integration of a tendon into bone involves the establishment of a structurally continuous gradient from

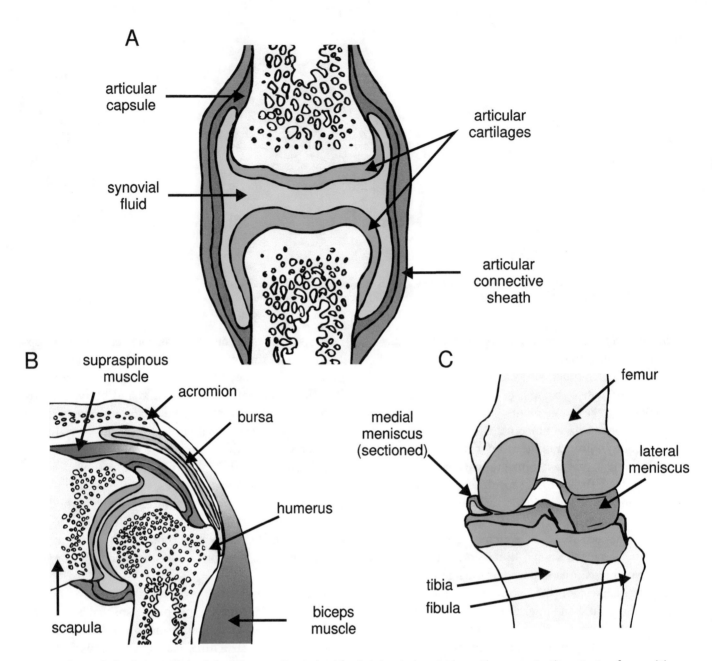

A.13. Joints. *A*, idealized joint (with the skeletal elements disarticulated for clarity); note the articular cartilages covering the contact surfaces and the articular capsule containing the synovial fluid. *B*, shoulder joint in a human; note the presence of a bursa, with synovial fluid, interposed between the tendons of two muscles; many tendons are surrounded by a similar structure, termed a *tendon sheath* (not shown). *C*, right knee joint in a human (seen from behind), showing the two menisci interposed between the femoral condyles and tibial plateau; on the *left* side of the joint, the medial meniscus is sectioned to show the articulation between the medial femoral condyle and the tibia. Modified from Levangie and Norkin (2005).

uncalcified tendon to calcified bone, generally including four distinct zones of varying cellular compositions, mechanical properties, and functions. Entheses frequently leave evident marks (generally termed *muscle scars*) on the bone that provide very useful information about the configuration of the muscular system of a fossil vertebrate.

In some entheses, the tendon or ligament anchors itself in the periosteum of the skeletal element by means of dense fibrous connective tissue; these are termed *fibrous* or *periosteal entheses*. In other entheses, the attachment is more complex, as the tendon-bone interface also includes fibrous cartilage; these are termed *fibrocartilaginous entheses*. Commonly, in mammals at least, entheses can become ossified as an individual becomes older or as a result of pathologies.

The tendons of some muscles regularly exhibit localized ossifications (sesamoid bones) that reinforce their structure and/or increase a muscle's moment arm in

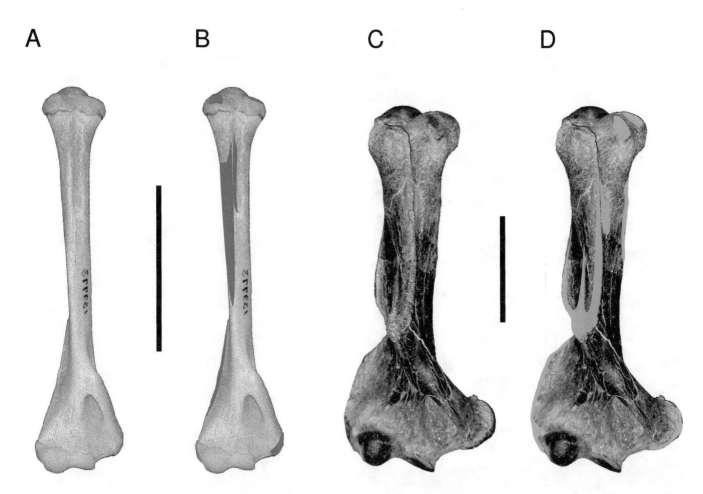

A.14. Bony remnants (commonly termed *muscle scars*) of the entheses of some muscles inserting on the right humerus (in anterior view) of extant and extinct sloths. *A*, humerus of a two-toed sloth *Choloepus*. *B*, muscle scars indicated in *red*. *C*, humerus of the Miocene sloth *Hapalops* from Patagonia. *D*, homologous muscle scars indicated in *blue*. Scale bars = 5 cm.

places where the tendon slides over skeletal elements (e.g., patella of the knee joint, an ossification in the tendon of the quadriceps muscle).

When a muscle contracts, its two ends experience the same pulling force (chap. 3, fig. 3.7). Generally, one of the ends is attached to a relatively mobile element; this end is called the insertion. The other attachment, on a relatively more fixed element, is called the origin. However, these terms should be applied carefully. Which end of a muscle experiences the greater degree of movement depends also on posture, the activity of other muscles, and the particular movement with respect to the environment. Another option is to consider the more proximal attachment site as the origin and the more distal site as the insertion (e.g., humeral origin and radial insertion of the brachialis biceps muscle). While this scheme works well for appendicular muscles (recall that proximal and distal are particularly useful in describing position along an appendage), it is not equally unambiguous for many other muscles.

Dentition

Teeth are characteristic elements of jawed vertebrates. In most vertebrates, they are functionally part of the system of food acquisition and mechanical digestion (chaps. 7 and 8). Indeed, they are instrumental in helping to obtain and grasp food and introduce it into the oral cavity. Through mastication (chewing), they also begin the process of reducing it into smaller portions to facilitate deglutition (swallowing) and to increase the surface area available for enzymatic action. Although phylogeny certainly plays a role, the distribution, structure, and pattern of the replacement of teeth are also closely related to dietary habits. Teeth are often thought of as being located only along the margins of the jaw, as is the condition in humans and other mammals, but in many vertebrates (fish, amphibians and reptiles) they may also be present on elements of the palate, inner (or lingual) margin of the lower jaw, and even on elements of the pharynx, such as branchial arches.

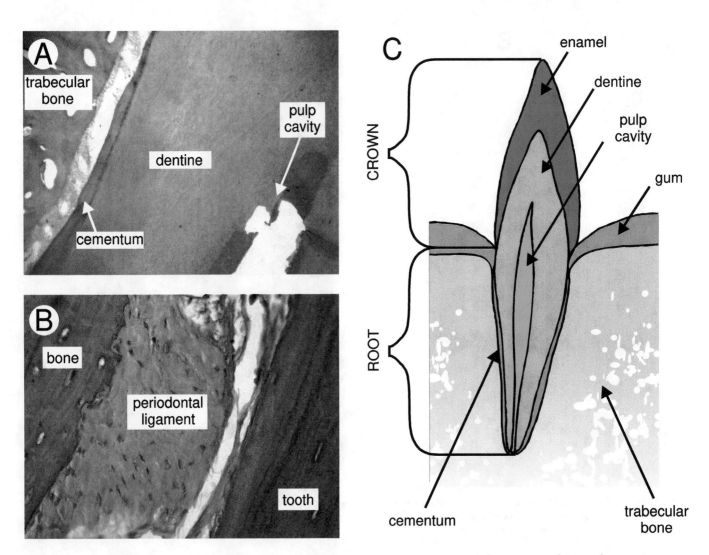

A.15. Teeth. *A*, longitudinal section of a mammalian tooth (4×, haematoxylin-eosin stain). *B*, higher magnification image of the interface zone between the root and the alveolus, showing the connective tissue forming the periodontal ligament (40×). *C*, diagram of a typical mammalian tooth (modified from Liem et al. 2001). Photos courtesy of the Cátedra de Histología y Embriología Animal, Facultad de Ciencias Naturales y Museo, Universidad Nacional de La Plata, Argentina.

An adult tooth (fig. A.15) includes a part, the crown, typically exposed beyond the gum and a part, the root, anchored to the jaw through connective tissue fibers that form a periodontal ligament (see gomphosis, above) and sometimes by an osteoid substance called cementum. The root of the tooth includes the pulp cavity, containing nerves, vessels, and odontoblasts, which are cells producing the ground tissue of the tooth, the dentine. This tissue is more mineralized than the bone itself, and is formed 70% by inorganic hydroxyapatite crystals and 30% by organic matter—collagen and other proteins that provide it with elasticity (making dentine less hard but also less brittle than enamel)—and water. It is harder and more resistant to wear than the cementum that may cover it.

Like bone, the cementum has cellular and acellular regions. It grows in layers on the dentine, but in many herbivorous mammals, it extends onto the crown between the enamel (see below) folds and helps form the occlusal surface. The cementum cells (cementocytes) build the matrix in seasonal pulses, and the cementum thus increases irregularly with age. The appearance of the concentric rings that result from this deposition is affected by factors related to stresses due to the mechanical properties (hardness) of the food, the nutritional state of the animal (e.g., during periods of food shortage), and seasonality. Hence, the aspects of cementum in fossil teeth can provide information on diet or the time of the year that the animal died.

Enamel is the mineralized sheath that covers the crown of the teeth of most vertebrates, with few exceptions including mammals such as xenarthrans and aardvarks. It constitutes the hardest and most friction-resistant substance that has evolved among vertebrates. It develops from the ectodermal ameloblasts that surround the dental primordium. Enamel is an acellular biomaterial,

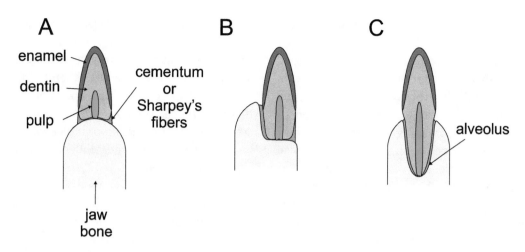

A.16. Main types of tooth attachment in vertebrates. *A,* acrodont. *B,* pleurodont. *C,* thecodont.

96% of which is formed by hydroxyapatite crystals. Shark teeth have a vitreous cover, enameloid; it is very similar to enamel (indeed, some authorities consider it a form of enamel) but includes substances of mesodermal origin.

Based on their attachment to the bones of the jaws (fig. A.16), teeth or dentitions are classified as acrodont, fixed to the edge of the mandible in shallow, superficial pits (fishes, amphibians, squamates); pleurodont, fixed to the shelflike inner edge of the jaw bones (squamates); and thecodont, in which each tooth is set in a deep alveolus or socket of the jaw bones (synapsids and archosaurians), enabling them to support greater mechanical tensions.

Dentitions may also be classified based on the number of tooth replacements. Polyphyodont dentitions exhibit multiple sets or generations of tooth replacement, as occurs in almost all fishes, amphibians, and reptiles. Oligophyodont dentitions exhibit few generations, and there is variation on the number of tooth replacements. Mammals and some other vertebrates, such as crocodiles, are considered to have oligophyodont replacement. Mammals are typically diphyodont, with two generations of teeth, with the first set being deciduous or milk teeth that typically begin to erupt soon after birth. Most of the teeth of this first wave are shed and replaced by a permanent or adult dentition during early stages of ontogeny. However, as the last of the milk teeth (i.e., the more distal, at the back of the jaws) complete their development and erupt during the later part of juvenile stages, as a consequence of lengthening of the facial region, they are not replaced. That is, they function during the adult life of an individual, but are not shed and replaced; they are the molars of mammals, and the term *hemiphyodont* is generally preferred for such teeth. The replacement of milk teeth and the eruption of more posterior permanent teeth (see below) occur in a relatively fixed and specific sequence for species, often making it possible to estimate the age of a young animal through the presence of particular teeth of one or the other set. Differences also occur among groups of mammals; for example, the basal pattern in marsupials is the presence of four molars, whereas three molars are present in placentals. The dentition of some mammals, as well as a few other vertebrates (such as holocephalians), is monophyodont, with only one set of teeth (e.g., almost all xenarthrans); in others, the milk dentition erupts and is replaced by the permanent dentition in utero (odontocete cetaceans, some rodents); and in still others, teeth seem to be completely lacking, as in anteaters (although transient tooth buds may appear during early development, as in mysticete cetaceans).

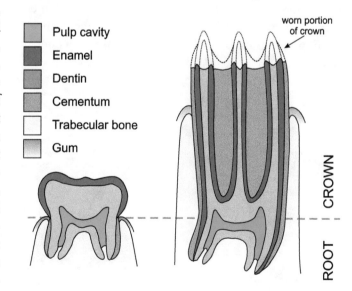

A.17. Comparative molar crown morphology in mammalian teeth. *Left* image, brachyodont molar (of a human, *Homo*). *Right* image, hypsodont molar (of a horse, *Equus*). Note that in hypsodont teeth a good part of the expanded crown is contained within the alveolus. Modified from Steel and Harvey (1979).

As the tooth grows, the radicular root canal closes and the pulp cavity is sealed, except for a small opening to admit vessels and nerves. These teeth, with well-developed roots and limited growth, are termed *brachydont*. By contrast, in many mammals the root canal remains open and the tooth continues to grow for an extended or even an undefined period of time, allowing the crown to recover continuously from wear. Such teeth are generally termed *hypsodont*, although variations in their precise nature occur among different groups. The teeth of xenarthran, incisor teeth of rodents, and cheek teeth of many rodents and ungulates often exhibit this sort of growth (fig. A.17).

The crowns of teeth exhibit considerable morphological diversity depending on taxon, diet, and age, among other variables. There are homodont dentitions, consisting of similarly shaped teeth along the jaws (e.g., fish, amphibians, most reptiles, cetaceans), and heterodont dentitions, in which the teeth differ in shape and function (catshark, some reptiles, and most mammals). Heterodonty is particularly widespread among mammals (although a few groups exhibit nearly homodont dentitions, such as odontocete cetaceans and armadillos), with three main morphological patterns recognized (fig. A.18). Incisor teeth, at the mesial or front part of the jaws, function mainly in cutting and grasping food, but may also be involved in nonfeeding behaviors such as grooming, digging, or fighting. On either side of the incisors is a canine tooth, usually pointed and suited for piercing; in some mammals (e.g., sheep) the canine may become incisiform (modified into an incisor-like tooth). Beyond the canine teeth, toward the distal or back part of the jaws, are the cheek teeth, which may function to crush and grind, or slice, or some combination of these functions (see below). Among cheek teeth, premolars and molars are distinguished. As noted

above, the latter are not replacement teeth for deciduous precursors (that is, there are no molar milk teeth; the teeth that function as molars in juveniles are actually premolar teeth) and they erupt late, when the facial region has lengthened sufficiently during ontogeny to accommodate them and/or the premolars have worn (e.g., the specialized form of tooth eruption in elephants, where a tooth becomes worn and another tooth of the same series erupts and occupies its place).

The number of each type of tooth may differ among groups, and indeed even within species of mammals. Dental formulae are defined using tooth initials for each tooth type and a number for each type. For example, humans have two incisors, one canine, two premolars, and three molars in each jaw quadrant (each half of the upper and of the lower jaws). The dental formula, expressing the number of teeth in one half of the oral cavity, is written as I 2/2 C 1/1 P 2/2 M 3/3. The dental formula of basal eutherians is I 3/3 C 1/1 P 4/4 M 3/3, and that of basal marsupials is I 5/4 C 1/1 P 3/3 M 4/4. In many lineages, a reduction in the number of teeth has occurred, whereas an increase has occurred in only a few cases (e.g., odontocete cetaceans, giant armadillo). For example, compared to basal eutherians, humans have lost I3 (the most distal incisor, next to the canine) and P1 and P2, and the eruption of M3 is delayed in both the upper and lower jaws.

Cheek teeth have complex crowns and carry out specific food-processing functions. These varied functions, not only among teeth in a particular individual but also of teeth among groups of mammals, are due to different patterns of protuberances or cusps and the ridges and crests (when present) that connect them. The basic patterns and names of such occlusal features have a long and complex history, with paleontologists attempting to

A

B

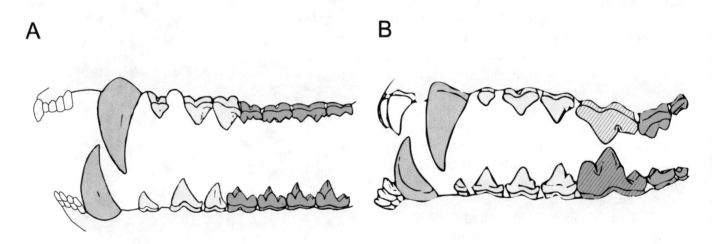

A.18. Heterodont dentition of mammals: incisors (*white*), canines (*orange*), premolars (*light blue*), and molars (*blue*). *A*, marsupial (opossum, *Didelphis*; dental formula = I 5/4 C 1/1 P 3/3 M 4/4). *B*, placental (dog, *Canis*; dental formula = I 3/3 C 1/1 P 4/4 M 2/3; carnassial pair, P4/M1, indicated by *hatching*). Modified from Thenius (1989).

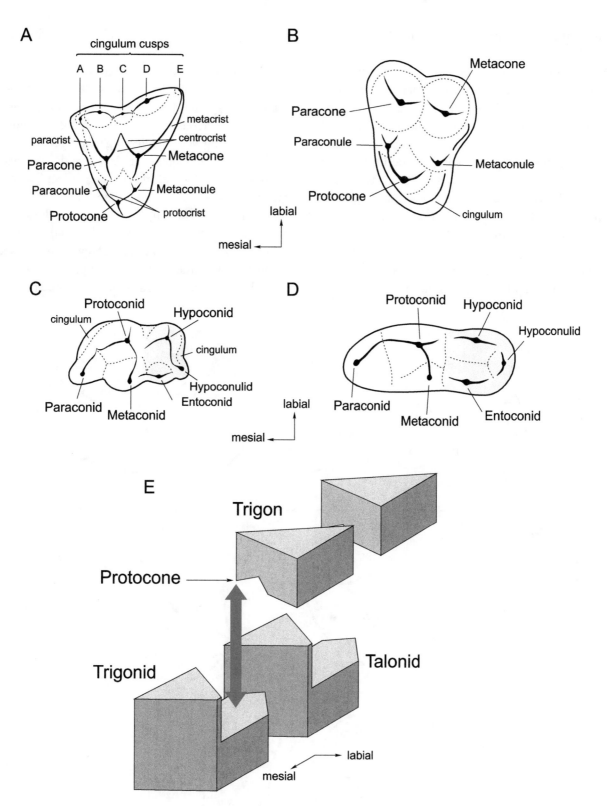

A.19. Cusps of upper (*top* images) and lower (*bottom* images) tribosphenic molars, in occlusal view. *A* and *C*, opossum (*Didelphis*). *B* and *D*, dog (*Canis*). Trigon and trigonid in *white*; talonid in *yellow*. *E*, diagrammatic molar teeth from the right upper and lower jaws, showing occlusion of the protocone into the talonid basin (modified from Luo et al. 2007 and Davis 2011).

reconcile anatomical features with phylogenetic history. It has become increasingly apparent, however, that the development of complex cheek teeth in mammals and their earlier kin is not straightforward, with many convergences in morphology having occurred (see Davis 2011 for a detailed review of this topic). As the phylogenetic aspect is beyond our immediate concern, our descriptions here of the occlusal features are intended to provide a basis for understanding functional aspects, rather than phylogenetic history.

To this end, it is useful first to consider the terminology of these features. In upper teeth, each cusp is called a cone (fig. A.19). The different cones are identified by particular prefixes, the main ones being proto-, para-, meta-, hypo-, and entocone. For smaller cusps, the suffix -ule is added (e.g., hypoconule). To distinguish occlusal features in upper and lower teeth, the suffix -id is added to the cusp name in lower teeth; for example, protoconid and hypoconid for main cusps and hypoconulid for smaller cusps. Additional features may also be present. For example, a crest, the cingulum (upper teeth) or cingulid (lower teeth), is often developed around the periphery of a tooth. Frequently, the cingulum and cingulid are expanded into a shelflike ridge that bears small cusps, each of which is denoted by a particular term.

The terminology for cheek teeth was established mainly based on a particular tooth morphology considered basal among crown mammals termed the *tribosphenic molar* (from the Greek τρίβος, meaning rubbing or grinding, and σφήν, meaning wedge), referring to wedge-shaped and basin-shaped components of the upper and lower teeth that, as they pass each other in occlusion, effect frictional and grinding actions. The wedge-shaped parts are triangular, with main cusps forming the apices, in occlusal view, and are termed the *trigon* in upper molars and the *trigonid* in lower molars. These triangular shapes are reversed with respect to each other in occluding upper and lower teeth, as explained in more detail below. In upper teeth, the apex

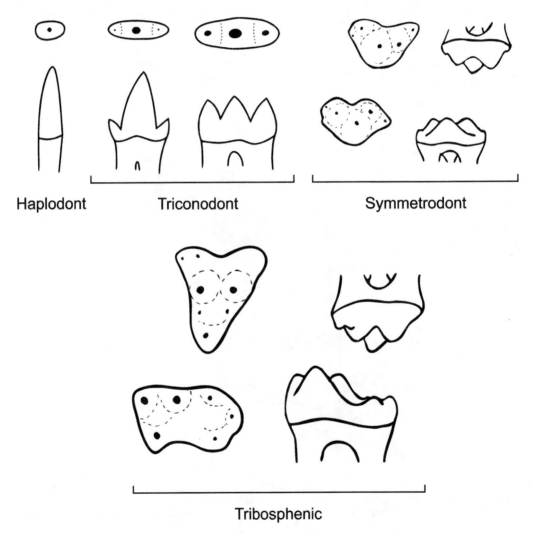

Haplodont   Triconodont   Symmetrodont

Tribosphenic

A.20. Evolution of dental morphology. The illustrated crown morphologies represent stages in the elaboration of crown features among synapsids and are not meant to indicate a phylogenetic sequence. Beginning with a simple tooth crown (haplodont) among early synapsids, additional cusps and their triangular arrangement (triconodont and symmetrodont) developed among several groups of therapsids and basal mammaliaforms, and extension of the distal end of lower molars to form a basined talonid to produce the tribosphenic tooth pattern, typical of therians. Modified from Thenius (1989) and Davis (2011).

is lingual (toward the tongue or internally in the oral cavity) in position and defined by the protoconid, whereas the base is labial (or vestibular, toward the lips or outer margin of the oral cavity) and defined by the paraconid mesially (toward the front) and the metaconid distally (toward the back). The terms coined for these cusps were based originally on an early interpretation of the homology of the cusps, with the protocone (first cusp) interpreted as representing the original (and often only) cusp present in reptilelike teeth. However, our interpretations have changed, and it is now recognized that the "first cusp" is represented by the paracone, but the names for the cusps have been retained.

As mentioned above, in the lower teeth the triangular wedge, the trigonid, is reversed. The base, defined by the paraconid and metaconid, is lingual in position, and the apex, defined by the protoconid, is labial in position. The presence of only these two wedge-like components, the trigon and trigonid in early stages leading to the tribosphenic pattern, produces a shearing action owing to the reversed configuration of the upper and lower triangles; the teeth interlock with each other as they occlude, and as their surfaces move past each other produce the shearing effect. However, a tribosphenic pattern involves another component, this being an extension (principally developed in the lower teeth) at the distal (or back) part of the

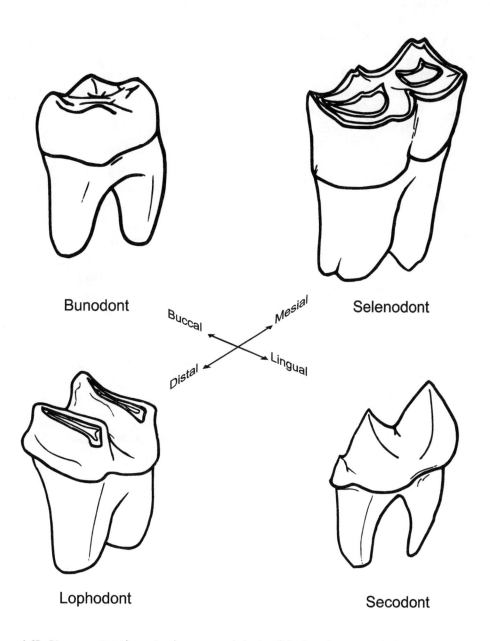

A.21. Diagrammatic tridimensional crown morphologies of cheek teeth in mammals: bunodont, typical of animals with omnivorous diets; selenodont and lophodont, typical of herbivorous diets; and secodont, typical of carnivorous diets. Modified from Martin and von Koenigswald (2021).

triangular component that effects the grinding action. The extension is termed a *talon* in upper teeth and a *talonid* in lower teeth. As noted, the talonid (at least originally) is the more prominent component. When the upper and lower teeth make contact during mastication, the upper tooth protocone occludes within a valley, termed the *basin*, on the talonid. Tribosphery refers to this particular arrangement of occlusal features of teeth—that is, wedge-shaped trigon and trigonid, with the protocone occluding against a basin-shaped talonid—and it is often also referred to as a tribosphenic molar complex (fig. A.19).

This basal condition is present unaltered or slightly modified in some extant mammals, such as opossums (*Didelphis*). From this pattern, however, the morphological evolution of most of the other teeth in mammals can be interpreted (figs. A.20, A.21). Early in the history of mammals, for example, a fourth main cusp, the hypocone, was added to the lingual side, behind the protocone, of the upper molar. As a result, the outline of the tooth assumes a nearly squared appearance, termed *quadritubercular* or *euthemorphic*. Elaboration of cusps and the ridges and crests connecting them has resulted during evolution in several morphological patterns, some of which are described below. Although this discussion has principally referred to molar teeth, in some groups, such as artiodactyls and perissodactyls, the morphology of some of the premolars has evolved to become similar to that of the molars, a process termed *molarization*.

> **Bunodont:** derived from the Greek βουνός, meaning hill, and ὀδούς, meaning tooth, these teeth are fairly quadrangular teeth with cusps in the shape of rounded peaks. Examples of such teeth are found among omnivores, such as raccoons (Procyonidae), bears (Ursidae), pigs (Suidae), and many primates (Hominidae, Cercopithecidae, Cebidae).

> **Selenodont:** from the Greek σελήνη, meaning moon, and ὀδούς, meaning tooth, these are teeth in which number and length of the enamel cutting surfaces are increased by lengthening of primary cusps in a mesiodistal (front-to-back) direction; with wear, the cusps adopt a crescent shape. Selenodont teeth are characteristic of herbivores, such as deer (Cervidae) and bovids (Bovidae), which have to grind hard and abrasive food.

> **Lophodont:** derived from the Greek λόφος, meaning crest, and ὀδούς, meaning tooth, the crown has crests or lophs that extend between cusps and can be oriented along mesiodistal or transverse directions. In the least elaborated examples, such as the bilophodont (two-crested) teeth of tapirs, the cones and conids can still be readily identified. The lophodont tooth constitutes another morphological strategy among herbivores for thoroughly grinding hard and abrasive foods. These teeth are present in tapirs (Tapiridae), manatees (Trichechidae), lagomorphs, and many rodents. Polylophodontia (an increase in the number of lophs beyond the basal condition) is observed in elephants (Elephantidae) and capybaras (Hydrochaeriidae). A lopho-selenodont condition is present in equids.

> **Secodont:** derived from the Latin *secare*, meaning to cut, and the Greek ὀδούς, meaning tooth, secodont teeth are specialized for cutting; the cusps are mesiodistally arranged and linked by sharply edged crests. Particularly within Carnivora, the fourth upper premolar and the first lower molar are prominent and especially developed in this morphology in the most carnivorous forms; these specialized teeth are termed *carnassials* and are often referred to as a carnassial pair. The plagiaulacoid tooth of some marsupials is a type of secodont cheek tooth present in both herbivores and carnivores.

References

Abba, A. M., Cassini, G. H., Valverde, G., Tilak, M., Vizcaíno, S. F., Superina, M., and Delsuc, F. 2015. Systematics of hairy armadillos and the taxonomic status of the Andean hairy armadillo (*Chaetophractus nationi*). *Journal of Mammalogy* 96(4): 673–689.

Abdala, F., and Damiani, R. 2004. Early development of the mammalian superficial masseter muscle in cynodonts. *Palaeontologia Africana* 40: 23–29.

Abel, O. 1911. *Grundzüge der Palaeobiologie der Wirbeltiere.* E. Schweizerbart, Stuttgart.

Ackerly, D. D. 2000. Taxon sampling, correlated evolution, and independent contrast. *Evolution* 54: 1480–1492.

Ackermans, N. L. 2020. The history of mesowear: A review. *PeerJ* 8: e8519.

Alexander, D. 2017. *Nature's Machines. An Introduction to Organismal Biomechanics.* 1st ed. Academic, London.

Alexander, R. McN. 1976. Estimates of speeds of dinosaurs. *Nature* 261: 129–130.

Alexander, R. McN. 1977. Allometry of the limbs of antelopes (Bovidae). *Journal of Zoology, London* 183: 125–146.

Alexander, R. McN. 1983. *Animal Mechanics.* 2nd ed. Blackwell, Oxford.

Alexander, R. McN. 1985. Mechanics of posture and gait of some large dinosaurs. *Zoological Journal of the Linnean Society* 83: 1–25.

Alexander, R. McN. 1989. *Dynamics of Dinosaurs and Other Extinct Giants.* Columbia University Press, New York.

Alexander, R. McN. 1990. *Animals.* Cambridge University Press, Cambridge.

Alexander, R. McN. 1996. *Optima for Animals.* Princeton University Press, Princeton, NJ.

Alexander, R. McN., Fariña, R. A., and Vizcaíno, S. F. 1999. Tail blow energy and carapace fractures in a large glyptodont (Mammalia, Xenarthra). *Zoological Journal of the Linnean Society* 126: 41–49.

Alexander, R. McN., Jayes, A. S., Maloiy, G. M. O., and Wathuta, E. M. 1979. Allometry of the limb bones of mammals from shrews (*Sorex*) to elephant (*Loxodonta*). *Journal of Zoology, London* 189: 305–314.

Alexander, R. McN., and Pond, C. M. 1992. Locomotion and bone strength of the White Rhinoceros, *Ceratotherium simum. Journal of Zoology* 227: 63–69.

Amador, L. I., Simmons, N. B., and Giannini, N. P. 2019. Aerodynamic reconstruction of the primitive fossil bat *Onychonycteris finneyi* (Mammalia: Chiroptera). *Biology Letters* 15(3):20180857.

Ameghino, F. 1894. Enumération synoptique des espèces de mammifères fossiles des formations éocènes de Patagonie. Coni é Hijos, Buenos Aires.

Ameghino, F. 1898. Sinopsis geológico-paleontológica de la Argentina. Pp. 113–255 in *Censo Nacional de la República Argentina*, Tomo I (Territorio), Capítulo I, Parte 3, Buenos Aires.

Amundson, R., and Lauder, G. V. 1994. Function without purpose: The uses of causal role function in evolutionary biology. *Biology and Philosophy* 9: 443–469.

Anderson, J. F., Hall-Martin, A., and Russell, D. A. 1985. Long bone circumference and weight in mammals, birds and dinosaurs. *Journal of Zoology, London* 207: 53–61.

Andrews, P., Lord, J. M., and Evans, E. M. N. 1979. Patterns of ecological diversity in fossil and modern mammalian faunas. *Biological Journal of the Linnean Society* 11: 177–205.

Aramayo, S. A., and Manera de Bianco, T. 1987. Hallazgo de una icnofauna continental (Pleistoceno Tardío) en la localidad de Pehuén-Co (partido de Coronel Rosales) provincia de Buenos Aires, Argentina. Parte I: Edentata, Litopterna, Prosboscidea. Parte II: Carnivora, Artiodactyla y Aves. *IV Congreso Latinoamericano de Paleontología*, Actas 1: 532–547.

Aramayo, S. A., Manera de Bianco, T., Bastianelli, N. V., and Melchor, R. N. 2015. Pehuén Co: Updated taxonomic review of a late Pleistocene ichnological site in Argentina. *Palaeogeography, Palaeoclimatology, Palaeoecology* 439: 144–165.

Arbour, V. M., and Zanno, L. E. 2018. The evolution of tail weaponization in amniotes. *Proceedings of the Royal Society B: Biological Sciences* 285: 20172299.

Argot, C. 2001. Functional-adaptive anatomy of the forelimb in the Didelphidae, and the paleobiology of the Paleocene marsupials *Mayulestes ferox* and *Pucadelphys andinus. Journal of Morphology* 247: 51–79.

Argot, C. 2002. Functional-adaptive anatomy of the hindlimb in the Didelphidae, and the paleobiology of the Paleocene marsupials *Mayulestes ferox* and *Pucadelphys andinus*. *Journal of Morphology* 253: 76–108.

Argot, C. 2004. Evolution of South American mammalian predators (Borhyaenoidea): Anatomical and palaeobiological implications. *Zoological Journal of the Linnean Society* 140: 487–521.

Arnold, S. J. 1983. Morphology, performance and fitness. *American Zoologist* 23: 347–361.

Bakker, R. T. 1971. Dinosaur physiology and the origin of mammals. *Evolution* 25: 636–658.

Barbero, S., Teta, P., and Cassini, G. H. 2020. An ecomorphological comparative study of extant and late Holocene Sigmodontinae (Rodentia, Cricetidae) assemblages from Central-Eastern Argentina. *Journal of Mammalian Evolution* 27: 697–711.

Bargo, M. S. 2001. The ground sloth *Megatherium americanum*: Skull shape, bite forces, and diet. *Acta Paleontologica Polonica* 46: 41–60.

Bargo, M. S., De Iuliis, G., and Toledo, N. 2019. Early Miocene sloths (Xenarthra, Folivora) from the Río Santa Cruz valley (Southern Patagonia, Argentina). Ameghino, 1887 revisited. *Publicación Electrónica de la Asociación Paleontológica Argentina* 19(2): 102–137.

Bargo, M. S., Toledo, N., and Vizcaíno, S. F. 2006. Muzzle of South American ground sloths (Xenarthra, Tardigrada). *Journal of Morphology* 267: 248–263.

Bargo, M. S., Toledo, N., and Vizcaíno, S. F. 2012. Paleobiology of the Santacrucian sloths and anteaters (Xenarthra, Pilosa). Pp. 216–242 in Vizcaíno, S. F., Kay, R. F., and Bargo M. S., eds. *Early Miocene Paleobiology in Patagonia: High-Latitude Paleocommunities of the Santa Cruz Formation*. Cambridge University Press, Cambridge.

Bargo, M. S., and Vizcaíno, S. F. 2008. Paleobiology of Pleistocene ground sloths (Xenarthra, Tardigrada): Biomechanics, morphogeometry and ecomorphology applied to the masticatory apparatus. *Ameghiniana* 45:175–196.

Bargo, M. S., Vizcaíno, S. F., Archuby, F., and Blanco, R. E. 2000. Limb bone proportions, strength and digging in some Lujanian (Late Pleistocene-Early Holocene) mylodontids ground sloths (Mammalia, Xenarthra). *Journal of Vertebrate Paleontology* 20: 601–610.

Bargo, M. S., Vizcaíno, S. F., and Kay, R. F. 2009. Predominance of orthal masticatory movements in the early Miocene *Eucholaeops* (Mammalia, Xenarthra, Tardigrada, Megalonychidae) and other megatherioid sloths. *Journal of Vertebrate Paleontology* 29: 870–880.

Barone, R. 1976. *Anatomie comparée des mammifères domestiques. Tome I: Osteologie*. Vigot Frères, Paris.

Barrick, R. E., Showers, W. J., and Fischer, A. G. 1996. Comparison of thermoregulation of four ornithischian dinosaurs and a varanid lizard from the Cretaceous Two Medicine Formation: Evidence from oxygen isotopes. *PALAIOS* 11: 295–305.

Behrensmeyer, A. K. 1988, Vertebrate preservation in fluvial channels. *Palaeogeography, Palaeoclimatology, Palaeoecology* 63: 183–199.

Benton, M. J. 2010. Studying function and behavior in the fossil record. *PLoS Biology* 8(3): e1000321.

Bernard, A. Lécuyer, C., Vincent, P., Amiot, R., Bardet, N., Cuny, G., Buffetaut, E., Fourel, F., Martineau, F., Mazin, J. M., and Prieur, A. 2010. Regulation of body temperature by some Mesozoic marine reptiles. *Science* 328: 1379.

Berthaume, M. A., Lazzari, V., and Guy, F. 2020. The landscape of tooth shape: Over 20 years of dental topography in primates. *Evolutionary Anthropology* 29(5): 245–262.

Bertram, J. E. A. 1988. The biomechanics of bending and its implications for terrestrial support. PhD dissertation, University of Chicago.

Bertram, J. E. A. 2016a. Concepts through time: Historical perspectives on mammalian locomotion. Pp. 1–25 in Bertram, J. E. A., ed. *Understanding Mammalian Locomotion: Concepts and Applications*. Wiley-Blackwell, Hoboken, NJ.

Bertram, J. E. A. 2016b. Design for prodigious size without extreme body mass: Dwarf elephants, differential scaling and implications for functional adaptation. Pp. 349–368 in Bertram. J. E. A., ed. *Understanding Mammalian Locomotion: Concepts and Applications*. Wiley-Blackwell, Hoboken, NJ.

Bertram, J. E. A. 2016c. The most important feature of an organism's biology: Dimension, similarity and scale. Pp. 193–227 in Bertram, J. E. A., ed. *Understanding Mammalian Locomotion: Concepts and Applications*. Wiley-Blackwell, Hoboken, NJ.

Betz, O. 2006. Ecomorphology: Integration of form, function, and ecology in the analysis of morphological structures. *Mitteilungen der Deutschen Gesellschaft für Allgemeine und Angewandte Entomologie* 15: 409–416.

Bhullar, B. A. S., Manafzadeh, A. R., Miyamae, J. A., Hoffman, E.A, Brainerd, E. L., Musinsky, C., and Crompton, A. W. 2019. Rolling of the jaw is essential for mammalian chewing and tribosphenic molar function. *Nature* 566: 528–532.

Biewener, A. A. 2005. Biomechanical consequences of scaling. *Journal of Experimental Biology* 208: 1665–1676.

Biewener, A. A., and Taylor, C. R. 1986. Bone strain: A determinant of gait and speed? *Journal of Experimental Biology* 123: 383–400.

Billet, G., Blondel, C., and de Muizon, C. 2009. Dental microwear analysis of notoungulates (Mammalia) from Salla (Late Oligocene, Bolivia) and discussion on their precocious hypsodonty. *Palaeogeography, Palaeoclimatology, Palaeoecology* 274: 114–124.

Black, K. H., Camens, A. B., Archer, M., and Hand, S. J. 2012. Herds overhead: *Nimbadon lavarackorum* (Diprotodontidae), heavyweight marsupial herbivores in the Miocene forests of Australia. *PLoS ONE* 7(11): e48213.

Blanco, R. E., and Czerwonogora, A. 2003. The gait of *Megatherium. Senckenbergiana Biologica* 83: 61–68.

Blanco, R. E., Jones, W. W., and Rinderknecht, A. 2009. The sweet spot of a biological hammer: The centre of percussion of glyptodont (Mammalia: Xenarthra) tail clubs. *Proceedings of the Royal Society B: Biological Sciences* 276: 3971–3978.

Blazevich, A. J. 2007. *Sports Biomechanics: The Basics. Optimising Human Performance.* Bloomsbury, New York.

Blomberg, S. P., and Garland, T., Jr. 2002. Tempo and mode in evolution: Phylogenetic inertia, adaptation and comparative methods. *Journal of Evolutionary Biology* 15(6): 899–910.

Blomberg, S. P., Rathnayake, S. I., and Moreau, C. M. 2020. Beyond Brownian motion and the Ornstein-Uhlenbeck process: Stochastic diffusion models for the evolution of quantitative characters. *American Naturalist* 195(2): 145–165

Bock, W. J. 1977. Toward an ecological morphology. *Vogelwarte* 29: 127–135.

Bock, W. J. 1980. The definition and recognition of biological adaptation. *American Zoologist* 20: 217–227.

Bock, W. J. 1988. The nature of explanations in morphology. *American Zoologist* 28: 205–215.

Bock, W. J. 1990. From Biologische Anatomie to Ecomorphology. *Journal of Zoology* 40: 254–277.

Bock, W. J. 2009. Design—An inappropriate concept in evolutionary theory. *Journal of Zoological Systematics and Evolutionary Research* 47(1): 7–9.

Bock, W. J., and von Wahlert, Y. G. 1965. Adaptation and the form-function complex. *Evolution* 19: 269–299.

Bodine, S. C., Roy, R. R., Meadows, D. A., Zernicke, R. F., Sacks, R. D., Fournier, M., and Edgerton, V. R. 1982. Architectural, histochemical, and contractile characteristics of a unique biarticular muscle: The cat semitendinosus. *Journal of Neurophysiology* 48: 192–201.

Bona, P., Degrange, F. J., and Fernández, M. S. 2013. Skull anatomy of the bizarre Crocodylian *Mourasuchus nativus* (Alligatoridae, Caimaninae). *Anatomical Record* 296(2): 227–239.

Bonaparte, J. 1965. Nuevas icnitas de la Quebrada del Yeso (La Rioja) y consideraciones acerca de la edad de los afloramientos. *Acta Geológica Lilloana* 7: 5–16.

Bond, M. 1986. Los ungulados fósiles de Argentina: Evolución y paleoambientes. En Simposio "Evolución de los Vertebrados Cenozoicos," *IV Congreso Argentino de Paleontología y Bioestratigrafía*, Actas 2: 187–190.

Bond, M., Cerdeño, E. P., and López, G. 1995. Los ungulados nativos de América del Sur. Pp. 259–275 in Alberdi, M. T., Leone, G., and Tonni, E. P., eds. *Evolución climática y biológica de la región pampeana durante los últimos cinco millones de años: Un ensayo de correlación con el Mediterráneo occidental.* Consejo Superior de Investigaciones Científicas, Monografías, Madrid.

Bond, M., Perea, D., Ubilla, M., and Tauber, A. A. 2001. *Neolicaphrium recens* Frenguelli, 1921, the only surviving Proterotheriidae (Litopterna, Mammalia) into the South American Pleistocene. *Palaeovertebrata* 30: 37–50.

Bookstein, F. L. 1986. Size and shape spaces for landmark data in two dimensions: Comment. *Statistical Science* 1: 181–222.

Bookstein, F. L. 1989. "Size and shape": A comment on semantics. *Systematic Zoology* 38(2): 173–180.

Bookstein, F. L., Sampson, P. D., Streissguth, A. P., and Barr, H. M. 1990. Measuring "dose" and "response" with multivariate data using partial least squares techniques. *Communication in Statistics: Theory and Methods* 19: 765–804.

Boscaini, A., Iurino, D. A., Sardella, R., Tirao, G., Gaudin, T. J., and Pujos, F. 2020. Digital cranial endocasts of the extinct sloth *Glossotherium robustum* (Xenarthra, Mylodontidae) from the Late Pleistocene of Argentina: Description and comparison with the extant sloths. *Journal of Mammalian Evolution* 27: 55–71.

Botha, J., and Angielczyk, K. 2007. An integrative approach to distinguishing the Late Permian dicynodont species *Oudenodon baini* and *Tropidostoma microtrema* (Therapsida: Anomodontia). *Palaeontology* 50: 1175–1209.

Botha-Brink, J., and Angielczyk, K. 2010. Do extraordinarily high growth rates in Permo-Triassic dicynodonts (Therapsida, Anomodontia) explain their success before and after the end-Permian extinction? *Zoological Journal of the Linnean Society* 160: 341–365.

Boué, C. 1970. Morphologie fonctionnelle des dents labiales chez les ruminants. *Mammalia* 34: 696–711.

Bramwell, C. D., and Whitfield, G. R. 1974. Biomechanics of *Pteranodon. Philosophical Transactions of the Royal Society B: Biological Sciences* 267: 503–581.

Brassey, C. 2017. Body-mass estimation in paleontology: A review of volumetric techniques. *Paleontological Society Papers* 22: 133–156.

Brassey, C. A., Maidment, S. C. R., and Barrett, P. M. 2015. Body mass estimates of an exceptionally complete *Stegosaurus* (Ornithischia: Thyreophora): Comparing volumetric and linear bivariate mass estimation methods. *Biology Letters* 11: 20140984.

Brody, S. 1945. *Bioenergetics and Growth.* Reinhold, New York.

Broeckhoven, C., du Plessis, A., and Hui, C. 2017. Functional trade-off between strength and thermal capacity of dermal armor: Insights from girdled lizards. *Journal of the Mechanical Behavior of Biomedical Materials* 74: 189–194.

Bromley, R. G. 1996. *Trace Fossils: Biology, Taphonomy and Applications*. Chapman and Hall, London.

Brooks, D. R., and McLennan, D. A. 1991. *Phylogeny, Ecology, and Behavior: A Research Program in Comparative Biology*. University of Chicago Press, Chicago.

Brown, J. H., Gillooly, F., Allen, A. P., Savage, V. M., and West, G. B. 2004. Toward a metabolic theory of ecology. *Ecology* 85: 1771–1789.

Brown, J. H., Gupta, V. K., Li, B. L., Milne, B. T., Restrepo, C., and West, G. B. 2002. The fractal nature of nature: Power laws, ecological complexity, and biodiversity. *Proceedings of the Royal Society B: Biological Sciences* 357: 619–626.

Brown, J. H., and West, G. B. 2000. *Scaling in Biology*. Oxford University Press, New York.

Bryant, H. N., and Russell, A. P. 1992. The role of phylogenetic analysis in the inference of unpreserved attributes of extinct taxa. *Philosophical Transactions of the Royal Society B: Biological Sciences* 337: 405–418.

Buatois, L. A., and Mángano, M. G. 2011. *Ichnology: Organism-Substrate Interactions in Space and Time*. Cambridge University Press, Cambridge.

Buckman, S. S. 1893. The Bajocian of the Sherborne District: Its relation to subjacent and superjacent strata. *Quarterly Journal of the Geological Society* 49: 479–522.

Bunn, J. M., Boyer, D. M., Lipman, Y., St. Clair, E. M., Jernvall, J., and Daubechies, I. 2011. Comparing Dirichlet normal surface energy of tooth crowns, a new technique of molar shape quantification for dietary inference, with previous methods in isolation and in combination. *American Journal of Physical Anthropology* 145(2): 247–261.

Bunn, J. M., and Ungar, P. S. 2009. Dental topography and diets of four old world monkey species. *American Journal of Primatology* 71(6): 466–477.

Calder, W. A. 1996. *Size, Function, and Life History*. Dover, New York.

Campione, N. E., and Evans, D. C. 2020. The accuracy and precision of body mass estimation in non-avian dinosaurs. *Biological Reviews of the Cambridge Philosophical Society* 95: 1759–1797.

Campo, D. H., Caraballo, D. A., Cassini, G. H., Lucero, S. O., and Teta, P. 2020. Integrative taxonomy of extant maras supports the recognition of the genera *Pediolagus* and *Dolichotis* within the Dolichotinae (Rodentia, Caviidae). *Journal of Mammalogy* 101(3): 817–834.

Canals, M., Figueroa, D., and Sabat, P. 2010. Symmorphosis in the proximal pathway for oxygen in the leaf-eared mouse *Phyllotis darwini*. *Biological Research* 43(1): 75–81.

Candela A. M., and Picasso, M. B. J. 2008. Functional anatomy of the locomotor behavior in Miocene porcupines. *Journal of Morphology* 269: 552–593.

Candela, A. M., Rasia, L. L., and Pérez, M. E. 2012. Paleobiology of Santacrucian Caviomorph rodents: A morphofunctional approach. Pp. 287–305 in Vizcaíno, S. F., Kay, R. F., and Bargo, M. S., eds. *Early Miocene Paleobiology in Patagonia: High-Latitude Paleocommunities of the Santa Cruz Formation*. Cambridge University Press, Cambridge.

Capra, F. 2007. *The Science of Leonardo: Inside the Mind of the Great Genius of the Renaissance*. Doubleday, New York.

Cardini, A., Nagorsen, D., O'Higgins, P., Polly, P. D., Thorington, R. W., Jr., and Tongiorgi, P. 2009. Detecting biological distinctiveness using geometric morphometrics: An example case from the Vancouver Island marmot. *Ethology, Ecology and Evolution* 21: 209–223.

Carrano, M. T. 1999. What, if anything, is a cursor? Categories versus continua for determining locomotor habit in mammals and dinosaurs. *Journal of Zoology, London* 247:29–42.

Carrano, M. T. 2006. Body-size evolution in the Dinosauria. Pp. 225–268 in Carrano, M. T., Blob, R. W., Gaudin, T. J., and Wible, J. R., eds. *Amniote Paleobiology: Perspectives on the Evolution of Mammals, Birds, and Reptiles*. University of Chicago Press, Chicago.

Carrano, M. T., and Hutchinson, J. R. 2002. Pelvic and hindlimb musculature of *Tyrannosaurus rex* (Dinosauria: Theropoda). *Journal of Morphology* 253: 207–228.

Carreño, C. A, and Nishikawa, K. C. 2010. Aquatic feeding in pipid frogs: The use of suction for prey capture. *Journal of Experimental Biology* 213: 2001–2008.

Carril, J., Degrange, F. J., and Tambussi, C. P. 2014. Jaw-muscle reconstruction of the Late Pliocene Psittaciform *Nandayus vorohuensis* from Argentina. *Ameghiniana* 51(4): 361–365.

Cartmill, M. 1985. Climbing. Pp. 73–88 in Hildebrand, M., Bramble, D. M., Liem, K. F., and Wake, D. B., eds. *Functional Vertebrate Morphology*. University of Chicago Press, Chicago.

Casamiquela, R. M. 1974. El bipedismo de los megaterioideos: Estudio de pisadas fósiles en la Formación Río Negro típica. *Ameghiniana* 11: 249–282.

Casamiquela, R. M. 1983. Pisadas del Pleistoceno (Superior?) del balneario Monte Hermoso, Buenos Aires: La confirmación del andar bipedal en los Megaterioides. *Cuadernos del Instituto Superior Juan XXIII* 4: 1–15.

Casinos, A. 1996. Bipedalism and quadrupedalism in *Megatherium*: An attempt at biomechanical reconstruction. *Lethaia* 29: 87–96.

Cassini, G. H. 2011. *Paleobiología de ungulados de la Formación Santa Cruz (Mioceno temprano-medio),*

Patagonia, Argentina. Una aproximación morfométrica y morfofuncional al estudio del aparato masticatorio. Unpublished PhD. Dissertation, Facultad de Ciencias Naturales y Museo, Universidad Nacional de La Plata, Argentina.

Cassini, G. H. 2013. Skull geometric morphometrics and Paleoecology of Santacrucian (late Early Miocene; Patagonia) native ungulates (Astrapotheria, Litopterna, and Notoungulata). *Ameghiniana* 50: 193–216.

Cassini, G. H., Cerdeño, M. E., Villafañe, A., and Muñoz, N. A. 2012a. Paleobiology of Santacrucian native ungulates (Meridiungulata; Astrapotheria, Litopterna and Notoungulata). Pp. 243–286 in Vizcaíno, S. F., Kay, R. F., and Bargo, M. S., eds. *Early Miocene Paleobiology in Patagonia: High-Latitude Paleocommunities of the Santa Cruz Formation.* Cambridge University Press, Cambridge.

Cassini, G. H., Hernández del Pino, S., Muñoz, N. A., Acosta, W. G., Fernández, M., Bargo, M. S., and Vizcaíno, S. F. 2017a. Teeth complexity, hypsodonty and body mass in Santacrucian (Early Miocene) notoungulates (Mammalia). *Earth and Environmental Science Transactions of the Royal Society of Edinburgh* 106:303–313.

Cassini, G. H., Mendoza, M., Vizcaíno, S. F., and Bargo, M. S. 2011. Inferring habitat and feeding behaviour of early Miocene notoungulates from Patagonia. *Lethaia* 44:153–165.

Cassini, G. H., Muñoz, N. A., and Vizcaíno, S. F. 2017b. Morphological integration of native South American ungulate mandibles: A tribute to D'Arcy Thompson in the centennial of "On growth and form." *Publicación Electrónica de la Asociación Paleontológica Argentina* 17(2):58–74.

Cassini, G. H., and Toledo, N. 2021. An ecomorphological approach to craniomandibular integration in Neotropical deer. *Journal of Mammalian Evolution* 28(1): 111–123.

Cassini, G. H., Toledo, N., and Vizcaíno, S. F. 2021a. Form-function correlation paradigm in mammalogy: A tribute to Leonard B. Radinsky (1937–1985). *Journal of Mammalian Evolution, Special Issue* 28(1): 1–5.

Cassini, G. H., Toledo, N., and Vizcaíno, S. F., eds. 2021b. Form-function correlation paradigm in mammalogy: A tribute to Leonard B. Radinsky (1937–1985). *Journal of Mammalian Evolution, Special Issue* 28(1).

Cassini, G. H., and Vizcaíno, S. F. 2012. An approach to the biomechanics of the masticatory apparatus of early Miocene (Santacrucian Age) South American ungulates (Astrapotheria, Litopterna, and Notoungulata): Moment arm estimation based on 3D landmarks. *Journal of Mammalian Evolution* 19: 9–25.

Cassini, G. H., Vizcaíno, S. F., and Bargo, M. S. 2012b. Body mass estimation in Early Miocene native South

American ungulates: A predictive equation based on 3D landmarks. *Journal of Zoology* 287: 53–64.

Cerda, I. A., and Chinsamy, A. 2012. Biological implications of the bone microstructure of the Late Cretaceous ornithopod dinosaur *Gasparinisaura cincosaltensis.* *Journal of Vertebrate Paleontology* 32(2): 355–368.

Cerda, I. A., Pereyra, M. E., Garrone, M., Ponce, D., Navarro, T. G., González, R., Militello, M., Luna, C. A., and Jannello, J. M. 2020. A basic guide for sampling and preparation of extant and fossil bones for histological studies. *Publicación Electrónica de la Asociación Paleontológica Argentina* 20(1): 15–28.

Chaffin, D. B., Andersson, G. B. J., and Martin, B. J. 2006. *Occupational Biomechanics.* 4th ed. Wiley, Hoboken, NJ.

Chatterjee, S., Templin, R. J., and Campbell, K. E. 2007. The aerodynamics of *Argentavis*, the world's largest flying bird from the Miocene of Argentina. *PNAS* 104(30): 12398–12403.

Chen, M., and Wilson, G. P. 2015. A multivariate approach to infer locomotor modes in Mesozoic mammals. *Paleobiology* 41: 280–312.

Christiansen, F., Sironi, M., Moore, M. J., Di Martino, M., Ricciardi, M., Warick, H. A., Irschick, D. J., Gutierrez, R., and Uhart, M. M. 2019. Estimating body mass of free-living whales using aerial photogrammetry and 3D volumetrics. *Methods in Ecology and Evolution* 10:2034–2044.

Christiansen, P., and Adolfssen, J. S. 2005. Bite forces, canine strength and skull allometry in carnivores (Mammalia, Carnivora). *Journal of Zoology, London* 266: 133–151.

Christiansen, P., and Harris, J. M. 2005. Body size of *Smilodon* (Mammalia: Felidae). *Journal of Morphology* 266: 369–384.

Ciccioli, P. L. 2008. *Evolución paleoambiental, estratigrafía y petrología sedimentaria de la Formación Toro Negro, Sierras Pampeanas Noroccidentales, provincia de La Rioja.* Unpublished PhD dissertation, Facultad de Ciencias Exactas y Naturales, Universidad de Buenos Aires.

Cifelli, R. L., and Guerrero Díaz, J. 1997. *Litopterns.* Pp. 289–302 in Kay, R. F., Madden, R. H., and Flynn, J. J., eds. *Vertebrate Paleontology in the Neotropics: The Miocene Fauna of La Venta, Colombia.* Smithsonian, Washington, DC.

Clack, J. A. 2012. *Gaining Ground: The Origin and Evolution of Tetrapods.* 2nd ed. Indiana University Press, Bloomington.

Clauss, M., Kaiser, T., and Hummel, J., 2008. The morpho-physiological adaptations of browsing and grazing mammals. Pp. 47–88 in Gordon, I. J., and Prins, H. H. T., eds. *The Ecology of Browsing and Grazing.* Springer, Heidelberg.

Clemente, C. J., and Wu, N. C. 2018. Body and tail-assisted pitch control facilitates bipedal locomotion in Australian agamid lizards. *Journal of the Royal Society Interface* 15: 20180276.

Clementz, M. T. 2012. New insight from old bones: Stable isotope analysis of fossil mammals. *Journal of Mammalogy* 93(2): 368–380.

Clementz, M. T., Holroyd, P. A., and Koch, P. L. 2008. Identifying aquatic habits of herbivorous mammals through stable isotope analysis. *PALAIOS* 23(9): 574–585.

Cloutier, R., Clement, A. M., Lee, M. S. Y., Noël, R., Béchard, I., Roy, V., and Long, J. A. 2020. *Elpistostege* and the origin of the vertebrate hand. *Nature* 579: 549–554.

Cock, A. G. 1966. Genetical aspects of metrical growth and form in animals. *Quarterly Review of Biology* 41: 131–190.

Corner, B. D., and Richtsmeier, J. T. 1991. Morphometric analysis of craniofacial growth in *Cebus paella*. *American Journal of Physical Anthropology* 84: 323–342.

Cornwell, W., and Nakawaga, S. 2017. Phylogenetic comparative methods. *Current Biology* 27(9): R333–R336.

Costa, R. L., Jr., and Greaves, W. S. 1981. Experimentally produced tooth wear facets and the direction of jaw motion. *Journal of Paleontology* 55: 635–638.

Cressie, N. 1986. Size and shape spaces for landmark data in two dimensions: Comment. *Statistical Science* 1: 226.

Croft, D. A. 2000. Archaeohyracidae (Mammalia: Notoungulata) from the Tinguiririca Fauna, central Chile, and the evolution and paleoecology of South American mammalian herbivores. Unpublished PhD dissertation, University of Chicago.

Croft, D. A. 2001. Cenozoic environmental change in South America as indicated by mammalian body size distributions (cenograms). *Diversity and Distributions* 7: 271–287.

Croft, D. A., and Anderson, L. C. 2008. Locomotion in the extinct notoungulate *Protypotherium*. *Palaeontologia Electronica* 11.1.1A.

Croft, D., Su, D. F., and Simpson, S. W., eds. 2018. *Methods in Paleoecology: Reconstructing Cenozoic Terrestrial Environments and Ecological Communities*. Vertebrate Paleobiology and Paleoanthropology Series, Delson, E., and Sargis, E., eds. Springer, Berlin.

Crompton, A. W., Lieberman, D. E., and Aboelela, S. 2006. Tooth orientation during occlusion and the functional significance of condylar translation in Primates and Herbivores. Pp. 367–388 in Carrano, M. T., Gaudin, T. J., Blob, R. W., and Wible, J. R., eds. *Amniote Paleobiology: Perspectives on the Evolution of Mammals, Birds, and Reptiles*. University of Chicago Press, Chicago.

Cuitiño, J. I., Vizcaíno, S. F., Bargo, M. S., and Aramendía, I. 2019. Sedimentology and fossil vertebrates of the Santa Cruz Formation (early Miocene) in Lago Posadas, southwestern Patagonia, Argentina. *Andean Geology* 46(2): 383–420.

Currey, J. D. 1984. *The Mechanical Adaptations of Bones*. Princeton University Press, Princeton, NJ.

Currie, A. 2013. Convergence as evidence. *British Journal for the Philosophy of Science* 64: 763–786.

Currie, A. 2015. Marsupial lions and methodological omnivory: Function, success and reconstruction in paleobiology. *Biology and Philosophy* 30: 187–209.

Czerwonogora, A., Fariña, R. A., and Tonni, E. P. 2011. Diet and isotopes of Late Pleistocene ground sloths: First results for *Lestodon* and *Glossotherium* (Xenarthra, Tardigrada). *Neues Jahrbuch für Geologie und Paläontologie: Abhandlungen* 262: 257–266.

Damuth, J. 1981a. Home range, home range overlap and energy use among animals. *Biological Journal of the Linnean Society* 15: 185–193.

Damuth, J. 1981b. Population density and body size in mammals. *Nature* 290:699–700.

Damuth, J. 1987. Interspecific allometry of population density in mammals and other animals: The independence of body mass and population energy use. *Biological Journal of the Linnean Society* 31: 193–246.

Damuth, J. 1990. Problems in estimating body masses of archaic ungulates using dental measurements. Pp. 229–253 in Damuth, J., and MacFadden, B. J., eds. *Body Size in Mammalian Paleobiology: Estimation and Biological Implications*. Cambridge University Press, Cambridge.

Damuth, J. 1991. Of size and abundance. *Nature* 351: 268–269.

Damuth, J. 1993. Cope's rule, the island rule and the scaling of mammalian population density. *Nature* 365: 748–750.

Damuth, J., and MacFadden, B. J., eds. 1990. *Body Size in Mammalian Paleobiology: Estimation and Biological Implications*. Cambridge University Press, Cambridge.

Dantas, M. A. T. 2022. Estimating the body mass of the Late Pleistocene megafauna from the South America Intertropical Region and a new regression to estimate the body mass of extinct xenarthrans. *Journal of South American Earth Sciences* 119: 103900.

Dantas, M. A. T., Cherkinsky, A., Lessa, C. M. B., Santos, L. V., Cozzuol, M. A., Omena, É. C., Silva, J. L. L., Sial, A. N., and Bocherens, H. 2020. Isotopic paleoecology ($\delta^{13}C$, $\delta^{18}O$) of a late Pleistocene vertebrate community from the Brazilian Intertropical Region. *Revista Brasileira de Paleontologia* 23(2): 138–152.

Davis, B. M. 2011. Evolution of the tribosphenic molar pattern in early mammals, with comments on the "dual-origin" hypothesis. *Journal of Mammalian Evolution* 18(4): 227–244.

Davit-Béal, T., Tucker A., and Sire, J. 2009. Loss of teeth and enamel in tetrapods: Fossil record, genetic data and morphological adaptations. *Journal of Anatomy* 214: 477–501.

Deban, S. M., and Olson, W. M. 2002. Suction feeding by a tiny predatory tadpole. *Nature* 420(6911): 41–42.

De Esteban-Trivigno, S. 2011. Ecomorfología de xenartros extintos a través del análisis de la mandíbula con métodos de morfometría geométrica. *Ameghiniana* 48(3): 381–398.

De Esteban-Trivigno, S., Mendoza, M., and De Renzi, M. 2008. Body mass estimation in Xenarthra: A predictive equation suitable for all quadrupedal terrestrial placentals? *Journal of Morphology* 269: 1276–1293.

Deevey, E. S. 1974. Book review of *Models in Paleobiology*, Schopf, T. J. M., ed. *Lymnology and Oceanography* 19(2): 375–376.

Degrange, F. J., Noriega, J. I., and Areta, J. I. 2012. Diversity and paleobiology of the Santacrucian birds. Pp. 138–155 in Vizcaíno, S. F., Kay, R. F., and Bargo, M. S., eds. *Early Miocene Paleobiology in Patagonia: High-Latitude Paleocommunities of the Santa Cruz Formation.* Cambridge University Press, Cambridge.

Degrange, F. J., Tambussi, C. P., Moreno, K., Witmer, L. M., and Wroe, S. 2010. Mechanical analysis of feeding behavior in the extinct "Terror Bird" *Andalgalornis steulleti* (Gruiformes: Phorusrhacidae). *PLoS ONE* 5(8): e11856.

De Iuliis, G., Bargo, M. S., and Vizcaíno, S. F. 2000. Variation in skull morphology and mastication in the fossil giant armadillos *Pampatherium* spp. and allied genera (Mammalia: Xenarthra: Pampatheriidae), with comments on their systematic and distribution. *Journal of Vertebrate Paleontology* 20: 743–754.

De Iuliis, G., and Pulerà, D. 2019. *The Dissection of Vertebrates: A Laboratory Manual.* 3rd ed. Academic, Cambridge.

Delsuc, F., Kuch, M., Gibb, G. C., Karpinski, E., Hackenberger, D., Szpak, P., Martínez, J. G., et al. 2019. Ancient mitogenomes reveal the evolutionary history and biogeography of sloths. *Current Biology* 29(12): 2031–2042.

Demar, R. E. 1973. Book review of *Models in Paleobiology*, Schopf, T. J. M., ed. *Journal of Geology* 81(4): 517.

Demes, B., Larson, S. G., Stern, J. T., Jungers, W. L., Biknevicius, A. R., and Schmitt, D. 1994. The kinetics of primate quadrupedalism: "Hindlimb drive" reconsidered. *Journal of Human Evolution* 26: 353–374.

de Muizon, C. 1998. *Mayulestes ferox*, a borhyaenoid (Metatheria, Mammalia) from the early Palaeocene of Bolivia: Phylogenetic and paleobiologic implications. *Geodiversitas* 20:19–142.

Depew, M. J., Lufkin, T., and Rubenstein, J. L. 2002. Specification of jaw subdivisions by Dlx genes. *Science* 298(5592): 381–385.

Derrickson, E. M., and Ricklefs, R. E. 1988. Taxon-dependent diversification of life-history traits and the perception of phylogenetic constraints. *Functional Ecology* 2: 417–423.

Desojo, J. B., and Vizcaíno, S. F. 2009. Jaw biomechanics in the South American aetosaur *Neoaetosauroides engaeus. Paläontologische Zeitschrift* 83: 499–510.

Diniz-Filho, J. A. F., Ramos de Sant'Ana, C. E., and Bini, L. M. 1998. An eigenvector method for estimating phylogenetic inertia. *Evolution* 52: 1247–1262.

Diogo, R., and Abdala, V. 2010. *Muscles of Vertebrates: Comparative Anatomy, Evolution, Homologies and Development.* CRC, Boca Raton, FL.

Diogo, R., Ziermann, J. M., Molnar, J., Siomava, N., and Abdala, V. 2018. *Muscles of Chordates—Development, Homologies, and Evolution.* CRC Press, Boca Raton, FL.

Dobzhansky, T. 1973. Nothing in biology makes sense except in the light of evolution. *American Biology Teacher* 35: 125–129.

Doe, B. R. 1983. The past is the key to the future. *Geochimica et Cosmochimica Acta* 47: 1341–1354.

Dressino, V., and Lamas, S. G. 2003. Teoría craneana funcional de Cornelis Jakob van der Klaauw: Una teoría sobre adaptación morfológica. *Episteme* 16: 99–110.

Drew, L. 2017. *I, Mammal: The Story of What Makes Us Mammals.* Bloomsbury, London.

Dryden, I. L., and Mardia, K. V. 1998. *Statistical Shape Analysis.* Wiley, Chichester.

Duellman, W. E., and Trueb, L. 1986. *Biology of Amphibians.* McGraw-Hill, New York.

du Plessis, A., Broeckhoven, C., Yadroitsev, I., Yadroitsava, I., and le Roux, S. G. 2018. Analyzing nature's protective design: The glyptodont body armor. *Journal of the Mechanical Behavior of Biomedical Materials* 82: 218–223

Efremov, I. A. 1940. Taphonomy: A new branch of paleontology. *Pan American Geologist* 74: 81–93.

Egi, N., Takai, M., Shigehara, N., and Tsubamoto, T. 2004. Body mass estimates for Eocene eosimiid and Amphipithecid primates using prosimian and anthropoid scaling methods. *International Journal of Primatology* 25: 211–236.

Elftman, H. O. 1929. Functional adaptations of the pelvis in marsupials. *Bulletin of the American Museum of Natural History* 58:189–232.

Elgin, R. A., Hone, D. W. E., and Frey, E. 2011. The extent of the pterosaur flight membrane. *Acta Palaeontologica Polonica* 56(1): 99–111.

Elissamburu, A. 2004. Morphometric and morphofunctional analysis of the appendicular skeleton of

Paedotherium (Mammalia, Notoungulata). *Ameghiniana* 41(3): 363–380.

Elissamburu, A. 2010. Estudio biomecánico y morfofuncional del esqueleto apendicular de *Homalodotherium* Flower 1873 (Mammalia, Notoungulata). *Ameghiniana* 47: 25–43.

Elissamburu, A., and Vizcaíno, S. F. 2004. Limb proportions and adaptations in caviomorph rodents (Rodentia: Caviomorpha). *Journal of Zoology* 262: 145–159.

Emerson, S. B. 1985. Jumping and leaping. Pp. 58–72 in Hildebrand, M., Bramble, D. M., Liem, K. F., and Wake, D. B., eds. *Functional Vertebrate Morphology*. Harvard University Press, Cambridge, MA.

Emlen, D. J. 2014. *Animal Weapons: The Evolution of Battle*. Picador, New York.

Ercoli, M. D. 2010. *Estudio de los hábitos locomotores en los Borhyaenoidea (Marsupialia, Sparassodonta) de la Formación Santa Cruz (Mioceno Inferior de la provincia de Santa Cruz) a partir de la diferenciación morfológica en depredadores vivientes*. Unpublished Lic. dissertation, Universidad de Buenos Aires.

Ercoli, M. D., Álvarez, A., Moyano, S. R., Youlatos, D., and Candela, A. M. 2020. Tracing the paleobiology of *Paedotherium* and *Tremacyllus* (Pachyrukhinae, Notoungulata), the latest sciuromorph South American Native Ungulates—Part I: Snout and masticatory apparatus. *Journal of Mammalian Evolution* 28: 377–409.

Ercoli, M. D., and Prevosti, F. J. 2011. Estimación de masa de las especies de Sparassodonta (Metatheria, Mammalia) de la Edad Santacrucense (Mioceno Temprano) a partir de tamaños de centroide de elementos apendiculares: Inferencias paleoecológicas. *Ameghiniana* 48(4): 462–479.

Ercoli, M. D., Prevosti, F. J., and Álvarez, A. 2012 Form and function within a phylogenetic framework: Locomotory habits of extant predators and some Miocene Sparassodonta (Metatheria). *Zoological Journal of the Linnean Society* 165: 224–251.

Ercoli, M. D., Prevosti, F. J., and Forasiepi, A. M. 2014. The structure of the mammalian predator guild in the Santa Cruz Formation (late Early Miocene). *Journal of Mammalian Evolution* 21(4): 369–381.

Eronen, J. T., Polly, P. D., Fred, M., Damuth, J., Frank, D. C., Mosbrugger, V., Scheidegger, C., Stenseth, N. C., and Fortelius, M. 2010. Ecometrics: The traits that bind the past and present together. *Integrative Zoology* 5(2): 88–101.

Evans, A. R. 2013. Shape descriptors as ecometrics in dental ecology. *Hystrix, the Italian Journal of Mammalogy* 24: 133–140.

Evans, A. R., and Pineda-Munoz, S. 2018. Inferring mammal dietary ecology from dental morphology. Pp. 37–51 in Croft, D., Su, D., and Simpson, S., eds. *Methods in Paleoecology: Reconstructing Cenozoic Terrestrial Environments and Ecological Communities*. Vertebrate Paleobiology and Paleoanthropology Series, Delson, E., and Sargis, E., eds.. Springer, New York.

Evans, D. C. 2010. Cranial anatomy and ontogeny of *Hypacrosaurus altispinus*, and a comparative analysis of skull growth in lambeosaurines (Ornithischia: Hadrosauridae). *Zoological Journal of the Linnean Society* 159: 398–434.

Falcon-Lang, H. J., Benton, M. J., and Stimson, M. 2007. Ecology of earliest reptiles inferred from basal Pennsylvanian trackways. *Journal of the Geological Society, London* 164: 1113–1118.

Falk, A. R., Kaye, T. G., Zhou, Z., and Burnham, D. A. 2016. Laser fluorescence illuminates the soft tissue and life habits of the Early Cretaceous bird *Confuciusornis*. *PLoS ONE* 11(12): e0167284.

Fariña, R. A. 1995. Bone strength and habits in large glyptodonts. *Lethaia* 28: 189–196.

Fariña, R. A. 1996. Trophic relationships among Lujanian mammals. *Evolutionary Theory* 11: 125–134.

Fariña, R. A., Vizcaíno, S. F., and Bargo, M. S. 1998. Body mass estimations in Lujanian (Late Pleistocene-Early Holocene of South America) mammal megafauna. *Mastozoología Neotropical* 5(2):87–108.

Fariña, R. A., Vizcaíno, S. F., and Blanco, E. 1997. Scaling of the indicator of athletic capability in fossil and extant land tetrapods. *Journal of Theoretical Biology* 185: 441–446.

Fariña, R. A., Vizcaíno, S. F., and De Iuliis, G. 2013. *Megafauna: Giant Beasts of Pleistocene South America*. Indiana University Press, Bloomington.

Farlow, J. O. 1976. A consideration of the trophic dynamics of a late Cretaceous large-dinosaur community (Oldman Formation). *Ecology* 57: 841–857.

Farlow, J. O., Hurlburt, G. R., Elsey, R. M., Britton, A. R. C., and Langston, W., Jr. 2005. Femoral dimensions and body size of *Alligator mississipiensis*: Estimating the size of extinct mesoeucrocodylians. *Journal of Vertebrate Paleontology* 25: 354–369.

Feilich, K. L., and López-Fernández, H. 2019. When does form reflect function? Acknowledging and supporting ecomorphological assumptions. *Integrative and Comparative Biology* 59(2): 358–370.

Felsenstein, J. 1985. Phylogenies and the comparative method. *American Naturalist* 125(1): 1–15.

Fernández, M. S., and Herrera, Y. 2009. Paranasal sinus system of *Geosaurus araucanensis* and the homology of the antorbital fenestra of metriorhynchids (Thalattosuchia: Crocodylomorpha). *Journal of Vertebrate Paleontology* 29: 702–714.

Fernández, M. E., Vasallo, A. I., and Zárate, M. A. 2000. Functional morphology and palaeobiology of the

pliocene rodent †*Actenomys* (Caviomorpha: Octodon-
tidae): The evolution to a subterranean mode of life.
Biological Journal of the Linnean Society 71: 71–90.

Fernicola, J. C., Toledo, N., Bargo, M. S., and Vizcaíno,
S. F. 2012. A neomorphic ossification of the nasal
cartilages and the structure of paranasal sinus system
of the glyptodont *Neosclerocalyptus* Paula Couto 1957
(Mammalia, Xenarthra). *Paleontologia Electronica* 15(3):
27A.

Ferreira-Cardoso, S., Fabre, P.-H., de Thoisy, B., Delsuc, F.,
and Hautier, L. 2020. Comparative masticatory myology
in anteaters and its implications for interpreting
morphological convergence in myrmecophagous placentals. *PeerJ* 8: e9690.

Fiorelli, L. E. 2010. Predation bite-marks on a peirosaurid
crocodyliform from the Upper Cretaceous of Neuquén
Province, Argentina. *Ameghiniana* 47(3):387–400.

Fisher, R. A. 1936. The use of multiple measurements in
taxonomic problems. *Annals of Eugenics* 7(2): 179–188.

Fleagle, J. G. 1979. *Primate adaptation and evolution.*
Academic, New York.

Flessa, K. W., Erben, H. K., Hallam, A., Hsü, K. J.,
Hüssner, H. M., Jablonsky, D., Raup, D. M., et al. 1986.
Causes and consequences of extinctions. Pp. 235–257
in Raup, D. M., and Jablonski, D., eds. *Patterns and
Processes in the History of Life.* Springer, Berlin.

Fletcher, T. M., Janis, C. M., and Rayfield, E. J. 2010.
Finite element analysis of ungulate jaws: Can mode
of digestive physiology be determined? *Palaeontologia
Electronica* 13:21A.

Fogarty, M. J., and Sieck, G. C. 2019. Evolution and
functional differentiation of the diaphragm muscle of
mammals. *Comprehensive Physiology* 9(2): 715–766.

Forasiepi, A. M., Goin, F. J., and Tauber, A. A. 2004.
Las especies de *Arctodictis* Mercerat 1891 (Metatheria,
Borhyaenidae), grandes carnívoros del Mioceno de
América del Sur. *Revista Española de Paleontología* 19:
1–22.

Fortelius, M., and Solounias, N. 2000. Functional characterization of ungulate molars using the abrasion-attri-
tion wear gradient: A new method for reconstructing
paleodiets. *American Museum Novitates* 3301: 1–36.

Fraser, D., and Theodor, J. M. 2011. Comparing ungulate
dietary proxies using discriminant function analysis.
Journal of Morphology 272: 1513–1526.

Frenguelli, J. 1950. Ichnites del Paleozoico Superior del
oeste argentino. *Revista de la Asociación Geológica Argentina* 5: 136–148.

Frey, R. W., and Seilacher, A. 1980. Uniformity in marine
invertebrate ichnology. *Lethaia* 13: 183–207.

Gagnon, M., and Chew, A. E. 2000. Dietary preferences
in extant African Bovidae. *Journal of Mammalogy* 81(2):
490–511.

Garland, T., Jr., Bennet, A. F., and Rezende, E. 2005.
Phylogenetic approaches in comparative physiology.
Journal of Experimental Biology 208: 3015–3035.

Garland, T., Jr., Dickerman, A. W., Janis, C. M., and
Jones, J. A. 1993. Phylogenetic analysis of covariance by
computer simulation. *Systematic Biology* 42: 265–292.

Garland, T., Jr., and Janis, C. M. 1993. Does metatarsal/
femur ratio predict maximal running speed in cursorial
mammals? *Journal of Zoology, London* 229:133–151.

Gastaldo, R. A., Savrda, C. E., and Lewis, R. D. 1996. *Deciphering Earth History: A Laboratory Manual with Internet
Exercises.* : Contemporary, Raleigh, NC.

Gaudin, T. J. 2004. Phylogenetic relationships among
sloths (Mammalia, Xenarthra, Tardigrada): The craniodental evidence. *Zoological Journal of the Linnean Society*
140: 255–305.

Gayon, J. 2000. History of the concept of allometry.
American Zoologist 40: 748–758.

Gebo, D. L. 1996. Climbing, brachiation, and terrestrial
quadrupedalism: Historical precursors of hominid
bipedalism. *American Journal of Physical Anthropology*
101(1): 55–92.

Giannini, N. P. 2003. Canonical phylogenetic ordination.
Systematic Biology 52(5): 684–695.

Giannini, N. P., and García-López, D. A. 2014. Ecomorphology of mammalian fossil lineages: Identifying
morphotypes in a case study of endemic South American ungulates. *Journal of Mammalian Evolution* 21:
195–212.

Giannini, N. P., Wible, J. R., and Simmons, N. B. 2006.
On the cranial osteology of Chiroptera: I, *Pteropus*
(Megachiroptera, Pteropodidae). *Bulletin of the American
Museum of Natural History* 295: 1–133.

Gifford, J. R., Garten, R. S., Nelson, A. D., Trinity, J. D.,
Layec, G., Witman, M. A. H., Weavil, J. C., et al. 2016.
Symmorphosis and skeletal muscle $\dot{V}_{O_{2max}}$: In vivo
and in vitro measures reveal differing constraints in
the exercise-trained and untrained human. *Journal of
Physiology* 594(6): 1741–1751.

Gingerich, P. D. 1990. Prediction of body mass in
mammalian species from long bone lengths and diameters. *Contributions from the Museum of Paleontology,
University of Michigan* 28: 79–92.

Gingerich, P. D. 2005. Aquatic adaptation and swimming
mode inferred from skeletal proportions in the Miocene
desmostylian *Desmostylus. Journal of Mammalian Evolution* 12: 183–194.

Gingerich, P. D., Smith, B. H., and Rossenberg, K. 1982.
Allometric scaling in the dentition of primates and
prediction of body weight from tooth size in fossils.
American Journal of Physical Anthropology 58: 81–100.

Goldbogen, J. A., Cade, D. E., Calambokidis, J., Fried-
laender, A. S., Potvin, J., Segre, P. S., and Werth, A. J.

2017. How baleen whales feed: The biomechanics of engulfment and filtration. *Annual Review of Marine Science* 9: 367–386.

Goodall, C. R. 1991. Procrustes methods in the statistical analysis of shape. *Journal of the Royal Statistical Society B* 53: 285–339.

Goodwin, M., and Horner, J. R. 2004. Cranial histology of pachycephalosaurs (Ornithischia: Marginocephalia) reveals transitory structures inconsistent with head-butting behavior. *Paleobiology* 30(2): 253–267.

Gordon, A. M., Huxley, A. F., and Julian, F. J. 1966. The variation in isometric tension with sarcomere length in vertebrate muscle fibres. *Journal of Physiology, London* 184:170–192.

Gordon, I. J., and Illius, A. W. 1988. Incisor arcade structure and diet selection in ruminants. *Functional Ecology* 2: 15–22.

Gould, S. J. 1966. Allometry and size in ontogeny and phylogeny. *Biological Reviews of the Cambridge Philosophical Society* 41: 587–640.

Gould, S. J. 1970. Evolutionary paleontology and the science of form. *Earth Science Reviews* 6(2): 77–119.

Gould, S. J. 1975. On the scaling of tooth size in mammals. *American Zoologist* 15: 351–362.

Gould, S. J. 1993. *Eight Little Piggies: Reflections on Natural History*. Norton, New York.

Gould, S. J., and Lewontin, R. C. 1979. The spandrels of San Marco and the panglossian paradigm: A critique of the adaptationist programme. *Proceedings of the Royal Society B: Biological Sciences* 205(1161): 581–598.

Grafen, A. 1995. The phylogenetic regression. *Philosophical Transactions of the Royal Society B: Biological Sciences* 326(1233): 119–157.

Grayson, D. K. 2016. *Giant Sloths and Sabertooth Cats: Extinct Mammals and the Archaeology of the Ice Age Great Basin*. University of Utah Press, Salt Lake City.

Greaves, W. S. 1973. The inference of jaw motion from tooth wear facets. *Journal of Paleontology* 47: 1000–1001.

Greaves, W. S. 1974. Functional implications of mammalian jaw joint position. *Forma et Functio* 7(4): 363–376.

Greaves, W. S. 1980. The mammalian jaw mechanism: The high glenoid cavity. *American Naturalist* 116(3): 432–440.

Green, J. L. 2009a. Dental microwear in the orthodentine of the Xenarthra (Mammalia) and its use in reconstructing the palaeodiet of extinct taxa: The case study of *Nothrotheriops shastensis* (Xenarthra, Tardigrada, Nothrotheriidae). *Zoological Journal of the Linnean Society* 156:201–222.

Green, J. L. 2009b. Intertooth variation in orthodentine microwear in armadillos (Cingulata) and tree sloths (Pilosa). *Journal of Mammalogy* 90: 768–778.

Grellet-Tinner, G., and Fiorelli, L. E. 2010. A new Argentinean nesting site showing neosauropod dinosaur reproduction in a Cretaceous hydrothermal environment. *Nature Communications* 1: 32.

Grellet-Tinner, G., Fiorelli, L. E., and Brincalepe Salvador, R. 2012. Water vapor conductance of the Lower Cretaceous dinosaurian eggs from Sanagasta, La Rioja, Argentina: Paleobiological and paleoecological implications for South American faveoloolithid and megaloolithid eggs. *PALAIOS* 27(1): 35–47.

Gröning, F., Jones, M. E. H., Curtis, N., Herrel. A., O'Higgins, P., Evans, S. E., and Fagan, M. J. 2013.The importance of accurate muscle modelling for biomechanical analyses: A case study with a lizard skull. *Journal of the Royal Society Interface* 10: 20130216

Grossnickle, D. M. 2017. The evolutionary origin of jaw yaw in mammals. *Scientific Reports* 7: 45094.

Grossnickle, D. M. 2020a. Feeding ecology has a stronger evolutionary influence on functional morphology than on body mass in mammals. *Evolution* 74(3):610–628.

Grossnickle, D. M. 2020b. Jaw roll and jaw yaw in early mammals. *Nature* 582: E6–E8.

Grossnickle, D. M., Chen, M., Wauer, J. G. A., Pevsner, S. K., Weaver, L. N., Meng, Q. J., Liu, D., Zhang, Y. G., and Luo, Z. X. 2020. Incomplete convergence of gliding mammal skeletons. *Evolution* 74: 2662–2680.

Grossnickle D. M., Weaver, L. N., Jäger, R. K., and Schultz, J. A. 2022. The evolution of anteriorly directed molar occlusion in mammals. *Zoological Journal of the Linnean Society* 194: 349–365.

Haas, G. 1973. Muscles of the jaws and associated structures in the Rhynchocephalia and Squamata. Pp. 285–490 in: *Biology of the Reptilia*, vol. 4.: Academic, London.

Hair, J. F., Jr., Black, W. C., Babin, B. J., and Anderson, R. E. 2019. *Multivariate Data Analysis*. 8th ed. Cengage, Hampshire.

Hanna, R. R. 2002. Multiple injury and infection in a sub-adult theropod dinosaur *Allosaurus fragilis* with comparisons to allosaur pathology in the Cleveland-Lloyd Dinosaur Quarry collection. *Journal of Vertebrate Paleontology* 22(1): 76–90.

Harvey, P. H., and Pagel, M. D. 1991. *The Comparative Method in Evolutionary Biology*. : Oxford University Press, Oxford.

Hasiotis, S. T. 2002. *Continental Trace Fossil Short Course Number 51*. SEPM, Tulsa, OK.

Hasiotis, S. T. 2007. Continental ichnology: Fundamental processes and controls on trace-fossil distribution. Pp. 268–284 in Miller, W., III, ed. *Trace Fossils: Concepts, Problems, Prospects*. Elsevier, Amsterdam.

Hasiotis, S. T., Platt, B. F., Hembree, D. I., and Everhart, M. 2007. The trace-fossil record of vertebrates. Pp.

196–218 in Miller, W., III, ed. *Trace Fossils: Concepts, Problems, Prospects.* Elsevier, Amsterdam.

Hastie, T., Tibshirani, R., and Buja, A. 1994. Flexible discriminant analysis by optimal scoring. *Journal of American Statistical Association* 89: 1–41.

Henderson, D. M. 1999. Estimating the masses and centres of mass of extinct animals by 3-D mathematical slicing. *Paleobiology* 25(1): 88–106.

Herrel, A., Aerts, P., and De Vree, F. 2000. Cranial kinesis in geckoes: Functional implications. *Journal of Experimental Biology* 203: 1415–1423.

Herrel, A., Aerts, P., Fret, J., and De Vree, F. 1999. Morphology of the feeding system in Agamid lizards: Ecological correlates. *Anatomical Record* 254: 496–507.

Herrel, A., De Smet, A., Aguirre, L. F., and Aerts, P. 2008. Morphological and mechanical determinants of bite force in bats: Do muscles matter? *Experimental Biology* 211: 86–91.

Herrera, Y. 2012. Análisis morfológico y paleobiológico de *Cricosaurus araucanensis* (Gasparini y Dellapé, 1976) (Crocodyliformes: Metriorhynchidae). Unpublished PhD dissertation, Facultad de Ciencias Naturales y Museo, Universidad Nacional de La Plata.

Herrera, Y., Fernández, M. S., and Gasparini, Z. N. 2013. The snout of *Cricosaurus araucanensis*: A case study in novel anatomy of the nasal region of metriorhynchids. *Lethaia* 46: 331–340.

Herring, S. W., and Herring, S. E. 1974. The superficial masseter and gape in mammals. *American Naturalist* 108: 561–576.

Hiiemae, K. M. 1978. Mammalian mastication: A review of the activity of the jaw muscles and movements they produce in chewing. Pp. 359–398 in Butler, P. M., and Joysey, K. A., eds. *Development, Function and Evolution of Teeth.* Academic,: New York.

Hiiemae, K. M., and Crompton, A. W. 1985. Mastication, food transport, and swallowing. Pp. 262–290 in Hildebrand, M., Bramble, D. M., Liem, K. F., and Wake, D. B., eds. *Functional Vertebrate Morphology.* Harvard University Press, Cambridge.

Hildebrand, M., and Goslow, G. E. 2001. *Analysis of Vertebrate Structure.* 5th ed. Wiley, Chichester.

Hinchliffe, J. R. 1994. Evolutionary developmental biology of the tetrapod limb. *Development 1994 Supplement*: 163–168.

Hofmann, R. R., and Stewart, D. R. M. 1972. Grazer or browser: A classification based on the stomach structure and feeding habits of East African ruminants. *Mammalia* 36: 226–240.

Hood, C. S. 2000. Geometric morphometric approaches to the study of sexual size dimorphism in mammals. *Hystrix* 11: 77–90.

Hoppe, K. A., Koch, P. L., Carlson, R. W., and Webb, S. D. 1999. Tracking mammoths and mastodons: Reconstruction of migratory behavior using strontium isotope ratios. *Geology* 27: 439–442.

Hoppeler, H., and Weibel, E. R. 2005. Scaling functions to body size: Theories and facts. *Journal of Experimental Biology* 208: 1573–1574.

Horner, J. R., de Ricqlès, A., and Padian, K. 2000. Long bone histology of the hadrosaurid *Maiasaura peeblesorum*: Growth dynamics and physiology based on an ontogenetic series of skeletal elements. *Journal of Vertebrate Paleontology* 20(1):109–123.

Horner, J. R., and Padian, K. 2004. Age and growth dynamics of *Tyrannosaurus rex. Proceedings of the Royal Society B: Biological Sciences* 271: 1875–1880.

Horner, J. R., Padian, K., and de Ricqlès, A. 2001. Comparative osteohistology of some embryonic and perinatal archosaurs: Developmental and behavioral implications for dinosaurs. *Paleobiology* 27(1): 39–58.

Howell, A. B. 1944. *Speed in Animals.* University of Chicago Press, Chicago.

Hubbe, A., and Machado, F. A. 2022. Comments on: "Estimating the body mass of the Late Pleistocene megafauna from the South America Intertropical Region and a new regression to estimate the body mass of extinct xenarthrans." *Journal of South American Earth Sciences* 103994.

Huey, R. B., and Kingsolver, J. 1989. Evolution of thermal sensitivity of ectotherm performance. *Trends in Ecology and Evolution* 4: 131–135.

Huneman, P. 2006. Naturalising purpose: From comparative anatomy to the "adventure of reason." *Studies in History and Philosophy of Biological and Biomedical Sciences* 37: 649–674.

Hunt, K. D., Cant, J. G. H., Gebo, D. L., Rose, M. D., Walker, S. E., and Youlatos, D. 1996. Standardized descriptions of primate locomotor and postural modes. *Primates* 37(4): 363–387.

Hutchinson, J. R. 2004. Biomechanical modeling and sensitivity analysis of bipedal running ability II: Extinct taxa. *Journal of Morphology* 262: 441–461.

Hutchinson, J. R., and García, M. 2002. *Tyrannosaurus* was not a fast runner. *Nature* 415: 1018–1021.

Huttenlocker, A. K., and Botha-Brink, J. 2013. Body size and growth patterns in the therocephalian *Moschorhinus kitchingi* (Therapsida: Eutheriodontia) before and after the end-Permian extinction in South Africa. *Paleobiology* 39: 253–277.

Huttenlocker, A. K., and Botha-Brink, J. 2014. Bone microstructure and the evolution of growth patterns in Permo-Triassic therocephalians (Amniota, Therapsida) of South Africa. *PeerJ* 2: e325.

Huxley, J. S. 1932. *Problems of Relative Growth*. : MacVeagh Dial, New York.

Huxley, J. S., and Teissier, G. 1936. Terminology of relative growth. *Nature* 137: 780–781.

Iriarte-Diaz, J., Terhune, C. E., Taylor, A. B., and Ross, C. F. 2017. Functional correlates of the position of the axis of rotation of the mandible during chewing in non-human primates. *Zoology* 124: 106–118.

Iwasaki, S. 2002. Evolution of the structure and function of the vertebrate tongue. *Journal of Anatomy* 201: 1–13.

Jackson, J. B. C., and Erwin, D. H. 2006. What can we learn about ecology and evolution from the fossil record? *Trends in Ecology and Evolution* 21: 322–328.

Janis, C. M. 1988. An estimation of tooth volume and hypsodonty indices in ungulate mammals, and the correlation of these factors with dietary preference. Pp. 367–387 in Russell, D. E., Santoro, J. P., and Sigoneau-Russell, D., eds. *Teeth Revisited: Proceedings of the VII International Symposium on Dental Morphology*. Mémoires du Muséum National d'Histoire Naturelle, Paris.

Janis, C. M. 1990. Correlation of cranial and dental variables with body size in ungulates and macropodoids. Pp. 255–299 in Damuth, J., and MacFadden, B. J., eds. *Body Size in Mammalian Paleobiology: Estimation and Biological Implications*. Cambridge University Press, Cambridge.

Janis, C. M. 1995. Correlations between craniodental morphology and feeding behavior in ungulates: Reciprocal illumination between living and fossil taxa. Pp. 76–98 in Thomason, J. J., ed. *Functional Morphology in Vertebrate Paleontology*. Cambridge University Press, Cambridge.

Janis, C. M. 2008. An evolutionary history of browsing and grazing ungulates. Pp. 21–45 in Gordon I. J., and Prins, H. H. T., eds. *The Ecology of Browsers and Grazers*. Springer, Berlin.

Janis, C. M., and Constable, E. 1993. Can ungulate craniodental features determine digestive physiology? *Journal of Vertebrate Paleontology* 13: 43A.

Janis, C. M., and Ehrhardt, D. 1988. Correlation of relative muzzle width and relative incisor width with dietary preference in ungulates. *Zoological Journal of the Linnean Society* 92: 267–284.

Jarman, P. J. 1974. The social organization of antelope in relation to their ecology. *Behaviour* 48: 215–267.

Jayne, B. C. 1988. Muscular mechanisms of snake locomotion: An electromyographic study of the sidewinding and concertina modes of *Crotalus cerastes, Nerodia fasciata* and *Elaphe obsoleta*. *Journal of Experimental Biology* 140: 1–33.

Jayne, B. C. 2020. What defines different modes of snake locomotion? *Integrative and Comparative Biology* 60(1):1–15.

Jernvall, J., and Selanne, L. 1999. Laser confocal microscopy and geographic information systems in the study of dental morphology. *Palaeontologia Electronica* 2(1): 1–18.

Jungers, W. L. 1990. Problems and methods in reconstructing body size in primates. Pp. 103–118 in Damuth, J., and MacFadden, B. J., eds. *Body Size in Mammalian Paleobiology: Estimation and Biological Implications*. Cambridge University Press, Cambridge.

Jungers, W. L., Godfrey, L. R., Simons, E. L., Wunderlich, R. E., Richmond, B. G., and Chatrath, P. S. 2002. Ecomorphology and behaviour of giant extinct lemurs from Madagascar. Pp 371–411 in Plavcan, J. M., Kay, R. F., Jungers, W. L., and van Schaik, C. P., eds. *Reconstructing Behavior in the Primate Fossil Record*. Kluwer/Plenum, New York.

Kaiser, T. M., and Fortelius, M. 2003. Differential mesowear in occluding upper and lower molars: Opening mesowear analysis for lower molars and premolars in hypsodont horses. *Journal of Morphology* 258(1): 67–83.

Kardong, K. 2018. *Vertebrates: Comparative Anatomy, Function, Evolution*. 8th ed. McGraw-Hill Education, New York.

Karr, J. R., and James, F. C. 1975. Eco-morphological configurations and convergent evolution in species and communities. Pp. 258–291 in Cody M. L,. and Diamond, J. M., eds. *Ecology and Evolution of Communities*. Harvard University Press, Cambridge.

Kay, R. F. 2021. Leonard B. Radinsky (1937–1985), Radical Biologist. *Journal of Mammalian Evolution* 28: 7–14.

Kay, R. F., and Madden, R. H. l997a. Mammals and rainfall: Paleoecology of the middle Miocene at La Venta (Colombia, South America). *Journal of Human Evolution* 32: 161–199.

Kay, R. F., and Madden, R. H. 1997b. Paleogeography and paleoecology. Pp. 520–550 in Kay, R. F., Madden, R. H., Cifelli, R. L., and Flynn, J. J., eds. *Vertebrate Paleontology in the Neotropics*. Smithsonian, Washington, DC.

Kay, R. F., Perry, J. M. G., Malinzak, M. D., Allen, K. L., Kirk, E. C., Plavcan, J. M., and Fleagle, J. G. 2012a. The paleobiology of Santacrucian primates. Pp. 306–330 in Vizcaíno, S. F., Kay, R. F., and Bargo, M. S., eds. *Early Miocene Paleobiology in Patagonia: High-Latitude Paleocommunities of the Santa Cruz Formation*. Cambridge University Press, Cambridge.

Kay, R. F., Vizcaíno, S. F., and Bargo, M. S. 2012b. A review of the paleoenvironment and paleoecology of the Miocene Santa Cruz Formation. Pp. 331–365 in Vizcaíno, S. F., Kay, R. F., and Bargo, M. S., eds. *Early*

Miocene Paleobiology in Patagonia: High-Latitude Paleo-communities of the Santa Cruz Formation. Cambridge University Press, Cambridge.

Kay, R. F., Vizcaíno, S. F., Bargo, M. S., Spradley, J. P., and Cuitiño, J. I. 2021. Paleoenvironments and paleoecology of the Santa Cruz Formation (early–middle Miocene) along the Río Santa Cruz, Patagonia (Argentina). *Journal of South American Earth Sciences* 109: 103296.

Kemp, E. S. 1979. The primitive Cynodont *Procynosuchus*: Functional anatomy of the skull and relationships. *Philosophical Transactions of the Royal Society B: Biological Sciences* 285(1005): 73–122.

Kendall, D. G. 1984. Shape-manifolds, Procrustean metrics and complex projective spaces. *Bulletin of the London Mathematical Society* 16: 81–121.

Kendall, D. G. 1986. Size and shape spaces for landmark data in two dimensions: Comment. *Statistical Science* 1: 222–226.

Kilbourne, B. M., and Hoffman, L. C. 2013. Scale effects between body size and limb design in quadrupedal mammals. *PLoS ONE* 8(11): e78392.

King, S. J., Arrigo-Nelson, S. J., Pochron, S. T., Semprebon, G. M., Godfrey, L. R., Wright, P. C., and Jernvall, J. 2005. Dental senescence in a long-lived primate links infant survival to rainfall. *Proceedings of the National Academy of Sciences of the USA* 102(46):16579–16583.

Kleiber, M. 1932. Body size and metabolism. *Hilgardia* 6: 315–353.

Kleinteich, T., Maddin, H. C., Herzen, J., Beckmann, F., and Summers, A. P. 2012. Is solid always best? Cranial performance in solid and fenestrated caecilian skulls. *Journal of Experimental Biology* 215: 833–844.

Klingenberg, C. P., and Ekau, W. 1996. A combined morphometric and phylogenetic analysis of an ecomorphological trend: Pelagization in Antarctic fishes (Perciformes: Nototheniidae). *Biological Journal of the Linnean Society* 59: 143–177.

Klukkert, Z. S., Dennis, J. C., M'kirera, F., and Ungar, P. S. 2012. Dental topographic analysis of the molar teeth of primates. *Methods in Molecular Biology* 915: 145–152.

Koch, P. L. 1998. Isotopic paleoecology and land vertebrates: Individuals, species, and ecosystems. *PALAIOS* 13(4): 309–310.

Koch, P. L., Fisher, D. C., and Dettman, D. L. 1989. Oxygen isotopes in the dentition of extinct proboscideans: A measure of season-of-death and seasonality. *Geology* 17: 515–519.

Koch, P. L., Hoppe, K. A., and Webb, S. D. 1998. The isotopic ecology of late Pleistocene mammals in North America, Part 1: Florida. *Chemical Geology* 52: 119–138.

Koehl, M. A. R. 1996. When does morphology matter? *Annual Reviews in Ecology and Systematics* 27: 501–542.

Kohn, M. J. 1996. Predicting animal $\delta^{18}O$: Accounting for diet and physiological adaptation. *Geochimica et Cosmochimica Acta* 60(23): 4811–4829.

Kowaleswski, M. 1996. Time-averaging, overcompleteness, and the geological record. *Journal of Geology* 104: 317–326.

Kramarz, A. 2009. Adiciones al conocimiento de *Astrapothericulus* (Mammalia, Astrapotheria): Anatomía cráneo-dentaria, diversidad y distribución. *Revista Brasileira de Paleontologia* 12(1): 55–66.

Kramarz, A. G., and Bond, M. 2005. Los Litopterna (Mammalia) de la Formación Pinturas, Mioceno temprano–medio de Patagonia. *Ameghiniana* 42: 611–625.

Kramarz, A. G., and Bond, M. 2009. A new Oligocene astrapothere (Mammalia, Meridiungulata) from Patagonia and a new appraisal of astrapothere phylogeny. *Journal of Systematic Palaeontology* 7: 117–128.

Krapovickas, V. 2010. El rol de las trazas fósiles de tetrápodos en los modelos de icnofacies continentales en ambientes de climas áridos-semiáridos. Unpublished PhD dissertation, Facultad de Ciencias Exactas y Naturales, Universidad de Buenos Aires.

Krapovickas, V., Ciccioli, P. L., Mángano, M. G., Marsicano, C. A., and Limarino, C. O. 2009. Paleobiology and paleoecology of an arid-semiarid Miocene South American ichnofauna in anastomosed fluvial deposits. *Palaeogeography, Palaeoclimatology, Palaeoecology* 284: 129–152.

Krapovickas, V., and Vizcaíno, S. F. 2016. South American Cenozoic Mammalian ichnology: Towards a global research program. Pp. 371–410 in Mangano, G., and Buatois, L., eds. *The Trace-Fossil Record of Major Evolutionary Events, Vol. 2: Mesozoic and Cenozoic.* Springer, Amsterdam.

Krause, D. W. 1982. Jaw movement, dental function, and diet in the Paleocene multituberculate Ptilodus. *Paleobiology* 8(3): 265–281.

Kubo, T. 2011. Estimating body weight from footprints: Application to pterosaurs. *Palaeogeography, Palaeoclimatology, Palaeoecology* 299: 197–199.

Kutschera, U. 2007. Palaeobiology: The origin and evolution of a scientific discipline. *Trends in Ecology and Evolution* 22: 172–173.

Lamm, E. T. 2007. Paleohistology widens the field of view in paleontology. *Microscopy and Microanalysis* 13(S02): 50–51.

Langer, P., and Chivers, D. J. 1994. Classification of foods for comparative analysis of gastro-intestinal tracts. Pp. 74–82 in Chivers, D. J., and Langer, P., eds. *The*

Digestive System in Mammals: Food, Form and Function. Cambridge University Press, New York.

Lauder, G. V. 1995. On the inference of form from structure. Pp. 1–18 in Thomason, J. J., ed. *Functional Morphology in Vertebrate Paleontology.* Cambridge University Press, New York.

Laurin, M. 2011. Limb origin and development. *Mémoires du Muséum National d'Histoire Naturelle* 201: 75–89.

Lautenschlager, S., Gill, P., Luo, Z. X., Fagan, M. J., and Rayfield, E. J. 2017. Morphological evolution of the mammalian jaw adductor complex. *Biological Reviews of the Cambridge Philosophical Society* 92(4): 1910–1940.

Lemell, P., Beisser, C. J., and Weisgram, J. 2000. Morphology and function of the feeding apparatus of *Pelusios castaneus* (Chelonia; Pleurodira). *Journal of Morphology* 244: 127–135.

Lessa, E. P., and Fariña, R. A. 1996. Reassessment of extinction patterns among the late Pleistocene mammals of South America. *Palaeontology* 39(3): 651–662.

Lessa, E. P., and Stein, B. R. 1992. Morphological constraints in the digging apparatus of pocket gophers (Mammalia: Geomydae). *Biological Journal of the Linnean Society* 47: 439–453.

Lessa, E. P., van Valkenburgh, B., and Fariña, R. A. 1997. Testing hypotheses of differential mammalian extinctions subsequent to the Great American biotic interchange. *Palaeogeography. Palaeoclimatology, Palaeoecology* 135: 157–162.

Levangie, P. K., and Norkin, C. C. 2005. *Joint Structure and Function: A Comprehensive Analysis.* 5th ed. Davis, Philadelphia.

Lieber, R. L., and Ward, S. R. 2011. Skeletal muscle design to meet functional demands. *Philosophical Transactions of the Royal Society B: Biological Sciences* 366: 1466–1476.

: Liem, K. F. 1991a. Functional morphology. Pp. 129–150 in Keenleyside, M. H. A., ed. *Cichlid Fishes: Behaviour, Ecology and Evolution.* Chapman & Hall, London.

Liem, K. F. 1991b. Toward a new morphology: Pluralism in research and education. *American Zoologist* 31: 759–767.

Liem, K. F., Bemis, W. E., Walker, W. F., Jr., and Grande, L. 2001. *Functional Anatomy of the Vertebrates: An Evolutionary Perspective.* Harcourt, New York.

Lindstedt, S. L., Miller, B. J., and Buskirk, S. W. 1986. Home range, time, and body size in mammals. *Ecology* 67: 413–418.

Lister, A. 2014. Behavioural leads in evolution: Evidence from the fossil record. *Biological Journal of the Linnean Society* 112: 315–331.

Lu, T. W., and Chang, C. F. 2012. Biomechanics of human movement and its clinical applications. *Kaohsiung Journal of Medical Sciences* 28: S13eS25.

Lucas, P. W. 2004. *Dental Functional Morphology: How Teeth Work.* Cambridge University Press, Cambridge.

Luo, Z. X. 2007. Transformation and diversification in early mammal evolution. *Nature* 450(7172): 1011–1019.

Luo, Z. X., Ji, Q., and Yuan, C. X. 2007. Convergent dental adaptations in pseudo-tribosphenic and tribosphenic mammals. *Nature* 450(7166): 93–97.

Lusk, G. 1928. *The Elements of the Science of Nutrition.* 4th ed. Saunders, Philadelphia.

Lyman, R. L. 2010. What taphonomy is, what it isn't, and why taphonomists should care about the difference. *Journal of Taphonomy* 8(1): 1–16.

Lyons, S. K., and Smith, F. A. 2013. Macroecological patterns of body size in mammals across time and space. Pp. 116–145 in Smith, F. A., and Lyons, S. K., eds. *Animal Body Size: Linking Pattern and Process across Space, Time and Taxonomic Group.* University of Chicago Press, Chicago.

MacFadden, B. J. 1998. Tale of two rhinos: Isotopic ecology, paleodiet, and niche differentiation of *Aphelops* and *Teleoceras* from the Florida Neogene. *Paleobiology* 24(2): 274–286.

MacFadden, B. J., and Hulbert, R. C., Jr. 1990. Body size estimates and size distribution of ungulate mammals from the Late Miocene Love Bone Bed of Florida. Pp. 337–363 in Damuth, J., and MacFadden, B. J., eds. *Body Size in Mammalian Paleobiology: Estimation and Biological Implications.* Cambridge University Press, Cambridge.

MacLeod, N. 2014. Concluding thoughts. Pp. 247–258 in Sánchez-Villagra, M. R., and MacLeod, N., eds. *Issues in Palaeobiology: A Global View: Interviews and Essays.* Scidinge Hall, Zürich.

Mahler, D. L., Ingram, T., Revell, L. J., and Losos, J. B. 2013. Exceptional convergence on the macroevolutionary landscape in island lizard radiations. *Science* 341(6143): 292–295.

Manera de Bianco, T., and Aramayo, S. A. 2004. Taphonomic features of Pehuén-Co palaeoichnological site (Late Pleistocene), Buenos Aires Province, Argentina. *1st International Congress on Ichnology (Trelew, Argentina)* Abstracts:49.

Marquet, P. A., Quiñones, R. A., Abades, S., Labra, F., Tognelli, M., Arim, M., and Rivadeneira, M. 2005. Scaling and power-laws in ecological systems. *Journal of Experimental Biology* 208: 1749–1769.

Marshall, L. G. 1977. Evolution of the carnivorous adaptive zone in South America. Pp. 709–722 in Hecht, M. K., Goody, P. C., and Hecht, B. M., eds. *Major Patterns in Vertebrate Evolution.* Plenum, New York.

Marshall, L. G. 1978. Evolution of the Borhyaenidae, extinct South American predaceous marsupials.

University of California Publications in Geological Sciences 117: 1–89.

Martin, T., and von Koenigswald, W., eds. 2021. *Mammalian Teeth: Form and Function*. Pfeil, Munich.

Martins, E. P., and Hansen, T. F. 1997. Phylogenies and the comparative method: A general approach to incorporating phylogenetic information into the analysis of interspecific data. *American Naturalist* 149(4): 646–667.

Massare, J. A. 1988. Swimming capabilities of Mesozoic marine reptiles: Implications for method of predation. *Paleobiology* 14(2): 187–205.

Maynard Smith, J., and Savage, R. J. G. 1959. The mechanics of mammalian jaws. *School Sciences Review* 141: 289–301.

McGowan, C. 1994. *Diatoms to Dinosaurs: The Size and Scale of Living Things*. Island, Washington, DC.

McGowan, C. P., and Collins, C. E. 2018. Why do mammals hop? Understanding the ecology, biomechanics and evolution of bipedal hopping. *Journal of Experimental Biology* 221(12): jeb161661.

McMahon, T. A. 1973. Size and shape in biology. *Science* 179: 1201–1204.

McNab, B. K. 1985. Energetics, population biology, and distribution of xenarthrans, living and extinct. Pp. 219–232 in Montgomery, G. G., ed. *The Evolution and Ecology of Armadillos, Sloths and Vermilinguas*. Smithsonian, Washington, DC.

McNab, B. K. 1988. Complications inherent in scaling the basal rate of metabolism in mammals. *Quarterly Review of Biology* 63: 25–54.

Meeh, K. 1879. Oberflächenmessungen des menschlichen Körpers. *Zeitschrift für Biologie* 15: 425–485.

Mendoza, M. 2005. Hacia una caracterización ecomorfológica compleja: Una revisión de la paleoautoecología de los ungulados. *Ameghiniana* 42: 233–248.

Mendoza, M., Janis, C. M., and Palmqvist, P. 2002. Characterizing complex craniodental patterns related to feeding behaviour in ungulates: A multivariate approach. *Journal of Zoology* 258: 223–246.

Mendoza, M., Janis, C. M., and Palmqvist, P. 2005. Ecological patterns in the trophic-size structure of large mammal communities: A "taxon-free" characterization. *Evolutionary Ecology Research* 7: 505–530.

Mendoza, M., Janis, C. M., and Palmqvist, P. 2006. Estimating the body mass of extinct ungulates: A study on the use of multiple regression. *Journal of Zoology* 270: 90–101.

Mendoza, M., and Palmqvist, P. 2006. Characterizing adaptive morphological patterns related to diet in Bovidae (Mammalia: Artiodactyla). *Acta Zoologica Sinica* 52:988–1008.

Mendoza, M., and Palmqvist, P. 2008. Hypsodonty in ungulates: An adaptation for grass consumption or for foraging in open habitat? *Journal of Zoology* 274: 134–142.

Meng, Q. J., Grossnickle, D. M., Liu, Y. G., Zhang, A. I., Neander, Q. Ji, and Luo, Z. X. 2017. New gliding mammaliaforms from the Jurassic. *Nature* 548: 291–296.

Miller, W., III., ed. 2007. *Trace Fossils: Concepts, Problems, Prospects*. Elsevier, Amsterdam.

Miljutin, A. 2009. Substrate utilization and feeding strategies of mammals: Description and classification. *Estonian Journal of Ecology* 58: 60–71.

Millien, V., and Bovy, H. 2010. When teeth and bones disagree: Body mass estimation of a giant extinct rodent. *Journal of Mammalogy* 91: 11–18.

Minter, N. J., and Braddy, S. J. 2009. Ichnology of an Early Permian tidal flat: The Robledo Mountains Formation of the Robledo Mountains, southern New Mexico, USA. *Special Papers in Palaeontology* 82.

Mitteroecker, P., and Huttegger, S. 2009. The concept of morphospaces in evolutionary and developmental biology: Mathematics and metaphors. *Biological Theory* 4(1): 54–67.

Mones, A. 1982. An equivocal nomenclature: What means hypsodonty? *Paläontologische Zeitschrift* 56: 107–111.

Montalvo, C. I., Melchor, R. N., Visconti, G., and Cerdeño, M. E. 2008. Vertebrate taphonomy in loess-palaeosol deposits: A case study from the late Miocene of central Argentina. *Geobios* 41: 133–143.

Montalvo, C. I., Raigemborn, M. S., Tomassini, R. L., Zapata, L., Bargo, M. S., Martínez Uncal, M. C., and Vizcaíno, S. F. 2019. Floodplain taphonomic mode of Early Miocene vertebrates of Southern Patagonia, Argentina. *PALAIOS* 34: 105–120.

Morgan Ernest, S. K. 2013. Using size distributions to understand the role of body size in mammalian community assembly. Pp. 146–167 in Smith, F. A., and Lyons, S. K., eds. *Animal Body Size: Linking Pattern and Process across Space, Time and Taxonomic Group*. University of Chicago Press, Chicago.

Motani, R. 2001. Estimating body mass from silhouettes: Testing the of elliptical body cross-sections. *Paleobiology* 27(4): 735–750.

Motani, R. 2002. Swimming speed estimation of extinct marine reptiles: Energetic approach revisited. *Paleobiology* 28: 251–262.

Motani, R., and Schmitz, L. 2011. Phylogenetic versus functional signals in the evolution of form-function relationships in terrestrial vision. *Evolution* 65(8): 2245–2257.

Münkemüller, T., Lavergne, S., Bzeznik, B., Dray, S., Jombart, T., Schiffers, K., and Thuiller, W. 2012. How to measure and test phylogenetic signal. *Methods in Ecology and Evolution* 3:743–756.

Muñoz, M. M. 2019. The evolutionary dynamics of mechanically complex systems. *Integrative Comparative Biology* 59(3): 705–715.

Muñoz, N. A. 2021. Locomotion in rodents and small carnivorans: Are they so different? *Journal of Mammalian Evolution* 28(1): 87–98.

Muñoz, N. A., Cassini, G. H., Candela, A. M., and Vizcaíno, S. F. 2017. Ulnar articular surface 3D landmarks and ecomorphology of small mammals: A case of study in two Early Miocene typotheres (Notoungulata) from Patagonia. *Earth and Environmental Science Transactions of the Royal Society of Edinburgh* 106: 315–323.

Muñoz, N. A., Toledo, N., Candela, A. M., and Vizcaíno, S. F. 2019. Functional morphology of the forelimb of Early Miocene caviomorph rodents from Patagonia. *Lethaia* 52: 91–106.

Myers, T. J. 2001. Prediction of marsupial body mass. *Australian Journal of Zoology* 49: 99–118.

Nabavizadeh, A. 2014. Hadrosauroid jaw mechanics and the functional significance of the predentary bone. Pp. 467–483 in Eberth, D. A., and Evans, D. C., eds. *Hadrosaurs*. Indiana University Press, Bloomington.

Nabavizadeh, A. 2016. Evolutionary trends in the jaw adductor mechanics of ornithischian dinosaurs. *Anatomical Record* 299(3): 271–294.

Nabavizadeh, A. 2020. New reconstruction of cranial musculature in ornithischian dinosaurs: Implications for feeding mechanisms and buccal anatomy. *Anatomical Record* 303(2): 347–362.

Nagy, K. A. 2005. Field metabolic rate and body size. *Journal of Experimental Biology* 208: 1621–1625.

Naples, V. L. 1999. Morphology, evolution and function of feeding in the giant anteater (*Myrmecophaga tridactyla*). *Journal of Zoology* 249(1): 19–41.

Nations, J. A., Heaney, L. R., Demos, T. C., Achmadi, A. S., Rowe, K. C., and Esselstyn, J. A. 2019. A simple skeletal measurement effectively predicts climbing behaviour in a diverse clade of small mammals. *Biological Journal of the Linnean Society* 128(2): 323–336.

Navalon, G., Bright, J. A., Marugan-Lobon, J., and Rayfield, E. J. 2019. The evolutionary relationship among beak shape, mechanical advantage, and feeding ecology in modern birds. *Evolution* 73: 422–435.

Nee, S., Read, A. F., Greenwood, J. J. D., and Harvey, P. H. 1991. The relationship between abundance and body size in British birds. *Nature* 351: 312–313.

Nesbitt, S. J. 2011. The early evolution of archosaurs: Relationships and the origin of major clades. *Bulletin of the American Museum of Natural History* 352: 1–292.

Niklas, K. 1992. *Plant Biomechanics: An Engineering Approach to Plant Form and Function*. University of Chicago Press, Chicago.

Norberg, U. M., and Rayner, J. M. V. 1987. Ecological morphology and flight in bats (Mammalia, Chiroptera): Wing adaptations, flight performance, foraging strategy and echolocation. *Philosophical Transactions of the Royal Society B: Biological Sciences* 316: 335–427.

Norman, D. B., Crompton, A. W., Butler, R. J., Porro, L. B., and Charig, A. J. 2011. The Lower Jurassic ornithischian dinosaur *Heterodontosaurus tucki* Crompton and Charig, 1962: Cranial anatomy, functional morphology, taxonomy, and relationships. *Zoological Journal of the Linnean Society* 163(1): 182–276.

Norman, D. B., and Weishampel, D. B. 1985. Ornithopod feeding mechanisms: Their bearing on the evolution of herbivory. *American Naturalist* 126(2): 151–164.

Nowak, R. M. 1999. *Walker's Mammals of the World*. 6th ed. Johns Hopkins University Press, Baltimore.

Ollier, S., Couteron, P., and Chessel, D. 2006. Orthonormal transform to decompose the variance of a life-history trait across a phylogenetic tree. *Biometrics* 62: 471–477.

Olsen, A. M. 2015. Exceptional avian herbivores: Multiple transitions toward herbivory in the bird order Anseriformes and its correlation with body mass. *Ecology and Evolution* 5(21): 5016–5032.

Olsen, A. M. 2017. Feeding ecology is the primary driver of beak shape diversification in waterfowl. *Functional Ecology* 31(10): 1985–1995.

Orlandi Laureto, L. M., Vinicius Cianciaruso, M., and Menezes Samia, D. S. 2015. Functional diversity: An overview of its history and applicability. *Perspectives in Ecology and Conservation* 13(2): 112–116.

Otero, A., Pérez Moreno, A., Falkingham, P. L., Cassini, G. H., Ruella, A., Militello, M., and Toledo, N. 2020. Three-dimensional image surface acquisition in vertebrate paleontology: A review of principal techniques. *Publicación Electrónica de la Asociación Paleontológica Argentina* 20(1): 1–14.

Owen-Smith, N. 1988. *Megaherbivores: The Influence of Very Large Body Size on Ecology*. Cambridge University Press, Cambridge.

Oxnard, C. 1984. *The Order of Man*. Hong Kong University Press, Hong Kong.

Padian, K. 1995. Form and function: The evolution of a dialectic. Pp. 264–277 in Thomason, J. J., ed. *Functional Morphology and Vertebrate Paleontology*. Cambridge University Press, Cambridge.

Padian, K., and Rayner, J. M. V. 1993. The wings of pterosaurs. *American Journal of Science* 293(A): 91–166.

Pagel, M. D. 1991. Constructing "every animal." Book review of *Body Size in Mammalian Paleobiology*, Damuth, J., and MacFadden, B., eds. *Nature* 351: 532–533.

Pagel, M. D. 1999. Inferring historical patterns of biological evolution. *Nature* 401: 877– 884.

Paradis, E., and Claude, J. 2002. Analysis of comparative data using generalized estimating equations. *Journal of Theoretical Biology* 218: 175–185.

Paul, G. S. 1988. *Predatory Dinosaurs of the World: A Complete Illustrated Guide.* Simon & Schuster, New York.

Paulina Carabajal, A., and Canale, J. I. 2010. Cranial endocast of the carcharodontosaurid theropod *Giganotosaurus carolinii* Coria and Salgado, 1995. *Neues Jahrbuch für Geologie und Paläontologie: Abhandlungen* 258(2): 249–256.

PC-MATLAB. 1990. PC-MATLABTM for MS-DOS Personal Computers. MathWorks, Natick, MA.

Pennell, M. W., and Harmon, L. J. 2013. An integrative view of phylogenetic comparative methods: Connections to population genetics, community ecology, and paleobiology. *Annals of the New York Academy of Sciences* 1289: 90–105.

Pennycuick, C. J. 1992. *Newton Rules Biology: A Physical Approach to Biological Problems.* Oxford University Press, Oxford.

Pérez, L. M., Toledo, N., De Iuliis, G., Bargo, M. S., and Vizcaíno, S. F. 2010. Morphology and function of the hyoid apparatus of fossil xenarthrans. *Journal of Morphology* 271(9): 1119–1133.

Pérez-Barbería, F. J., and Gordon, I. J. 2001. Relationships between oral morphology and feeding style in the ungulata: A phylogenetically controlled evaluation. *Proceedings of the Royal Society B: Biological Sciences* 268: 1023–1032.

Perry, J. M. G., and Prufrock, K. A. 2018. Muscle functional morphology in paleobiology: The past, present, and future of "paleomyology." *Anatomical Record* 301: 538–555.

Peters, R. H. 1983. *The Ecological Implications of Body Size.* Cambridge University Press, Cambridge.

Pineda-Munoz S., and Alroy, J. 2014. Dietary characterization of terrestrial mammals. *Proceedings of the Royal Society B: Biological Sciences* 281: 20141173.

Pineda-Munoz, S., Lazagabaster, I. A., Evans, A. R., and Alroy, J. 2017. Inferring diet from dental morphology in terrestrial mammals. *Methods in Ecology and Evolution* 8: 481–491.

Plotnick, R., and Baumiller, T. 2000. Invention by evolution: Functional analysis in paleobiology. *Paleobiology* 26(4, Supplement):305–323.

Polly, P. D. 2007. Limbs in mammalian evolution. Pp. 245–268 in Hall, B. K., ed. *Fins into Limbs: Evolution, Development, and Transformation.* University of Chicago Press: Chicago.

Pough, F. H., and Janis, C. M. 2019. *Vertebrate Life.* 10th ed. Prentice Hall, New York.

Presslee, S., Slater, G. J., Pujos, F. R., Forasiepi, A. M., Fischer, R., Molloy, K., Mackie, M., et al. 2019. Palaeoproteomics resolves sloth relationships. *Nature Ecology and Evolution* 3: 1121–1130.

Prevosti, F. J., and Forasiepí, A. M. 2018. *Evolution of South American Mammalian Predators during the Cenozoic: Paleobiogeographic and Paleoenvironmental Contingencies.* Springer Geology, Cham.

Prevosti, F. J., Forasiepi, A. M., Ercoli, M. D., and Turazzini, G. F. 2012. Paleoecology of the mammalian carnivores (Metatheria, Sparassodonta) of the Santa Cruz Formation. Pp. 173–193 in Vizcaíno, S. F., Kay, R. F., and Bargo, M. S., eds. *Early Miocene Paleobiology in Patagonia: High-Latitude Paleocommunities of the Santa Cruz Formation.* Cambridge University Press, Cambridge.

Prevosti, F. J., Forasiepi, A. M., and Zimicz, N. 2013. The evolution of the Cenozoic terrestrial mammalian predator guild in South America: Competition or replacement? *Journal of Mammalian Evolution* 20: 3–21.

Prevosti, F. J., and Palmqvist, P. 2001. Análisis ecomorfológico del cánido hipercarnívoro *Theriodictis platensis* (Mammalia, Carnivora) basado en un nuevo ejemplar del Pleistoceno de Sudamérica. *Ameghiniana* 38(4):375–384.

Prevosti, F. J., and Vizcaíno, S. F. 2006. Paleoecology of the large carnivore guild from the Late Pleistocene of Argentina. *Acta Palaeontologica Polonica* 51:407–422.

Pujos, F., Gaudin, T., De Iuliis, G., and Cartelle, C. 2012. Recent advances on variability, morpho-functional adaptations, dental terminology, and evolution of sloths. *Journal of Mammalian Evolution* 19: 159–169.

Pujos, F., and Salas-Gismondi, R. 2020. Predation of the giant Miocene caiman *Purussaurus* on a mylodontid ground sloth in the wetlands of proto-Amazonia. *Biology Letters* 16: 20200239.

Quader, S., Isvaran, K., Hale, R. E., Miner, B. G., and Seavy, N. E. 2004. Nonlinear relationships and phylogenetically independent contrasts. *Journal of Evolutionary Biology* 17: 709–715.

Radinsky, L. B. 1987. *The Evolution of Vertebrate Design.* University of Chicago Press, Chicago.

Rayner, J. M. V. 1991. Wake structure and force generation in avian flapping flight. *Acta XX Congressus Internationalis Ornithologici* (II): 702–715.

Reed, K. E. 1998. Using large mammal communities to examine ecological and taxonomic structure and predict vegetation in extant and extinct assemblages. *Paleobiology* 24: 384–408.

Reguero, M. A., Candela, A. M., and Cassini, G. H. 2010. Hypsodonty and body size in rodent-like notoungulates. Pp. 362–371 in Madden, R., Carlini, A. A., Vucetich, M. G., and Kay, R. F., eds. *The Paleontology of Gran*

Barranca: Evolution and Environmental Change through the Middle Cenozoic of Patagonia. Cambridge University Press, Cambridge.

Reilly, S. M., and Wainwright, P. C. 1994. Conclusion: Ecological morphology and the power of integration. Pp. 339–354 in Wainwright, P. C., and Reilly, S. M., eds. *Ecological Morphology. Integrative Organismal Biology*. University of Chicago Press, Chicago.

Reiss, M. 1988. Scaling of home range size: Body size, metabolic needs and ecology. *Trends in Ecology and Evolution* 3: 85–86.

Rensberger, J. M. 1973. An occlusion model for mastication and dental wear in herbivorous mammals. *Journal of Paleontology* 47: 515–528.

Richardson, P. L. 2011. How do albatrosses fly around the world without flapping their wings? *Progress in Oceanography* 88: 46–58.

Richardson, P. L., Wakefield, E. D., and Phillips, R. A. 2018. Flight speed and performance of the wandering albatross with respect to wind. *Movement Ecology* 6: 3.

Riggs, E. S. 1935. A skeleton of *Astrapotherium*. *Field Museum of Natural History, Geological Series* 6(13): 167–177.

Rinehart, L. F., and Lucas, S. G. 2001. A statistical analysis of a growth series of the Permian nectridean *Diplocaulus magnicornis* showing two-stage ontogeny. *Journal of Vertebrate Paleontology* 21(4): 803–806.

Rodríguez-Gómez, G., Cassini, G. H., Palmqvist, P., Bargo, M. S., Toledo, N., Martín-González, J. A., Muñoz, N. A., Kay, R. F., and Vizcaíno, S. F. 2020. Testing the hypothesis of an impoverished predator guild in the Early Miocene ecosystems of Patagonia: An analysis of meat availability and competition intensity among carnivores. *Palaeogeography, Palaeoclimatology, Palaeoecology* 554: 109805.

Rodríguez-Gómez, G., Mateos, A., Martín-González, J. A., Blasco, R., Rosell, J., and Rodríguez, J. 2014. Discontinuity of human presence at Atapuerca during the early Middle Pleistocene: A matter of ecological competition? *PLoS ONE* 9: e101938.

Rodríguez-Gómez, G., Rodríguez, J., Martín-González, J. A., Goikoetxea, I., and Mateos, A. 2013. Modeling trophic resource availability for the first human settlers of Europe: The case of Atapuerca-TD6. *Journal of Human Evolution* 64: 645–657.

Rodríguez-Gómez, G., Rodríguez, J., Martín-González, J. A., and Mateos, A. 2017. Carnivores and humans during the Early and Middle Pleistocene at Sierra de Atapuerca. *Quaternary International* 433: 402–414.

Rohlf, F. J. 1999. Shape statistics: Procrustes superimpositions and tangent spaces. *Journal of Classification* 16(2): 197–223.

Romer, A. S. 1962. *The Vertebrate Body*. 3rd ed. Saunders, Philadelphia.

Rowell, A. J. 1973. Book review of *Models in Paleobiology*, Schopf, T. J. M., ed. *Systematic Biology* 22(1): 94–95.

Rudwick, M. J. S. 1964. The inference of function from structure in fossils. *British Journal for the Philosophy of Science* 15(57): 27–40.

Ruff, C. 1990. Body mass and hindlimb bone cross-sectional and articular dimensions in anthropoid primates. Pp. 119–150 in Damuth, J., and MacFadden, B. J., eds. *Body Size in Mammalian Paleobiology: Estimation and Biological Implications*. Cambridge University Press, Cambridge.

Rüttimann, B. 1992. A noteworthy meeting of the Society for Nature Research in Zurich: Two important precursors of Julius Wolff: Carl Culmann and Hermann von Meyer. Pp. 13–22 in Regling, G., ed. *Wolff's Law and Connective Tissue Regulation*. De Gruyter, New York.

Sage, R. F. 2016. A portrait of the C_4 photosynthetic family on the 50th anniversary of its discovery: Species number, evolutionary lineages, and Hall of Fame. *Journal of Experimental Botany* 67(14): 4039–4056.

Sakamoto, M. 2010. Jaw biomechanics and the evolution of biting performance in theropod dinosaurs. *Proceedings of the Royal Society B: Biological Sciences* 277(1698):3327–3333.

Samuels, J. X. 2009. Cranial morphology and dietary habits of rodents. *Zoological Journal of the Linnean Society* 156: 864–888.

Samuels, J. X., and van Valkenburgh, B. 2008. Skeletal indicators of locomotor adaptations in living and extinct rodents. *Journal of Morphology* 269: 1387–1411.

Sánchez-Villagra, M. R., and MacLeod, N., eds. 2014. *Issues in Palaeontology: A Global View. Interviews and Essays*. Scidinge Hall, Zürich.

Sanford, G. M., Lutterschmidt, W. I., and Hutchinson, V. H. 2002. The comparative method revisited. *BioScience* 52: 830–836.

Sargis, E. J. 2002a. Functional morphology of the forelimbs of tupaiids (Mammalia, Scandentia) and its phylogenetic implications. *Journal of Morphology* 253: 10–42.

Sargis, E. J. 2002b. Functional morphology of the hindlimbs of tupaiids (Mammalia, Scandentia) and its phylogenetic implications. *Journal of Morphology* 254: 149–185.

Savage, V. M., Gillooly, J. F., Woodruff, W. H., West, G. B., Allen, A. P., Enquist, B. J., and Brown, J. H. 2004. The predominance of quarter-power scaling in biology. *Functional Ecology* 18: 257–282.

Savazzi, E. 1999. Introduction to functional morphology. Pp. 3–14 in Savazzi, E., ed. *Functional Morphology of the Invertebrate Skeleton*. Wiley, Chichester.

Scanlon, V. C., and Sanders, T. 2007. *Essentials of Anatomy and Physiology*. Davis, Philadelphia.

Schaefer, S. A., and Lauder, G. V. 1986. Historical transformation of functional design: Evolutionary morphology of feeding mechanisms in loricarioid catfishes. *Systematic Zoology* 35: 489–508.

Schmidt, M. 2010. Locomotion and postural behavior. *Advances in Science and Research* 5: 23–39.

Schmidt-Nielsen, K. 1984. *Scaling: Why Is Animal Size So Important?* Cambridge University Press, Cambridge.

Schopf, T. J. M., ed. 1972. *Models in Paleobiology*. Papers from a symposium, Washington, DC, November 1971. Freeman Cooper, San Francisco.

Scholey, K. D. 1983. Developments in vertebrate flight: Climbing and gliding of mammals and reptiles, and the flapping flight of birds. Unpublished PhD dissertation, University of Bristol.

Schwenk, K. 2000. *Feeding: Form, Function and Evolution in Tetrapod Vertebrates*. Academic, San Diego.

Scott, E. C., and Matzke, N. J. 2007. Biological design in science classrooms. *Proceedings of the National Academy of Sciences of the USA* 104(1):8669–8676.

Scott, K. M. 1983. Prediction of body weight of fossil Artiodactyla. *Zoological Journal of the Linnean Society* 27:199–215.

Scott, K. M. 1990. Postcranial dimensions of ungulates as predictors of body mass. Pp. 301–335 in Damuth, J., and MacFadden, B. J., eds. *Body Size in Mammalian Paleobiology: Estimation and Biological Implications*. Cambridge University Press, Cambridge.

Scott, W. B. 1913. *A History of Land Mammals in the Western Hemisphere*. MacMillan, New York.

Scott, W. B. 1928. Astrapotheria. Pp. 301–351 in Scott, W. B., ed. *Mammalia of the Santa Cruz Beds: Part IV, Reports of the Princeton University Expeditions to Patagonia, 1896–1899, Vol. VI: Paleontology*. Princeton University, Princeton, NJ.

Sepkoski, D. 2012. *Rereading the Fossil Record: The Growth of Paleobiology as an Evolutionary Discipline*. University of Chicago Press, Chicago.

Sepkoski, D., and Ruse, M., eds. 2009. *The Paleobiological Revolution: Essays on the Growth of Modern Paleontology*. University of Chicago Press, Chicago.

Shanahan, T. 2011. Phylogenetic inertia and Darwin's higher law. *Studies in History and Philosophy of Biological and Biomedical Sciences* 42: 60–68.

Shingleton, A. 2010. Allometry: The study of biological scaling. *Nature Education Knowledge* 3(10): 2.

Shipman, P. 1981. *Life History of a Fossil: An Introduction to Taphonomy and Paleoecology*. Harvard University Press, Cambridge.

Shu, D.-G., Conway Morris, S., Han, J., Zhang, Z.-F., Yasui, K., Janvierk, P., Chen, L., Zhang, X.-L., Liu, J.-N., Li, Y., and Liu, H.-Q. 2003. Head and backbone of the Early Cambrian vertebrate *Haikouichthys. Nature* 421(30): 526–529.

Shu, D.-G., Luo, H.-L., Conway Morris, S., Zhang, X.-L., Hu, S.-X., Chen, L., Han, J., Zhu, M., Li, Y., and Chen, L.-Z. 1999. Lower Cambrian vertebrates from south China. *Nature* 402:42–46.

Shubin, N. H., Daeschler, E. B., and Jenkins, F. A., Jr. 2014. Pelvic girdle and fin of *Tiktaalik roseae. Proceedings National Academy of Sciences of the USA* 111: 893–899.

Shuker, D., and Simmons, L., eds. 2014. *The Evolution of Insect Mating Systems*. Oxford University Press, Oxford.

Simpson, G. G. 1980. *Splendid Isolation: The Curious History of South American Mammals*. Yale University Press, New Haven, CT.

Skedros, J. G., and Baucom, S. L. 2007. Mathematical analysis of trabecular "trajectories" in apparent trajectorial structures: The unfortunate historical emphasis on the human proximal femur. *Journal of Theorical Biology* 244(1): 15–45.

Slater, G. J. 2013. Phylogenetic evidence for a shift in the mode of mammalian body size evolution at the Cretaceous–Palaeogene boundary. *Methods in Ecology and Evolution* 4(8): 734–744.

Slater, G. J., and Harmon, L. J. 2013. Unifying fossils and phylogenies for comparative analyses of diversification and trait evolution. *Methods in Ecology and Evolution* 4: 699–702.

Slice, D. E., ed. 2006. *Modern Morphometrics in Physical Anthropology*. Springer Science & Business, Berlin.

Smith, A. P., and Ganzhorn, J. U. 1996. Convergence and divergence in community structure and dietary adaptation in Australian possums and gliders and Malagasy lemurs. *Australian Journal of Ecology* 21: 31–46.

Smith, F. A., and Lyons, S. K., eds. 2013. *Animal Body Size: Linking Pattern and Process across Space, Time and Taxonomic Group*. University of Chicago Press, Chicago.

Smith, R. J. 1984. Allometric scaling in comparative biology: Problems of concept and method. *American Journal of Physiology* 246: R152–R160.

Smith, R. J. 1993. Logarithmic transformation bias in allometry. *American Journal of Physical Anthropology* 90: 215–228.

Smits, A. J. 2019. Undulatory and oscillatory swimming. *Journal of Fluid Mechanics* 874: P1.

Snelling, E. P., Seymour, R. S., Runciman, S., Matthews, P. G., and White, C. R. 2012. Symmorphosis and the insect respiratory system: A comparison between flight and hopping muscle. *Journal of Experimental Biology* 215(18): 3324–3333.

Snowdon, P. 1991. A ratio estimator for bias correction in logarithmic regression. *Canadian Journal of Forest Research* 21: 720–724.

Solounias, N., and Dawson-Saunders, B. 1988. Dietary adaptations and palaeocology of the late Miocene ruminants from Pikermi and Samos in Greece. *Palaeogeography, Palaeoclimatology, Palaeoecology* 65: 149–172.

Solounias, N., Fortelius, M., and Freeman, P. 1994. Molar wear rates in ruminants: A new approach. *Annales Zoologici Fennici* 31: 219–227.

Solounias, N., and Moelleken, S. M. C. 1993. Dietary adaptation of some extinct ruminants determined by premaxillary shape. *Journal of Mammalogy* 74: 1059–1071.

Solounias, N., Rivals, F., and Semprebon, G. M. 2010. Dietary interpretation and paleoecology of herbivores from Pikermi and Samos (late Miocene of Greece). *Paleobiology* 36: 113–136.

Solounias, N., and Semprebon, G. 2002. Advances in reconstruction of ungulate ecomorphology with application to early fossil equids. *American Museum Novitates* 3366: 1–49.

Spencer, L. M. 1995. Morphological correlates of dietary resource partitioning in the African Bovidae. *Journal of Mammalogy* 76: 448–471.

Steel, R., and Harvey, A. P. 1979. *The Encyclopedia of Prehistoric Life.* McGraw-Hill, New York.

Stein, B. R., and Casinos, A. 1997. What is a cursorial mammal? *Journal of Zoology, London* 242: 185–192.

Steinthorsdottir, M., Coxall, H. K., de Boer, A. M., Huber, M., Barbolini, N., Bradshaw, C. D., Burls, N. J., et al. 2021. The Miocene: The future of the past. *Paleoceanography and Paleoclimatology* 36: e2020PA00403.

Strömberg, C. A. E., Dunn, R. E., Madden, R. H., Kohn, M. J., and Carlini, A. A. 2013. Decoupling the spread of grasslands from the evolution of grazer-type herbivores in South America. *Nature Communications* 4: 1478.

Summers, A. P., and O'Reilly, J. C. 1997. A comparative study of locomotion in the caecilians *Demophis mexicanus* and *Typhlonectes natans* (Amphibia: Gymnophiona). *Zoological Journal of the Linnean Society* 121: 65–76.

Sutton, M. D., Rahman, I. A., and Garwood, R. J. 2014. *Techniques for Virtual Palaeontology.* Wiley, Chichester.

Swihart, R. K., Slade, N. A., and Bergstrom, B. J. 1988. Relating body size to the rate of home range use in mammals. *Ecology* 69: 393–399.

Talevi, M., and Fernandez, M. S. 2012. Unexpected skeletal histology of an ichthyosaur from the Middle Jurassic of Patagonia: Implications for evolution of bone microstructure among secondary aquatic tetrapods. *Naturwissenschaften* 99: 241–244.

Talevi, M., and Fernandez, M. S. 2015. Remodelling of skeletal tissue bone and structural specializations in an elasmosaurid (Sauropterygia: Plesiosauroidea) from the Upper Cretaceous of Patagonia, Argentina. *Historical Biology* 27: 60–67.

Talevi, M., Fernández, M. S., and Salgado, L. 2012. Variación ontogenética en la histología ósea de *Caypullisaurus bonapartei* Fernández, 1997 (Ichthyosauria: Ophthalmosauridae). *Ameghiniana* 49: 38–46.

Tambusso, P. S., and Fariña, R. A. 2015. Digital cranial endocast of *Pseudoplohophorus absolutus* (Xenarthra, Cingulata) and its systematic and evolutionary implications. *Journal of Vertebrate Paleontology* 30(5): e967853.

Tauber, A. A. 1997. Paleoecología de la Formación Santa Cruz (Mioceno inferior) en el extremo sudeste de la Patagonia. *Ameghiniana* 34: 517–529.

Taylor, B. K. 1978. The anatomy of the forelimb in the anteater (*Tamandua*) and its functional implications. *Journal of Morphology* 157: 347–368

Texera, W. A. 1974. Algunos aspectos de la biología del huemul (*Hippocamelus bisulcus*) (Mammalia: Artiodactyla, Cervidae) en cautividad. *Anales del Instituto Patagónico, Punta Arenas (Chile)* 5: 155–188.

Thenius, E. 1989. *Zähne und Gebiß der Säugetiere.* De Gruyter, Berlin.

Thiery, G., Guy, F., and Lazzari, V. 2019. A comparison of relief estimates used in three-dimensional dental topography. *American Journal of Biological Anthropology* 170: 260–274.

Thomason, J. J., ed. 1995. *Functional Morphology in Vertebrate Paleontology.* Cambridge University Press, Cambridge.

Thorington, R. W., Jr., and Santana, E. M. 2007. How to make a flying squirrel: *Glaucomys* anatomy in phylogenetic perspective. *Journal of Mammalogy* 88: 882–896.

Tobalske, B. W. 2007. Biomechanics of bird flight. *Journal of Experimental Biology* 210: 3135–3146.

Tobalske, B. W. and Dial, K. P. 1996. Flight kinematics of black-billed magpies and pigeons over a wide range of speeds. *Journal of Experimental Biology* 199: 263–280.

Toledo, N. 2016. Conceptual and methodological approaches for a paleobiological integration: The Santacrucian sloths (early Miocene of Patagonia) as a study case. *Ameghiniana* 53(2): 100–141.

Toledo, N., Bargo, M. S., Cassini, G. H., and Vizcaíno, S. F. 2012. The forelimb of Early Miocene sloths (Mammalia, Xenarthra, Folivora): Morphometrics and functional implications for substrate preferences. *Journal of Mammalian Evolution* 19: 185–198.

Toledo, N., Bargo, M. S., and Vizcaíno, S. F. 2013. Muscular reconstruction and functional morphology of the forelimb of Early Miocene sloths (Xenarthra, Folivora) of Patagonia. *Anatomical Record* 296: 305–325.

Toledo, N., Bargo, M. S., and Vizcaíno, S. F. 2015. Muscular reconstruction and functional morphology of the hind limb of Santacrucian (Early Miocene) sloths

(Xenarthra, Folivora) of Patagonia. *Anatomical Record* 298: 842–864.

Toledo, N., Cassini, G. H., Vizcaíno S. F., and Bargo, M. S. 2014. Mass estimation of Santacrucian sloths from the Early Miocene Santa Cruz Formation of Patagonia, Argentina. *Acta Palaeontologica Polonica* 59(2): 267–280.

Tomassini, R., Montalvo, C. I., Bargo, M. S., Vizcaíno, S. F., and Cuitiño, J. I. 2019. Sparassodonta (Metatheria) coprolites from the early-mid Miocene (Santacrucian Age) of Patagonia (Argentina): Evidence of exploitation by coprophagous insects. *PALAIOS* 31(12): 639–651.

Toon, S., and Toon, A. 2004. Koalas (Phascolarctidae). In *Grzimek Animals Life Encyclopedia*. Gale, Farmington Hills, MI.

Townsend, K. E. B., and Croft, D. A. 2005. Low-magnification microwear analyses of South American endemic herbivores. *Journal of Vertebrate Paleontology* 25: 123A.

Townsend, K. E. B., and Croft, D. A. 2008. Diets of notoungulates from the Santa Cruz Formation, Argentina: New evidence from enamel microwear. *Journal of Vertebrate Paleontology* 28: 217–230.

Trayler, R. B., Kohn, M. J., Bargo, M. S., Cuitiño, J. I., Kay, R. F., Strömberg, C. A. E., and Vizcaíno, S. F. 2020. Patagonian aridification at the onset of the mid-Miocene Climate Optimum. *Paleoceanography and Paleoclimatology* 35(9): e2020PA003956.

Tucker, V. 1977. Scaling and avian flight. Pp. 497–509 in Pedley, T. J., ed. *Scale Effects in Animal Locomotion*. Academic, New York.

Ungar, P. S. 2006. *Evolution of the Human Diet: The Known, the Unknown, and the Unknowable*. Oxford University Press, New York.

Ungar, P. S., and M'Kirera, F. 2003. A solution to the worn tooth conundrum in primate functional anatomy. *Proceedings of the National Academy of Sciences of the USA* 100(7): 3874–3877.

Urey, H. C. 1947. The thermodynamic properties of isotopic substances. *Journal of the Chemical Society*: 562–581.

van Bergen, Y., and Phillips, K. 2005. Size matters. *Journal of Experimental Biology* 208: i–iii.

van Couvering, J. A. H. 1980. Community evolution in Africa during the Cenozoic. Pp. 272–298 in Behrensmeyer, A. K., and Hill, A., eds. *Fossils in the Making*. University of Chicago Press, Chicago.

van der Klaauw, C. J. 1948. Ecological studies and reviews: IV. Ecological morphology. *Bibliographia Biotheoretica* 4(2): 23–111.

van der Zwaard, S., Jo de Ruiter, C., Noordhof, D. A., Sterrenburg, R., Bloemers, F. W., de Koning, J. J., Jaspers, R. T., and van der Laarse, W. J. 2016. Maximal oxygen uptake is proportional to muscle fiber oxidative capacity, from chronic heart failure patients to professional cyclists. *Journal of Applied Physiology* 121: 636–645.

van Drongelen, W., and Dullemeijer, P. 1982. The feeding apparatus of *Caiman crocodilus*: A functional-morphological study. *Anatomischer Anzeiger* 151(4): 337–366.

van Valkenburgh, B. 1985. Locomotor diversity within past and present guilds of large predatory mammals. *Paleobiology* 11: 406–428.

van Valkenburgh, B. 1989. Carnivore dental adaptations and diet: A study of trophic diversity within guilds. Pp. 410–436 in Gittleman, J. L., ed. *Carnivore Behaviour, Ecology and Evolution*.: Chapman & Hall, London.

van Valkenburgh, B. 1990. Skeletal and dental predictors of body mass in carnivores. Pp. 181–205 in Damuth, J., and MacFadden, B. J., eds. *Body Size in Mammalian Paleobiology: Estimation and Biological Implications*. Cambridge University Press, Cambridge.

van Valkenburgh, B. 1991. Iterative evolution of hypercarnivory in canids (Mammalia: Carnivora): Evolutionary interactions among sympatric predators. *Paleobiology* 17: 340–362.

van Valkenburgh, B. 1999. Major patterns in the history of carnivorous mammals. *Annual Review of Earth and Planetary Sciences* 27: 463–93.

van Valkenburgh, B. 2007. Déjà vu: The evolution of feeding morphologies in the Carnivora. *Integrative and Comparative Biology* 47(1): 147–163.

van Valkenburgh, B., and Wayne, R. K. 1994. Shape divergence associated with size convergence in sympatric East African jackals. *Ecology* 75: 1567–1581.

van Wassenbergh, S., and Baeckens, S. 2019. Digest: Evolution of shape and leverage of bird beaks reflects feeding ecology, but not as strongly as expected. *Evolution* 73(3): 621–622.

Vassallo, A. I., Becerra, F., Echeverría, I. A., Díaz, A. O., Longo, M. V., Cohen, M., and Buezas, G. N. 2021 Analysis of the form-function relationship: Digging behavior as a case study. *Journal of Mammalian Evolution* 28: 59–74.

Verde Arregoitia, L. D., and D'Elía, G. 2021. Classifying rodent diets for comparative research. *Mammal Review* 51: 51–65.

Vermillion, W., Polly, P. D., Head, J., Eronen, J., and Lawing, A. M. 2018. Ecometrics: A trait-based approach to paleoclimate and paleoenvironmental reconstruction. Pp. 373–394 in Croft, D. A., Su, D. F., and Simpson, S. W., eds. *Methods in Paleoecology: Reconstructing Cenozoic Terrestrial Environments and Ecological Communities*. Vertebrate Paleobiology and Paleoanthropology Series, Delson, E., and Sargis, E., eds. Springer, New York.

Villafañe, A. L., Ortiz-Jaureguizar, E., and Bond, M. 2006. Cambios en la riqueza taxonómica y en las tasas de primera y última aparición de los Proterotheriidae

(Mammalia, Litopterna) durante el Cenozoico. *Estudios Geológicos* 62: 155–166.

Vizcaíno, S. F. 2009. The teeth of the "toothless": Novelties and key innovations in the evolution of xenarthrans (Mammalia, Xenarthra). *Paleobiology* 35(3): 343–366.

Vizcaíno, S. F. 2014. Interview on paleobiology. Pp. 181–192 in Sánchez-Villagra, M. R., and MacLeod, N., eds. *Issues in Palaeontology: A Global View. Interviews and Essays*. Scidinge Hall, Zürich.

Vizcaíno, S. F., and Bargo, M. S. 1998. The masticatory apparatus of *Eutatus* (Mammalia, Cingulata) and some allied genera: Evolution and paleobiology. *Paleobiology* 24: 371–383.

Vizcaíno, S. F., and Bargo, M. S. 2014. Loss of ancient diversity of xenarthrans and the value of protecting extant armadillos, sloths and anteaters. *Edentata* 15: 27–38.

Vizcaíno, S. F., and Bargo, M. S. 2021. Views on the form-function correlation and biological design. *Journal of Mammalian Evolution* 28: 15–22.

Vizcaíno, S. F., Bargo, M. S., and Cassini, G. H. 2006a. Dental occlusal surface area in relation to body mass, food habits and other biologic features in fossil Xenarthrans. *Ameghiniana* 43: 11–26.

Vizcaíno, S. F., Bargo, M. S., Cassini, G. H., and Toledo, N. 2016. *Forma y función en paleobiología de vertebrados*. Editorial Universidad Nacional de La Plata, La Plata. http://sedici.unlp.edu.ar/handle/10915/55101.

Vizcaíno, S. F., Bargo, M. S., and Fariña, R. A. 2008. Form, function and paleobiology in xenarthrans. Pp. 86–99 in Vizcaíno, S. F., and Loughry, W. L., eds. *The Biology of the Xenarthra*. University Press of Florida, Gainesville, FL.

Vizcaíno, S. F., Bargo, M. S., Kay, R. F., Fariña, R. A., Di Giacomo, M., Perry, J. M., Prevosti, F. J., Toledo, N., Cassini, G. H., and Fernicola, J. C. 2010. A baseline paleoecological study for the Santa Cruz Formation (late–early Miocene) at the Atlantic coast of Patagonia, Argentina. *Palaeogeography, Palaeoclimatology, Palaeoecology* 292: 507–519.

Vizcaíno, S. F., Bargo, M. S., Kay, R. F., and Milne, N. 2006b. The armadillos (Mammalia, Xenarthra) of the Santa Cruz Formation (early–middle Miocene): An approach to their paleobiology. *Palaeogeography, Palaeoclimatology, Palaeoecology* 237: 255–269.

Vizcaíno, S. F., Bargo, M. S., Pérez, M. E., Aramendía, I., Cuitiño, J. I., Monsalvo, E. S., Vlachos, E., Noriega, J. I., and Kay, R. A. 2022. Fossil vertebrates of the early–middle Miocene Cerro Boleadoras Formation, Northwestern Santa Cruz Province, Patagonia, Argentina. *Andean Geology* 49(3): 382–422.

Vizcaíno, S. F., Bargo, M. S., and Toledo, N. 2017. Revaluación crítica de los mamíferos actuales como indicadores paleoambientales: Ejemplos de la mastofauna neotropical. *XXX Jornadas Argentinas de Mastozoología* Abstract book: 41.

Vizcaíno, S. F., Bargo, M. S., Toledo, N., and De Iuliis, G. 2023. Conceptual challenges for the paleoecological reconstruction of the Pleistocene Pampean megafauna and the consequences of its extinction. *Publicación Electrónica de la Asociación Paleontológica Argentina* 23(1): 317–330.

Vizcaíno, S. F., Blanco, R. E., Bender, J. B., and Milne, N. 2011a. Proportions and function of the limbs of glyptodonts. *Lethaia* 44: 93–101.

Vizcaíno, S. F., Cassini, G. H., Fernicola, J. C., and Bargo, M. S. 2011b. Evaluating habitats and feeding habits through ecomorphological features in glyptodonts (Mammalia, Xenarthra). *Ameghiniana* 48: 305–319.

Vizcaíno, S. F., Cassini, G. H., Toledo, N., and Bargo, M. S. 2012a. On the evolution of large size in mammalian herbivores of Cenozoic faunas of southern South America. Pp. 76–101 in Patterson, B., and Costa, L., eds. *Bones, Clones and Biomes: The History and Geography of Recent Neotropical Mammals*. University of Chicago Press, Chicago.

Vizcaíno, S. F., and De Iuliis, G. 2003. Evidence for advanced carnivory in fossil armadillos (Mammalia: Xenarthra: Dasypodidae). *Paleobiology* 29: 123–138.

Vizcaíno, S. F., De Iuliis, G., and Bargo, M. S. 1998. Skull shape, masticatory apparatus, and diet of *Vassallia* and *Holmesina* (Mammalia: Xenarthra: Pampatheriidae): When anatomy constrains destiny. *Journal of Mammalian Evolution* 5: 291–-322.

Vizcaíno, S. F., and Fariña, R. A. 1997. Diet and locomotion in *Peltephilus*: A new view. *Lethaia* 30: 79–86.

Vizcaíno, S. F., and Fariña, R. A. 1999. On the flight capabilities and distribution of the giant Miocene bird *Argentavis magnificens* (Teratornithidae). *Lethaia* 32: 271–278.

Vizcaíno, S. F., and Fariña, R. A. 2000. El vuelo de un gigante. *Museo* (Revista de la Fundación Museo de La Plata) 3(14): 11–18.

Vizcaíno, S. F., Fariña, R. A., Bargo, M. S., and De Iuliis, G. 2004. Functional and phylogenetical assessment of the masticatory adaptations in Cingulata (Mammalia, Xenarthra). *Ameghiniana* 41: 651–664.

Vizcaíno, S. F., Fariña, R. A., and Mazzetta, G. 1999. Ulnar dimensions and fossoriality in armadillos and other South American mammals. *Acta Theriologica* 44: 309–320.

Vizcaíno, S. F., Fernicola, J. C., and Bargo, M. S. 2012b. Paleobiology of Santacrucian glyptodonts and armadillos (Xenarthra, Cingulata). Pp. 194–215 in Vizcaíno, S. F., Kay, R. F., and Bargo, M. S., eds. *Early Miocene*

Paleobiology in Patagonia: High-Latitude Paleocommunities of the Santa Cruz Formation. Cambridge University Press, Cambridge.

Vizcaíno, S. F., Kay, R. F., and Bargo, M. S. 2012c. Background for a palaeoecological study of the Santa Cruz Formation (late Early Miocene) on the Atlantic coast of Patagonia. Pp. 1–22 in Vizcaíno, S. F., Kay, R. F., and Bargo, M. S., eds. *Early Miocene Paleobiology in Patagonia: High-Latitude Paleocommunities of the Santa Cruz Formation.* Cambridge University Press, Cambridge.

Vizcaíno, S. F., Kay, R. F., and Bargo, M. S., eds. 2012d. *Early Miocene Paleobiology in Patagonia: High-Latitude Paleocommunities of the Santa Cruz Formation,* Cambridge University Press, Cambridge.

Vizcaíno, S. F., and Milne, N. 2002. Structure and function in armadillo limbs (Mammalia: Xenarthra: Dasypodidae). *Journal of Zoology* 257: 117–27.

Vizcaíno, S. F., Milne, N., and Bargo, M. S. 2003. Limb reconstruction of *Eutatus seguini* (Mammalia: Dasypodidae): Paleobiological implications. *Ameghiniana* 40: 89–101.

Vizcaíno, S. F., Toledo, N., and Bargo, M. S. 2018. Advantages and limitations in the use of extant xenarthrans (Mammalia) as morphological models for paleobiological reconstruction. *Journal of Mammalian Evolution* 25: 495–505.

Vizcaíno, S. F., Zárate, M. A., Bargo, M. S., and Dondas, A. 2001. Pleistocene burrows in the Mar del Plata area (Buenos Aires Province, Argentina) and their probable builders. *Acta Paleontologica Polonica* 46: 157–169.

von Koenigswald, W., Anders, U., Engels, S., Schultz, J. A., and Kullmer, O. 2013. Jaw movement in fossil mammals: Analysis, description and visualization. *Paläontologische Zeitschrift* 87: 141–159.

von Meyer, G. H. 1867. Die Architekur der Spongiosa. *Archiv für Anatomie, Physiologie und Wissenschaftliche Medicin* 34: 615–628.

Wagner, D. O., and Aspenberg, P. 2011. Where did bone come from? An overview of its evolution. *Acta Orthopaedica* 82(4): 393–398.

Wainwright, P. C. 1991. Ecomorphology: Experimental functional anatomy for ecological problems. *American Zoologist* 31: 680–693.

Wainwright, P. C. 2007. Functional versus morphological diversity in macroevolution. *Annual Review of Ecology, Evolution, and Systematics* 38: 381–401.

Wainwright, P. C., Alfaro, M. E., Bolnick, D. I., and Hulsey, C. D. 2005. Many-to-one mapping of form to function: A general principle in organismal design? *Integrative and Comparative Biology* 45(2): 256–262.

Wainwright, P. C., and Reilly, S. M. 1994. *Ecological Morphology: Integrative Organismal Biology.* University of Chicago Press, Chicago.

Wake, M. H. 1993. The skull as a locomotor organ. Pp. 197–240 in Hanken, J., and Hall, B. K., eds. *The Vertebrate Skull,* Vol. 3. University of Chicago Press, Chicago.

Walmsley, C. W., Smits, P. D., Quayle, M. R., McCurry, M. R., Richards, H. S., Oldfield, C. C., Wroe, S., Clausen, P. D., and McHenry, C. R. 2013. Why the long face? The mechanics of mandibular symphysis proportions in crocodiles. *PLoS ONE* 8(1): e53873.

Warton, D. I., Wright, I. J., Falster, D. S., and Westoby, M. 2006. Bivariate line-fitting methods for allometry. *Biological Reviews of the Cambridge Philosophical Society* 81: 259–291.

Weibel, E. R., Taylor, R., and Bolis, L. 1998. *Principles of Animal Design: The Optimization and Symmorphosis Debate.* Cambridge University Press, Cambridge.

Werner, C. F. 1961. *Wortelemente lateinisch-griechischer Fachausdrucke in den biologischen Wissenschaften,* 2 Aufl. Akademische Geest & Portig, Leipzig.

Werth, A. J. 2000. A kinematic study of suction feeding and associated behavior in the long-finned pilot whale, *Globicephala melas* (Traill). *Marine Mammal Science* 16(2): 299–314.

Werth, A. J. 2004. Functional morphology of the sperm whale (*Physeter macrocephalus*) tongue, with reference to suction feeding. *Aquatic Mammals* 30(3): 405–418.

West, G. B., Brown, J. H., and Enquist, B. J. 1997. A general model for the origin of allometric scaling laws in biology. *Science* 276: 122–126.

West, G. B., Brown, J. H., and Enquist, B. J. 1999a. The fourth dimension of life: Fractal geometry and allometric scaling of organisms. *Science* 284: 1677–1679.

West, G. B., Brown, J. H., and Enquist, B. J. 1999b. A general model for the structure and allometry of plant vascular systems. *Nature* 400: 664–667.

Westneat, M. W. 2003. A biomechanical model for analysis of muscle force, power output and lower jaw motion in fishes. *Journal of Theoretical Biology* 223: 269–281.

White, C. R., and Seymour, R. S. 2003. Mammalian basal metabolic rate is proportional to body mass$^{2/3}$. *Proceedings of the National Academy of Sciences of the USA* 100: 4046–4049.

White, C. R., and Seymour, R. S. 2005. Allometric scaling of mammalian metabolism. *Journal of Experimental Biology* 208: 1611–1619.

Whitenack, L. B., and Motta, P. J. 2010. Performance of shark teeth during puncture and draw: Implications for the mechanics of cutting. *Biological Journal of the Linnean Society* 100: 271–286.

Whitney, M. R., and Pierce, S. E. 2021. Osteohistology of *Greererpeton* provides insight into the life history of an early Carboniferous tetrapod. *Journal of Anatomy* 239(6): 1256–1272.

Wilga, C. D., and Motta, P. J. 2000. Durophagy in sharks: Feeding mechanics of the hammerhead *Sphyrna tiburo*. *Journal of Experimental Biology* 203: 2781–2796.

Wilga, C. D., Wainwright, P. C., and Motta, P. J. 2000. Evolution of jaw depression mechanics in aquatic vertebrates: Insights from Chondrichthyes. *Biological Journal of the Linnean. Society* 71: 165–185.

Wilson, G. P., Evans, A. R., Corfe, I. J., Smits, P. D., Fortelius, M., and Jernvall, J. 2012. Adaptive radiation of multituberculate mammals before the extinction of dinosaurs. *Nature* 483: 457–460.

Wilson, J., Marsicano, C., and Smith, R. M. H. 2009. Dynamic locomotor capabilities revealed by early dinosaur track makers from southern Africa. *PlosONE* 4(10): e7331.

Winters, T. M., Takahashi, M., Lieber, R. L., and Ward, S. R. 2010. Whole muscle length-tension relationships are accurately modeled as scaled sarcomeres in rabbit hind limb muscles. *Journal of Biomechanics* 44: 109–115.

Withers, P. C., Cooper, C. E., Maloney, S. K., Bozinovic, F., and Cruz-Neto, A. P. 2016. *Ecological and Environmental Physiology of Mammals*. Ecological and Environmental Physiology Series. Oxford University Press, Oxford.

Witmer, L. M. 1995. The Extant Phylogenetic Bracket and the importance of reconstructing soft tissues in fossils. Pp. 19–33 in Thomason, J., ed. *Functional Morphology in Vertebrate Paleontology*. Cambridge University Press, Cambridge.

Witton, M. P. 2008. A new approach to determining pterosaur body mass and its implications for pterosaur flight. *Zitteliana, Reihe B* 28: 143–158.

Wolff, J. 1869. Über die Bedeutung der Architectur der spongiösen Substanz für die Frage vom Knochenwachsthum. *Centralblatt für die medicinschen Wissenschaften:* 54: 849–851.

Wolff, J. 1870. Über die innere Architectur der Knochen und ihre Bedeutung für die Frage vom Knochenwachstum. *Virchow's Archiv* 50(3): 389–453.

Wolff, J. 1892. *Das Gesetz der Transformation der Knochen*. Hirschwald, Berlin.

Wolff, J. 1986. *The Law of Bone Remodeling*. Springer, Berlin.

Woodward, G., Ebenman, B., Emmerson, M., Montoya, J. M., Olesen, J. M., Valido, A., and Warren, P. H. 2005. Body size in ecological networks. *Trends in Ecology and Evolution* 20(7): 402–409.

Wooller, M. J., Bataille, C., Druckenmiller, P., Erickson, G. M., Groves, P., Haubenstock, N., Howe, T., et al. 2021. Lifetime mobility of an Arctic woolly mammoth. *Science* 373(6556): 806–808.

Xu, L., Mei, T., Wei, X., Cao, K., and Luo, M. 2013a. A bio-inspired biped water running robot incorporating the Watt-I planar linkage mechanism. *Journal of Bionic Engineering* 10(4): 415–422.

Xu, L., Mei, T., Wei, X., Cao, K., and Luo, M. 2013b. Development of lifting and propulsion mechanism for biped robot inspired by basilisk lizards. *Advances in Mechanical Engineering* 976864.

Zelditch, M. L., Swiderski, D. L., and Sheets, H. D. 2012. *Geometric Morphometrics for Biologists: A Primer*. 2nd ed. Academic, London.

Zhao, Q., Benton, M. J., Hayashi, S., and Xu, X. 2019. Ontogenetic stages of ceratopsian dinosaur Psittacosaurus in bone histology. *Acta Palaeontologica Polonica* 64(2): 323–334.

Zuccotti, L. F., Williamson, M. D., Limp, W. F., and Ungar, P. S. 1998. Technical note: Modeling primate occlusal topography using geographic information systems technology. *American Journal of Physical Anthropology* 107: 137–142.

Index

Page numbers in italics indicate illustrations.

diggers, 22, 24, 37, 128, 129, 135, 136, 208, 216

digging, 9, 10, 22, 23, 24, 37, 41, 68, 76, *112*, 113, 118, 127, 128, *128*, 132, 133, *135*, 136, 137, 142, 149, *149*, 195, 209, 211, 218, 223, 256

digging, types, *128*, 129

digitigrade. *See* posture

dik-dik. *See Madoqua*

dimensionality ratio, 54

dinosaurian and dinosaur(s), 7, 32, 42, *43*, *54*, 58, 110, 112, 119, *120*, 130, 132, 159, 173, *190*, 191, *192*, *193*, 200

Diomedea, 96, 103, 105

Diphylla, 138

diphyodonty. *See* tooth replacement

Diplocaulus, *54*

diprotodont, 209, 213

Diprotodontia, 179

direction, 16, 17, *18*, 19, 20, *23*, 33, 48, 77, *77*, 83, *100*, *118*, *127*, *131*, *163*, 237, *237*

discrete categories, 37

discriminant analysis (DA), 34, 37, 38, 135, 180, 184, *225*

discriminant function analysis (DFA), 37

discriminant functions, 37, 38, 225

disseminula, 139

distal(ly). *See* anatomical directions

diving, 74, 88, 89, 106

docodont, 112

Doedicurus, 177, *178*

dorsal fin. *See* fin

drag, 83, 84, 86, 93, 95, *96*, 96, 108; coefficient, 83, 86, 90

duty factor, 110, 116, *117*, 118, 134, 136

durodentine, 161, 164

durophage, 138

dynamics, 15, 83, *151*, 168

dynamic equilibrium, 66; friction forces, 74, pressure, 94

dynamic soaring. *See* flight

eagle. See *Heliaeetus*

eccentric isotonic contraction, 45, 46

ecology, 34, *35*, 36, 38, 50, 57, 138, 187, 200, 201, *202*, 235

ecological analogs, 66; categories, 38, 39; groups, 211, *211*, 212; morphology, 34

ecometric analysis, 203

ecometrics, 189, 203,

ecomorphological analyses, 137, 209; approaches, 36, 135, 177, 216; studies, 34, *35*, 36, 74, 132, 138; traits, 203

ecomorphology, xi, 12, 26, 29, 34, 36, 141, 189; quantitative, 29

ecosystem(s), 8, 9, 10, 50, *57*, 57, 58, 76, 140, 141, 187, 199, 200, 223, 231, 233

Efremov, Ivan A., 191, *192*, 235, 236

eggshell(s), *190*, 191, 192

elapids, 144

elastic, 40; deformation, 40, 55; energy, 45, 115, 121, 122, *123*; failure, 56; modulus, 40, 55; similarity, 55, 56

elastic fibers. See fibers

elastic connective tissue, 239

elastin, 40, 239, 241

elephant(s), *7*, 42, *43*, 51, 53, *55*, 56, 67, 113, 132, 149, *149*, *154*, 155, 161, 177, 201, 256, 260

Elephantidae and elephantid, *177*, 200, 260

elevator. *See* muscles

Elpistostege, 108

enamel. *See* tooth

enameloid. *See* tooth

enantiometric relationships, *55*

enantiometry, 55

endochondral bone. *See* bone (types)

endomysium, 244, *245*

energetic requirements, 177, 222

endoglyphous snakes, 144

endoskeleton, 238, 239, 245, 246

endurance, 88, 119

Enhydra, 50

enthesis(es), 155, 164, 209, 251, 252, *253*

environment, 4, 6, 10, 34, 35, 37, 73, 77, 80, 84, 89, 108, 137, 140, 143, 187, 189, 193, 194, 198, 203, 207, 217, 222, 231, 235. *See also* trait-environment dyad

Eocardia, 226

Eosclerocalyptus, *178*

epaxial musculature. *See* muscles

epicondylar index (EI). *See* index

epimysium, 244

epiphyseal plate, *243*

epiphysis(es). *See* bone (structure)

epithelial tissue, 238

epithelium, 112, 152, *240*

Equidae and equid(s), *70*, 176, 177, *180*, 181, 182, *183*, 203, 260

Equus, 175, 181, 203, *255*

Eremotherium, 203, 207

Erethizontidae, 130, 211, *232*

Ernanodon, 189

Eschrichtius, 150

Eubalaena, 60

Eucholoeops, 28, *28*, 69, *135*, 169, *170*, *178*, 207, *208*, 209, *209*, 210–212, 226

Eucinepeltus, *178*

Euclid, 51, *52*

Euphractus, 172

euhypsodont, 161,

Eumetopias, 50

euthemorphic. *See* quadritubercular

Eutheria and eutherian(s), 55, *56*, 147, 177, 218, 219, 222–226, 256

evolution, x, 3–10, 34, 77, *79*, 80, 81, 111, 112, 121, 129, 144, 147, 154, 159, 187, 233, 235, 244, *258*, 260

evolutionary constraint, 10, 24

exaptation, 5

Exocoetidae, *91*, 92

exoskeleton, 138, 192, 238, 245

extant phylogenetic brackets (EPB), 32, *33*

extension, 9, 10, 22, 24, 45, 46, *110*, 115, 122, *122*, *123*, 125, 129, 131–133, 209, 210

extensor. *See* muscles

extensor digitorum longus. *See* muscles

extinction, 6, 12, 50, 186, 191, 235

exudativore, 139

faculty, 9, *10*, 11, 13, 132, 209; faculties, 24, 32, 34, 118, 119, 124, 126, 142, 210

failure, 22, 40, 41, 56. *See also* spontaneous failure

Fariña, Richard A., ix, x, *x*, xiii

fascia(e), 239, 241, 251

fatigue, 15, 22

faunivore, 138, 139

faunivory, 138

feathers, *92*, 92, 93, 105, 192

feature, 9, 10, 11, 29, 32, 35, 142, 166, 209, 218, 244

feeding, 11, *11*, 50, 51, 57, 73, 76, 137, 138, 140–144, *143*, 149, 150, *151*, 157–159, 164, 179, 182, 194, 199, 211, 224, 227–229, 244; habits, 137, 139, 140,152, 154, *156*, 173, 182, 184, 204, 211, *220*; behavior, 38, 216, 217; resources, 212

Felidae and felid(s), 32, 53, 114, 161, 168, 169, 213, 223

Felis, 189

femur. *See* bone (skeletal element)

femur robustness index (FRI). *See* index

fibers: collagen(ous), 41, 84, 239–244, *242*, 249–251; elastic, 239, *240*, 241, *241*, 243; reticular, 239

fibroblast, 240

fibrous cartilage (or fibrocartilage). *See* cartilage (types)

fibrous connective tissue. *See* connective tissue

fibrous joint. *See* joint, types: synarthrosis

fibula. *See* bone (skeletal element)

fin(s), *78*, 86; anal, 77; caudal, *86*, 88, 89, 108; dorsal, 33, 77, *78*; median, 84, 86, 246; paired, 77, 78, 79, 86, *86*, 89, 92, 246; pectoral, *78*, 249; pelvic, 77, *78*, 79

finite elements, 15, *16*, 24, *25*, *26*

finite element analysis (FEA), 24

fitness, 35, *35*, 191

flexible discriminant analysis, 34, 38

flexion, 20, 22, *23*, 42, 44, 45, 88, 110, *110*, 115, 116, 122, 131, 132, 133, 196, 197, 209

flexor. *See* muscles

flier(s), 91, 106, *112*

flight, 6, 29, *30*, 88, 90, 92, 96, 103, 104, 113; aerodynamics, 15; dynamic soaring, *97*, 99, *100*, 103; flapping, 32, 99, 103; flapping-gliding, *98*; flapping-missile, *98*; gliding, 104; hovering, 100, *101*, 103; soaring, 99, *100*; stationary, 100, *101*;

photogrammetry, 58, 60
phalanges, 81, 92, 93, 249
phylogenetic ANOVA, 39
phylogenetic comparative methods (PCM), 39
phylogenetic effect, 63; flexible discriminant analysis, 34, 38; inertia, 10, 11; signal, 11, 38, 63
phylogenetic eigenvector regression analysis (PVR), 63
phylogeny, 11, 12, 34, 38, 63, 69, 179, 139, 207, *207*, 235, 253
Physeter, 149
physical scale models, 58, *59*, *66*; system(s), 15, *18*, 19
physiological cross-sectional area (PCSA), 46, *48*
physiology, 27, 13, 50, 52, 89, 150, 174, 203, 216, 222
phytoliths, 141
phytophage, 139
Picidae, *146*
pig(s), 70, 149, 161, 162, 218, 260
plagiaulacoid, *See* teeth (types)
planes of section, 237, *237*; frontal, 237; sagittal, 66, 110, 237; midsagittal, 46, 237; transverse, 237
pleurodont. *See* tooth attachment
pinnate. *See* muscles
pinniped(s), 50, 84, 88
Pinnipedia, 149
pipid, 152
piscivore, 138, 139
pitch, 86, *87*, 161, 168, *169*
pivot. *See also* fulcrum
Placodermi and placoderm(s), 143, 238, 245
plains viscacha, 29
planktonivore, 138, 139
Planops, 207, *208*
plantigrade. *See* posture
plastic, 40, 41
platyrrhine, 207
Platyrrhini, *232*
Pleistocene, 24, *59*, 67–69, 105, 131, 140, 155–157, *155*, *156*, *175*, 177, *178*, 180, *181*, 184, *184*, 189, 196, *199*, 201, 203, 207, 211
Plesiosaur(s), 89, 90, *90*, 159
pleurodont(y). *See* tooth attachment
Pliocene, *165*, 173, *178*, 196
Poaceae, 139
poëphage, 139
pöephagous, 140
pollinivores, 139
Polylophodontia, 260
Polyphage(s), 138–140
polyphyodont(y). *See* tooth replacement
Pongo, 125
population density, 11, 50, 55, 212, 233, 260
porcupine(s), 29, 130, 206, 211, 212, *212*, 227, *232*
possums, 213
postcranial skeleton, 76, 246

posterior(ly). *See* anatomical directions
posture, 108, 110, 114, 116, 119, 126, 130, 131, *131*, 211, 238, 244, 251, 253; bipedal, 119, 131, 132, 134, 196, 216; digitigrade, 114, *114*, 116, 134; parasagittal, 110, 111, *111*, *112*, 134; plantigrade, *114*, 134, 210; quadrupedal, 119, 134, 196; sprawling, 108, *108*, 110, *110*, 134; unguligrade, *114*
power coefficient, 54; equations, 51–54, *64*, 65; formula, 52
prearticular. *See* bone (skeletal element)
predator(s), 45, 50, *72*, 74, *75*, 138, 140, 143, 150, *192*, *193*, 194, 212, 213, 222, 225, 230, 233; richness, 229
predator/prey ratio, 228–230, *229*
prehensile hands, 126; feet, 126; foot, *127*; limb(s), *112*, 126, 127; tail, 126
premolar. *See* teeth (types)
Prepotherium, 69, 207, 209–212
primates, 55, 56, 62, 68, *112*, 113, 119, 124–127, 131, 135, *135*, 136, 150, 161, 203, 204, 206, 207, 213, 226, 227, 260
prime mover. *See* muscles
principal components analysis (PCA), 37, 132, 135, *135*, 136, 141, 179, 180, *180*, 181, *181*, 209, 229, 231
Priodontes, *24*, 157, 207, *209*
Proboscidea and proboscidean(s), 67, 200, 201, *202*
proboscis, 154, 213
Procrustes superimposition method, 37
Procyonidae and procyonid(s), 162, 223, 260
Proeutatus, 212, 226
pronation, 125, 131, 133, 250
pronator. *See* muscles
Propalaehoplophorus, 68, *178*, 212
propalaehoplophorids, 156
propalinal, 161
Propraopus, 198
Protamandua, 211, 227
Proterotheriidae and proterotheriid(s), 69, 196, 212–214, *215*, 216, 231, 218, 219, 221
Prothylacynus, 211, 212, 222, 223, *223*–225, 226
protocone and protoconid. *See* tooth, occlusal feature
protohypsodonty, 214
protractor. *See* muscles
protrusible mouth, 143
Protypotherium, 69, 134, 135, *175*, *180*, 214, *218*–220
proximal(ly). *See* anatomical directions
Pseudonotictis, 222–225, *224*, *225*
Pseudoprepotherium, 194
Psittacidae, *91*, 146
Pteranodon, 105
Pteropodidae and pteropodid(s), 70, *70*, *91*, 103
pterosaur(s), 29, *30*, 31–33, 91, 92, *92*, 103, 105, 106, *106*, 107
pterygoid. *See* muscles. *See also* bone (skeletal element)

pterygia(um), 247
Ptilodus, *185*
Pucadelphys, 29
pulp cavity. *See* tooth
puncture-crushing cycle, 162

quadrate. *See* bone (skeletal element)
quadritubercular, 260
quadrupedal. *See* posture
quantitative ecomorphology. *See* ecomorphology
Quetzalcoatlus, 105

raccoon. See *Procyon*
radicular root canal. *See* tooth
Radinsky, Leonard B., x, 8, *8*, 11, 25, 27, 34, 76, 80, 141, 142, 235
radioactive disintegration, 198; isotope(s), 198, 199
radioactivity, 198
radius. *See* bone (skeletal element)
rainfall, 194, 227, *228*, 228, 229, *229*, 230, *230*, 231, 233
raphe, 251
ratio estimator (RE) 66, *68*
rattlesnake, 144
rauisuchiid, 119
ray-finned fish, 142
rays 79, 84, 88, 142, 143, 247
realized niche, 35, *35*, 36, 38, 58, 60, *62*, 63–65, 68–70, *70*, *72*, 72, 103, 176,176–180, *179*, 216, 218, 219, 221, *221*, 222, 230, 231
rectilinear movement. *See* locomotion without appendages
rectus femoris. *See* muscles
red bone marrow. *See* bone (structure): myeloid marrow
red-legged seriema. See *Cariama cristata*
red panda. See *Ailurus*
regression(s), 24, 37
Reig, Osvaldo, ix, xi, 235
relative grinding area (RGA), 222, 225
relative mandible length (JAW) 217, *233*
relative muzzle width (RMW), 156
relative warps, 224, *224*
reproductive success, 35, 36
reptile(s), 89, *90*, 91, 116, 147, 150, 152, 154, 159, 171, 191, 200, 204, 238, 244, 253, 255, 256
resistance. *See* drag. *See also* load
reticular fibers. *See* fibers
reticular tissue. *See* connective tissue
retractor. *See* muscles
reverse engineering, 15
Reynolds number, 83, 84, 86, 90
ricochetal, 121
rhinoceros(es), 29, 51, 53, 70, 130, *154*, 155, 156; rhinos, 213
Rodentia and rodent(s), 24, 37, 68, 70, 113, 126, 127, 129, 131, 132, 135, 136, 140, 148, 150, 160, 161, 169, 176, 181, 186,

Contributors

SERGIO F. VIZCAÍNO is Professor of Vertebrate Zoology at Universidad Nacional de La Plata, Researcher of the Consejo Nacional de Investigaciones Científicas y Técnicas (CONICET), and Curator of Vertebrate Paleontology at the Museo de La Plata, Argentina. He has coauthored the books *Megafauna: Giant Beasts of Pleistocene South America* (IUP, 2013) and *Forma y Función en Paleobiología de Vertebrados* (EDULP, 2016) and coedited the volumes *The Biology of the Xenarthra* (University Press of Florida, 2008) and *Early Miocene Paleobiology in Patagonia: High-Latitude Paleocommunities of the Santa Cruz Formation* (Cambridge University Press, 2012).

M. SUSANA BARGO is Researcher of the Comisión de Investigaciones Científicas de la Provincia de Buenos Aires (CIC PBA) and Associate Curator of Vertebrate Paleontolgy at the Museo de La Plata, Argentina. She has coauthored the book *Forma y Función en Paleobiología de Vertebrados* (EDULP, 2016) and coedited the volume *Early Miocene Paleobiology in Patagonia: High-Latitude Paleocommunities of the Santa Cruz Formation* (Cambridge University Press, 2012).

GUILLERMO H. CASSINI is Professor of structural and functional adaptations of vertebrates at the Universidad Nacional de Luján, Researcher of the Consejo Nacional de Investigaciones Científicas y Técnicas (CONICET) at the Museo Argentino de Ciencias Naturales "Bernardino Rivadavia" (MACN-BR), Argentina, and Associate Curator from Colección Nacional de Mastozoología at the MACN-BR. He has coauthored the book *Forma y Función en Paleobiología de Vertebrados* (EDULP, 2016).

NÉSTOR TOLEDO teaches Comparative Anatomy at the Universidad Nacional de La Plata and is Researcher of the Consejo Nacional de Investigaciones Científicas y Técnicas (CONICET) at the Museo de La Plata, Argentina. He has coauthored the book *Forma y Función en Paleobiología de Vertebrados* (EDULP, 2016).

GERARDO DE IULIIS is Lecturer of Comparative Vertebrate Anatomy and Vertebrate Palaeontology at the University of Toronto, Professor of Anatomy and Physiology at George Brown College, and Research Associate at the Royal Ontario Museum, Canada. He has coauthored the books *Megafauna: Giant Beasts of Pleistocene South America* (IUP, 2013) and *The Dissection of Vertebrates* (Elsevier, 2019).

FOR INDIANA UNIVERSITY PRESS

Tony Brewer *Artist and Book Designer*

Gary Dunham *Acquisitions Editor and Director*

Anna Garnai *Editorial Assistant*

Brenna Hosman *Production Coordinator*

Katie Huggins *Production Manager*

David Miller *Lead Project Manager/Editor*

Dan Pyle *Online Publishing Manager*

Pamela Rude *Senior Artist and Book Designer*

Stephen Williams *Assistant Director of Marketing*